T0136325

CHROMOSOME TECHNIQUES

CHROMOSOME TECHNIQUES

A MANUAL

Arun Kumar Sharma and Archana Sharma

Centre for Advanced Study in Cell
and Chromosome Research
University of Calcutta
Calcutta 700019, India

CRC Press is an imprint of the
Taylor & Francis Group, an **informa** business

CRC Press
Taylor & Francis Group
6000 Broken Sound Parkway NW, Suite 300
Boca Raton, FL 33487-2742

First issued in paperback 2019

ISBN-13: 978-0-367-44947-6 (pbk)
ISBN-13: 978-3-7186-5513-7 (hbk)

Visit the Taylor & Francis Web site at
http://www.taylorandfrancis.com

and the CRC Press Web site at
http://www.crcpress.com

CONTENTS

PREFACE

Chromosome study in eukaryota covers a wide spectrum ranging from different groups of plants and animals to the human system. The last two decades have witnessed a tremendous spurt in activity in chromosome research because of availability of refined methods of biotechnology and their application to all facets of life sciences including agriculture, horticulture and medicine.

Notwithstanding such advances, no uptodate manual is available covering the representative protocols for studying the chromosomes of major groups of eukaryotes. The authors' earlier book on "Chromosome Techniques - Theory and Practice" (third edition 1980) is already more than a decade old. Lately, several new and good treatises have been published, which mainly cover human chromosomes and to some extent, mammalian systems. It was therefore desired to undertake the preparation of a laboratory manual to cover the study of chromosomes in plants, animal and human systems, mainly dealing with the protocols and the principles involved. An attempt has been made to cater to the requirements of scientists working in a simple laboratory and also of those engaged in the most sophisticated ones. Details of instrumentation such as electron microscope, image analyser, micromanipulator and such others have been avoided for obvious reasons. Only the principles of their operation have been included, together with the experimental part. Similarly in describing the methods involving details of molecular biology, such as chromosome walking, yeast artificial chromosomes and polymerase chain reaction, the principles and operational mechanism have been outlined with that much of detail as would be necessary just for their use at the chromosome level. Our task was made easier by the availability of very good treatises on molecular biology by Sambrooke, Fritsch and Maniatis (1989) and by Brown (1991). Information available in these books has been extensively utilized.

Moreover, the authors are extremely grateful to the following scientists in the preparation of this treatise : (1) Professor S. M. Albini, School of Biological Sciences, University of Birmingham (UK). (2) Professor M.D. Bennett and Dr Simon Bennett, Jodrell Laboratory, Kew Gardens (UK). (3) Professor G.D. Burkholder, Department of Anatomy, University of Saskatchewan (Canada). (4) Professor John E. Dille, Winthrop University, Rockhill, SC (USA). (5) Professor K. Fukui, Department of Molecular Biology, National Institute of Agrobiological Resources, Tsukuba (Japan). (6) Professor B.S. Gill, Department of Plant Biology, Kansas State University (USA). (7) Professor C.B. Gillies, School of Biological Sciences, University of Sydney (Australia). (8) Dr J.P. Gustafsson, University of Missouri (USA). (9) Dr A.L. Rayburn, Agronomy Department, University of Illinois (USA). (10) Dr T. Schwarzacher and Dr J.S. Heslop-Harrison, Karyology Group, John Innes Institute, Norwich (UK). Their timely despatch of the reprints, including protocols wherever available, have been of much use to us. The authors would consider their effort successful, if it meets the need of a laboratory manual for regular use in chromosome research involving the application of molecular and cytological methods in plants, animals and man.

INTRODUCTION

The progress in the study of chromosome structure had been due principally to refined techniques from different branches — cytology, biochemistry and biophysics. The uninemic, fibrillar and multirepliconic constitution of the chromosomes has been fully substantiated, specially through pulse labelling and autoradiography. A linear array of independent replicons, which are equivalent to supercoiled DNA loops, is present in chromosome structure, where the subsets of dispersed repeats serve as sites of initiation or termination. High resolution autoradiography, spreading technique and ultrastructural analysis have resolved the structure of nucleosomal subunits, to be histone octamers surrounded by nucleic acid fibres, of which a part forms a linker with the protein. The fibre is coiled to give a supercoiled solenoid structure, containing several nucleosomes. The presence of multiple copies of similar sequences in chromosomes has been shown by extraction procedures as well as *in situ* estimation. Even the occurrence of mobile sequences in chromosomes has been detected through hybridization with specific probes in chromosomes from different genotypes and in different phases of development.

The ultrastructural analysis has revealed a central protein core with DNA loops termed the *scaffold*, in chromosomes after extraction of histones. The presence of nonhistone protein in the core is established. Silver staining, coupled with analysis under electron microscopy, shows the core as a compact network of fibres.

Chromosome study, in the earlier decades, had shifted completely from a purely cytogenetic level to analysis of biochemical patterns from extracts. The development of elegant molecular techniques utilizing microbes, had permitted the study of molecular events in detail. However, in eukaryotes, the specialized nature of the chromosome and the gradual realization of the complex multichromosomal control of differentiation led to the invention of novel techniques in recent years for the study of chromosome dynamics and its functioning *in situ*. The identification of gene loci at the cytological level of chromosomes requires the perfection of chromosome spread under the microscope and clarification of the minute details. As a consequence of this requirement, there has been a tremendous upsurge and revival of chromosome research in different groups in eukaryotes. Molecular products, so far restricted to extracted DNA, have been used extensively on chromosomes for identification of the topography of sequences *in vivo*. There has been a remarkable merger of molecular and cytological techniques, enabling a deeper understanding of chromosome dynamics.

Refinements in banding techniques have led to the determination of sequence complexity with marked accuracy. Application of banding techniques, along with *in situ* hybridization, has enabled the accurate localization of functional segments. Ultrastructural analysis of the bands has opened up a new vista in structural analysis of chromosomes with molecular delineation. Application of restriction enzymes in banding, which cut at discrete sequences of DNA, has further aided the analysis of sequence complexity. Advances in micrurgy are responsible for rapid strides in chromosome manipulation including excision of specific fragments, their analysis and use in hybridization. Micromanipulation and image analysis have enabled argon-laser dissection of specific regions of chromosomes and site specific DNA sequences.

Molecular hybridization *in situ* at the chromosome level has emerged as a powerful tool in the identification of finer segments, even reaching up to a few copy or single copy DNA sequences in chromosomes. This study has been specially facilitated through the availability of numerous probes of specific gene sequences responsible for normal and abnormal metabolism. The chromosome mapping as such has reached a very high level of perfection.

Refinements in molecular hybridization *in situ* and their wide application are due principally to the development of a large number of nonradioactive probes, in addition to radioactive ones. These fluorescent nonradioactive probes yield differential fluorescence, depending on the nature of the tagged fluorochromes. Such fluorochrome labelling has led to the invention of methods ranging from hybridization with the entire genome on the one hand to a few copy sequences on the other. These methods have opened up new possibilities of analysis of even multiparental genomes at the chromosome level, using different fluorochromes for different genome probes. Chromosome *painting* is now a powerful aid in genome analysis and affinities, without undertaking any breeding programme.

Just as application of DNA probes *in situ* has helped in identifying gene loci, antibody protein probes against specific chromosomal proteins have unravelled alterations in the DNA protein pattern in chromosomes during differentiation in normal and abnormal states. The application of such antinuclear autoantibodies in the study of chromosomes, especially of nucleosomes, indicates the conservation of some nucleosomal constituents in evolution.

Progress in chromosome research, in its different facets, geared towards the basic objective of identification of gene loci has, therefore, undergone a very high level of perfection. The advancements in banding technology, culminating in precision in reverse banding, have facilitated the mapping of gene loci responsible for human genetic disorders. Improvements of cell fusion techniques and subsequent chromosome elimination have provided a new approach for chromosome mapping. Refinements in the methods of probe preparation and their hybridization *in situ* at the chromosome level are outstanding developments, enabling identification even of a single gene at the site on the chromosome. The principal developments which have helped in the evolution of the techniques of *in situ* molecular hybridization and the interpretation of results are: the availability of restriction fragments, amplified DNA through polymerase chain reaction, yeast artificial chromosome for cloning large chromosome segments and fluorescent probes, as mentioned earlier. The signal achievements in identification and localization of genes on chromosomes are responsible for launching the Human Genome Programme, a global programme in biology with immense potential. A remarkable synthesis of molecular and microscopic details of chromosomes is now possible. The techniques presented in the successive chapters of this book, it is hoped, will assist in obtaining an idea of such a synthesis.

CHAPTER 1

PRE-TREATMENT, HYPOTONIC TREATMENT, FIXATION AND STAINING

PRE-TREATMENT AND HYPOTONIC TREATMENT

Pre-treatment for the study of chromosomes is generally performed for several special reasons. It may be carried out for: (a) clearing the cytoplasm, (b) separation of the middle lamella causing softening of the tissue, or (c) bringing about scattering of chromosomes with clarification of constriction regions. Pre-treatment may also be needed to achieve rapid penetration of the fixative by removing undesirable deposits on the tissue as well as for the study of the spiral structure of chromosomes. The first two applications involve removal of extranuclear contaminants, whereas the third and most important one exerts a direct effect on the chromosomes.

Pre-Treatment for Clearing the Cytoplasm and Softening the Tissue

In order to clear the cytoplasm from its heavy contents, acid treatment has often been found to be very effective. Short treatment in normal hydrochloric acid or other acids brings about transparency of the cytoplasmic background. Such treatments require thorough washing for the removal of excess acid and acid soluble materials. However since acid treatment affects the basophilia of the chromosomes mordanting may become necessary after such treatment. It hampers banding patterns as well.

In addition to acids, several enzymes have been applied for clearing the cytoplasm and cell separation through digestion. Most common ones used are pectinase, cytase from snail stomach and also ribonuclease. A complex enzyme preparation 'Clarase' gave brilliant preparations following staining. In the authors' laboratory, treatment with cellulase has given very satisfactory preparations in difficult dicotyledonous materials.

The use of enzymes for cell separation has been dealt with in the chapters on tissue culture and mammalian chromosomes.

Alkali solutions as pre-treatment agents are effectively employed for materials with heavy oil content in the cytoplasm since alkalis remove the oil by saponification. Such solutions commonly employed are sodium hydroxide or sodium carbonate. As in acid treatment, thorough washing in water is necessary after the tissue has been kept in alkali solution.

In certain cases, pre-treatment may be necessary to remove deposits of secretory and excretory substances from the surface of the tissue which may hinder the access of the fixing fluid. The best example is the application of hydrofluoric acid to remove siliceous deposits prior to fixation in bamboos; or a very short treatment in Carnoy's fluid, containing chloroform, to remove oily or other secretory deposits on the cell walls before fixation, in a number of plant materials.

Separation of Chromosome and Clarification of Constrictions

The underlying principle is the viscosity change in the cytoplasm. As spindle formation is dependent on the viscosity balance between cytoplasmic and spindle constituents, a change in cytoplasmic viscosity brings about a destruction of the spindle mechanism with the chromosomes remaining free or, more precisely, not attached to any binding force within the cell. Pressure applied during squash or smear scatters chromosomes throughout the cell surface. Changes in cytoplasmic viscosity simultaneously affect the chromosome too, which undergoes differential hydration in its segments, and due to this differential effect, constriction regions in chromosomes appear well clarified. A number of pre-treatment chemicals are available, colchicine being the most important one.

Pre-treatment also aids in securing a high frequency of metaphase stages through spindle inhibition. Colchicine is the most active substance known, and also compounds having similar properties, such as chloral hydrate, gammexane, acenaphthene, vinblastine sulphate (Velban) and vincaleucoblastine, amongst others (see table 1 for list).

The concentration used and the period of treatment have to be strictly controlled, as also the temperature in most cases. Prolonged treatment leads to narcotic effects, including chromosome breaks. For further details, see Sharma and Sharma (1980).

Colchicine

($C_{22}H_{25}O_6N$) was first isolated from the roots of *Colchicum autumnale* by Zeisel in 1883. The general method of extraction is through the use of ethanol and subsequent dilution with water; finally the aqueous solution is extracted with chloroform and crystals of colchicine are obtained along with the solvent. The chemical structure of colchicine shows that it has a three-ringed configuration.

Colchicine is the methyl ether of an enolone containing three additional methoxy groups and acetylated primary amino group and three non-benzenoid double bonds. The threshold regions of colchicine-mitotic activity are identical for both crystalline and amorphous forms. Colchicine is soluble in water (500.00 in 10^{-6} mol/1).

Though highly water-soluble, it is very active at an extremely low concentration. It falls under Ferguson's second category of compounds and the reaction is chemical. It brings about a change in the colloidal state of the cytoplasm, causing spindle disturbance.

With regard to the exact reactive groups in the colchine molecule (a) at least one methoxy group in ring A is necessary for its action (b) ring C must be 7-membered and the hydroxyl group should preferably be replaced by an amino group; (c) esterification of amino group in ring B increases the activity; and (d) isocolchicine and its derivatives are less active. A proper distance should be maintained between esterified side chains of rings B and C. A number of sulphydryl poisons, such as iodoacetamide, dimercaptopropanol (BAL), mercaptoethanol, sodium diethyldithiocarbamate act in the same way as colchicine.

In the study of chromosomes, without inducing polyploidy, colchicine has to be applied in a low concentration, such as 0.5 per cent for 1 h, thereby straightening the chromosome arms to allow a thorough study of the constriction regions. It is especially effective for long chromosomes.

Colchicine has the added advantage of being active within a very wide range of temperature. The period necessary for the manifestation of effect varies in different plant and animal groups. The range of concentrations is also wide, between 0.001 and 1%. In animal cells the method of application is preferably by injection or by addition to culture medium 2 h prior to harvesting, whereas in plant tissue it is applied through soaking, plugging and injection. It can also be used in lanolin paste and in agar. In artificial culture colchicine is added to the medium. To avoid toxicity affecting other metabolic processes, it is generally applied in a low concentration to animal cells. In general, the drug is exceptionally suitable for the study of chromosome structure and metaphase arrest, provided that strict control is maintained over the concentration and period of treatment.

Colcemid (Ciba) is de-acetyl methyl colchicine and is used more extensively in studies of mammalian chromosomes since it does not show certain toxic effects attributed to colchicine.

Acenaphthene, a naphthalene derivative has the same property as colchicine of arresting metaphase though its use is limited. As its structure is quite different from that of colchicine, several other aromatic compounds were tried and various derivatives of benzene and naphthalene were found to be effective.

Chloral hydrate

[$C\ Cl_3\ C\ H\ (OH)_2$] has been used as a pre-treatment chemical in cytology since 1935. It has the same effect as colchicine, and is less expensive.

Gammexane

γ-hexachlorocyclo-hexane has been effectively used in plants. In addition to metaphase arrest, polyploidy and fragmentation have also been recorded in some cases. Its o-isomer was found to be similarly effective. It is most active at low temperatures.

Coumarin and its derivatives

Coumarin

The importance of coumarin in chromosome analysis was pointed out by Sharma and Bal (1953). It was also used for the study of the gross chromosome number in cancer of the human cervix by Manna and Raychaudhuri (1953). It is the *o*-coumaric acid lactone found mostly in the glycoside form in plants. Several natural and synthetic coumarins are available and have been screened for their pretreatment effects in this laboratory. It is effective even at a temperature of 30 °C, though cold treatment causes better action. With chromosomes of the human cervix, even a short treatment of 5 min results in their perfect scattering. Being sparingly soluble in water, coumarin is used as a saturated solution.

Aesculine is a derivative of coumarin and is extracted from *Aesculus hippocastanum* Tourn. It is the direct derivative of aesculetin, in which the H from the –OH group in the 6-position is replaced by $C_6 H_{11} O_5$. This particular compound has been found to be suitable for different groups of plants having chromosome numbers ranging from very high to very low. Its approximate solubility is 0.04 % in water, and for chromosome analysis it is applied both in saturated and half-saturated solutions. It is effective only at very low temperatures, ranging between 4 and 16 °C.

Isopsoralene another natural derivative of coumarin, gives excellent results in different groups of plants. Psoralene, an isomer of isopsoralene was found to be unsuitable for chromosome study. In view of this differential action Chaudhuri, Chakravarty and Sharma (1962) inferred that the position of the furan ring is an important controlling factor in the manifestation of karyotype clarification.

Umbelliferone another derivative of coumarin shown to have the property of clarifying chromosomes is extracted from several species belonging to the family Umbelliferae. Its application is very limited.

The majority of the coumarin derivatives can be employed for chromosome analysis, but the best results have been obtained with the four mentioned above. Their applicability is restricted mostly to specific groups of plants, with the exception of aesculine which has a very wide application.

As viscosity change constitutes the principal basis of action of all of these compounds, atempts were made to find out the degree of change in viscosity induced by these derivatives. Sharma and Chaudhuri (1962) observed that, of coumarin and its derivatives aesculine, daphnetin, and aesculetin, daphnetin increases viscosity of the plasma within a very short period, the other three differing with respect to this property, the effect of coumarin being least. These results indicate the differential efficacy of different chemicals as pre-treatment agents.

Oxyquinoline (OQ)

Oxyquinoline (OQ) is a member of the most important groups of compounds investigated for their action in studying chromosome structure — the quinoline complex. Tjio and Levan (1950) fist worked out the importance of 8-hydroxyquinoline in chromosome analysis. It not only causes mitotic arrest, thus demonstrating its c-mitotic property, but it is also endowed with certain characteristics not shared by colchicine. The spindle is inactivated and, as such, does not cause any hindrance to the chromosomes being spread out during squashing; the chromosome arms contract equally. Unlike colchicine, OQ allows the metaphase chromosomes to maintain their relative arrangements at the equatorial plane. Phenols can be effectively employed for the study of chromosome morphology if applied at concentrations below the one causing chromosomal abnormalities. All the three types of phenols, mono-, di- and trihydric, can help in the clarification of chromosome morphology. Dihydric forms are the most effective ones, being active at a very low concentration. Even amongst the dihydric types, such as resorcinol and hydroquinone, the location of the –OH groups is possibly responsible for their differential activity.

p-Dichlorobenzene ($C_6 H_4 Cl_2$)

p-Dichlorobenzene ($C_6 H_4 Cl_2$) is the most useful one in chromosome work amongst the benzene derivatives. It is sparingly soluble in water. Similar to aesculine and other coumarin derivatives, pDB not only causes spindle inhibition but also leads to clarification of chromosome constricitions due to the contraction and differential hydration of chromosome segments. Of all the chemicals tried so far, it has a comparatively much wider application, and complements with both long and short chromosomes appear to respond equally, even though the period of treatment may require modification. The limitations of its use are the prolonged period of treatment (3h) and the specificity of the temperature needed (10-16°C) for optimal results.

Monobromonaphthalene

Even 10-15 min treatment at low temperature in saturated aqueous solution of monobromonaphthalene may bring about the desired representation of chromosome morphology in certain groups of plants. For aquatic angiosperms especially, pre-treatment with this chemical is effective. The principal cause is the rapid rate of penetration of this chemical into the tissues.

Table 1 Some common pre-treatment chemicals

Chemical	*Effective concentration of aqueous solution*	*Period of treatment needed for karyotype clarification*	*Temperature (°C)*	*Remarks*
Acenaphthene	Saturated	1-4 h	Room temperature	Application limited to specific organs, particularly pollen tubes
Aesculine	Saturated or half saturated	5 min-24 h	4-16	Widely applicable in both plants and animals
Aesculetin	Saturated	Variable	10-16	Limited to only a few plants
α-Bromonaphthalene	Saturated	10 min-4 h	10-16	Very effective for aquatic plants and wheat chromosomes
Chloral hydrate	0.5-1%	30 min04 h	10-16	Effective in a number of plants but results rather inconsistent
Colchicine	0.5-1%	30 min-1 h for plants but 5-30 min for animals	Preferably between 8 and 16	Very wide range of effects. equally suitable for plants and animals
Colcemid	0.5 μg ml to 0.05 μg/ml	1-6 h	Room temperature	Animals *in vivo* and *in vitro*, Human leucocyte culture
Coumarin	Saturated	3-6 h for plants, 5-60 min for animals	Both cold and room temperature	Effective for long chromosomes and for animals
Daphnetin	Saturated	Variable	10-6	Limited application
Vinblastine sulphate 0.1 μg/ml to 0.01 μg/ml	1 to 3 h up to 6 h	37	Peripheral blood culture	
Vincaleucoblastine	0.15 ml of stock soln of 0.5 g/ml /10 ml of culture	1 to 3 h	37	Peripheral blood culture
Hypotonic solutions (Nacl, KCl sodium citrate)	Variable	5-40 min	20 to 37	Animal and human
p-Dichlorobenzene	Saturated	3-5 h	12-16	Very effective, particularly on plants with a very high number of chromosomes
Gammexane	Saturated	1-3 h	12-16	Effective mainly on plant tissue
Hormones	0.0002-0.02%	$3\text{-}3\frac{1}{2}$ h	10-16	Effective on plants with both low and high number of chromosomes

Table 1 *(continued)*

Chemical	*Effective concentration of aqueous solution*	*Period of treatment needed for karyotype clarification*	*Temperature (°C)*	*Remarks*
Isopsoralene	Saturated	1-2 h	12-14	Very effective on mitotic chromosomes of a number of plants
Oxyquinoline	0.002 M	3-4 h	12-16	Very effective on plants, particularly with medium sized to long chromosomes
Phenols	Variable, usually low doses	3-6 h	10-16	Applicable to almost all groups of plants with changes in concentration
Umbelliferone	Saturated	1-3 h	12-16	Applicable to some plant groups having high chromosome number
Veratrine	0.05-4%	30-40 min	12-16	Applicable to several groups of plants
Miscellaneous fluids	(i) 0.6% NaCl in double-distilled water: 6 parts			Used for minced mouse fetus.
(a) Hungerford's fluid (1955)	(ii) Colchicine in modified Ringer's solution (0.82% NaCl, 0.02% K, 0.02% CaCl in double-distilled water) in proportion $1/10^6$			Incubated at 37 °C in covered depression slide for 30-60 min with agitation after treatment for 5-10 min
(b) Dilute isotonic salt solution can also be used				

Veratrine

Veratrine ($C_{32} H_{40} N O_9$) is another important alkaloid, shown to possess the property of clarifying chromosome structure. Cold treatment is found to be essential for its activity.

FIXATION

Fixation may be defined as the process by which tissues or their components are fixed selectively at a particular stage to a desired extent. The purpose is to kill the tissue without causing any distortion of the components to be studied, as far as practicable.

Although a number of fixing reagents have been devised by various workers, all of them possess certain common characteristics. Each fixing chemical is lethal in its action. The structural integrity of the chromosome must be maintained intact. Precipitation of the chromatin matter is essential to render the chromosome visible and to increase its basophilic nature in staining. Under living conditions, the phase difference between different components of the cell is not enough to permit them to be observed as distinct entities. Coagulation of protein and consequent precipitation cause a marked change in the refractive index of the chromosomes, helping them to appear as differentiated bodies within the cell. All fixatives, so far used, have the property of crosslinking proteins. The primary requiste of a fixative for the study of chromosomes is therefore the possession of the *property of precipitating chromatin.*

Another important requisite is that it should have the property of rapid penetration so that the tissue is killed instantaneously, the divisional figures being arrested at their respective phases. *Immediate killing* is essential as otherwise the nuclear division may proceed further and attain the so-called 'resting' or metabolic phase. With the death of the cells, certian consequent changes occur which are detrimental to the preservation of chromosome structure. The most important change is the autolysis of protein. A fixative should also be able to *check autolysis of proteins.* With the onset of lethality, bacterial action causes the tissue to decompose. Another prerequisite is the *prevent this decomposition* by maintaining an aseptic condition in which bacterial decay cannot take place. As the purpose of chromosome study is to observe the minute details of chromosome morphology, the staining should be perfect.

A proper fixative should, in general, *enhance the basophilia* of the chromosome.

A mixture which fulfils all the conditions detailed above can be considered to be a truly effective fixative for chromosome study, but since all these properties are rarely to be found within a single chemical, a fixative is generally a combination of several compatible fluids which jointly satisfy all the above requirements.

An alternative to chemical fixatives is fixing through freezing at low temperature, followed by drying the tissue-the principle involving rapid cooling of the tissue to a low temperature, followed by extraction of water in a vacuum. This method allows a life-like preservation of the tissue. The cooling process must be so rapid that the water cannot crystallise as this results in the distortion of cellular components, and consequent misinterpretation of chromosome structure. The initial water contents of the tissue, the shape and size of the material, the temperature of the tissue and of the cooling bath are all factors contributing to suitable fixation by freezing without forming ice crystals. Water can be frozen into amorphous ice in a cooling bath of −175°C, secured by condensation and liquid nitrogen.

After fixation by freezing, water is removed in vacuum at a low temperature. The material can also be dried by passing it through a stream of dry cool gas. The material can be cut directly or before sectioning, and may be infiltrated with paraffin or some other medium. Several freeze drying equipments and cryostats have now been developed.

The freezing method of fixation has a number of advantages namely, (a) minimum distortion of the tissue after its death, (b) least possibility of diffusion, and (c) no significant effect on the enzyme system (d) the tissue can be directly embedded in paraffin. Its inherent drawbacks include the distortion of cellular components during embedding or sectioning. Interference may occur with very small materials by crystallization of water while preparing the material in bulk, and disintegration of the tissue between developing ice crystals, and the relatively high cost in setting up the apparatus.

Some of these limitations, have been eliminated in the process of dehydration by freezing-substitution. The specimen is rapidly frozen, followed by dehydration at a very low temperature (−20 to −78°C) through any one of the following reagents: *n*-propanol, *n*-butanol, methanol, ethanol, methyl cellosolve, or the chemical fixatives. After complete dehydration, the material is brought back to room temperature slowly.

The freeze-drying method of fixation has numerous advantages over the chemical one. It is particularly accurate for the study of the effect of chemical and physical agents on the chromosome, where the immediate effect has to be analysed, but it is not very useful for the study of the structure and behaviour of chromosomes during the process of divisions under normal conditions. Other advantages of chemical fixation, such as increase in the basophilia of chromosomes, differential precipitation of chromatin matter in its different segments and so on, are obtained by means of the freezing-substitution technique. These factors, taken together, have contributed largely to the wide use of chemical fixatives in routine work on chromosome studies, and only under special circumstances, is freezing-substitution applied.

The fixing chemicals, in general, may be classified into two categories, based on their property of precipitating proteins within the cell. The best examples of precipitant fixatives are chromic acid, mercuric chloride, ethanol, etc. Among the non- precipitant fixatives can be included osmium tetroxide, potassium dichromate, etc. There are certain fixatives which undergo chemical combination with proteins, some of which precipitate out proteins, some of which precipitate out proteins, while others do not.

Whenever a fixative has a strong precipitating action, it is usually counteracted to a certain extent by the addition of other reagents.

The main chemicals which have been used as ingredients of a fixing mixture may be (a) non-metallic-ethanol, methanol, acetic acid, formaldehyde, propionic acid, picric acid, chloroform or (b) metallic-chromic acid, osmic acid, platinic chloride, mercuric chloride, uranium nitrate, lanthanum acetate in nature. Several of these compounds are also used as vapour fixatives, with the sole object of converting the soluble substances into insoluble ones, before coming into contact with water or other solvents, so that *in situ* preservation is maintained.

Most non-metallic fixatives, have one advantage over the metallic ones-that no washing in water is required after fixation. In materials where the cells are loosely scattered in a suspension, as for example, in mammalian peripheral blood cultures, fixation is improved by decreasing the amount of fluid surrounding the cells. The cells, therefore, are centrifuged into a small pellet and the supernatant liquid removed. The fixative is added and allowed to remain undisturbed for up to 30 min. The cells may be centrifuged into a pellet again, followed by the addition of fresh fixative. Cells on a coverglass or a slide are fixed usually for not less than 5 min or more than 24 h. The more common fixing fluids are mentioned below:

Non-metallic fixatives

Ethanol between is used extensively as a constituent of chromosome fixatives, in percentages 70 to 100%. An important advantage of ethanol is its capacity for immediate penetration. It precipitates nucleic acid. Its dehydrating property is well known and it causes an irreversibel denaturation of proteins. It also has an undesirable hardening effect on the tissue.

Being a reducing agent, it undergoes immediate oxidation to acetaldehyde and then to acetic acid in the presence of an oxidiser, and so cannot be used in combination with many metallic fixatives. Therefore it can be used principally in combination with acetic acid, formaldehyde or chloroform. For enzyme studies on chromosomes, chilled 80% ethanol fixation for 1 h or more is recommended and for monolayer cultures, absolute or 96% ethanol fixation for 1-15 min is often applied.

Acetic acid can be mixed in all proportions with water and ethanol or methanol from very low concentration to even glacial (100%). This acid has a remarkable penetrating property, even higher than alcohols.

Acetic acid is, in general, an ideal fixative for chromosomes, and in spite of Pischinger's finding that it dissolves histones, it has been observed to maintain the chromosome structure intact. One of the limitations of this fixative is the excessive swelling of the chromosome segements induced by it. Therefore where a study of detailed structure of chromosomes is needed, acetic acid should be used in combination with alcohols, or similar chemicals, which shrink and harden the tissue.

Acetic acid is also a good solvent for aniline dyes, and is a necessary component of staining-cum-fixing mixtures, like acetic-carmine, acetic-orcein, acetic-lacmoid, etc.

Formaldehyde, in its commercial form, known as 'formalin', contains a 40% solution of formaldehyde in water. It is a bifunctional compound capable of forming crosslinks between protein end groups. For the fixation of chromosomes (10-40%) solution of commercial formalin in water is used. It has a very low precipitating action on protein in high concentrations. The action is chiefly on proteins. It reacts with the amino groups of proteins.

Tissues, which are placed in fixatives containing formaldehyde, often show well scattered chromosomes, especially after sectioning from paraffin blocks. The cell volume increases considerably, resulting in spreading of the chromosomes over a larger area. The constriction regions appear slightly exaggerated due to contraction of the euchromatic segments. This effect is possibly due to the action of formaldehyde on the chromosome proteins. In cytochemical work, washing after formalin fixation is essential so that the reactive groups of proteins remain unmasked to combine with the reagents.

A serious disadvantage in using formaldehyde as a fixing agent is the fact that the tissue treated with this reagent is difficult to smear; the exceptional hardening, which is a result of its action on protein, being responsible for this

Methanol is occasionally used in chromosome studies of plants but is extensively employed in fixing animal chromosomes. While ethanol causes heavy shrinkage of chromosomes, methanol causes swelling and this property has been used advantageously in the preparation of fixatives where a swelling agent is often needed to compensate for the shrinking effect of other chemicals. Its effective concentrations are the same as ethanol.

Acrolein and glutaraldehyde have the property of crosslikning protein molecules more actively than formaldehyde.

Propionic acid has been used extensively in the fixation of chromosomes. It is miscible with water, ethanol and ether in all proportions, and is a good solvent for aniline dyes. It is generally used as a substitute for acetic acid. Its penetration is not as rapid as that of acetic acid but it causes much less swelling of the chromosomes. It can be used, in staining-cum- fixing mixtures, like propionic-carmine, propionic-orcein, etc.

Chloroform is miscible in all proportions with alcohols, ether and acetone. In the study of plant chromosomes, it is generally used in the fixative to dissolve the fatty and waxy secretions from the upper surface, facilitating the penetration of the fixative. In the study of chromosomes from animal tissues, chloroform is frequently used to dissolve the fats which are present as accessories. A judicious use of chloroform in fixing mixtures is recommended, as an excessive dose or long period of treatment may be toxic rapid penetration of the fluid, and also because of its ready miscibility with most solvents, it forms a good ingredient of fixing mixtures.

Commericial diethyl ether mixes well with alcohols or liquid hydrocarbons in all proportions, being a solvent for fats and oils and has been used in a few fixing mixtures, like Newcomer's fluid to clear the cytoplasm and the cell of the fatty and oily substances, so that the other components can reach the substrate easily.

Isopropanol has the same effect as ethanol on chromatin. It has been preferred in some materials due to its comparatively less drastic action.

Acetone is miscible with water, ethanol and ether in all proportions. It is a good solvent for cellulose acetate and for many organic compunds, and when acting in a fixing mixture serves the same purpose as chloroform or ether in clearing the cytoplasm by dissolving the organic matter.

Metallic fixatives

Osmium tetroxide is a strong oxidising agent and should never be mixed with formaldehyde or alcohol. In solution, it oxidises aliphatic and aromatic double bonds, alcoholic (OH) groups, amines, -SH groups and other nitrogenous groups as well, but generally the carboxyl and carbonyl groups are not affected. Initially the entire molecule combines with the amino groups of proteins. In the secondary phase, the compund formed undergoes oxidation, during which the residual part of osmium tetroxide is reduced to a lower oxide or hydroxide. Due to this, the tissue fixed in osmium tetroxide turns black. In general, it fixes homogeneously, maintaining a life-like preservation of the tissue. An advantage is that it does not cause much shrinkage of the tissue, but on the other hand , there is a slight swelling.

Table 2 Some common fixatives

Chemical	*Molecular weight*	*pH in fresh condition (Lassek and Lunetta, 1950)*	*Used alone or in mixture*	*Concentration commonly used*
Acetic acid	60.05	2.3 95% aq.)	Both	100% and 45% When used slone 10% soln. in effective
Acetic acid	60.05	2.3 (5% aq.)	Both	Also 60% alone for human chromosomes
Acetone	58	7.0 (abs.)	Mixture	100%
Chloroform	119.59	7.2 (abs.0	Mixture	100%
Chromic acid	100.01	1.2 (1% aq.)	Mixture	1 and 2%
Dioxane	88.11		Mixture	100%
Ethanol	46	8.4 (abs.)	Both	50-100%
Ether	74		Mixture	Absolute
Formaldehyde	30	3.4 (comm.)	Mixture	40%,5%
Formic acid	46	2.0 (1% aq.)	Mixture	1%
Hydrochloric acid	36.5		Both	Normal
Isopropanol	60		Mixture	100%
Methanol	32		Mixture	100 and 95%
Mercuiric chloride	272	3.0 (6.9%)	Mixture	—
N-butyl alcohol	74.12	6.2 (abs.)	Mixture	100%
Nitric acid	63		Mixture	10%
N-propyl alcohol	60	6.7 (abs.)	Mixture	100%
Osmic acid	255	6.1 (1% aq.)	Both	0.5-2%
Picric acid	229	1.3 9sat.)	Mixture	Sat.

The effect of osmium fixation depends to a significant extent on the pH, toxicity and temperature of the fixing mixture.

Although osmium fixation preserves chromosomes during the divisional cycle, it cannot be recommended for the study of the interkinetic nuclei. Moreover is it often results in protein loss.

The best result is obtained if the fixative is applied in the form of vapour. This can, however, be applied only on small materials, such as prothalli of ferns, unicellular objects or materials having no cellulose wall, such as smeared animal tissues. Rapid penetration without any deformation of the tissue is the special advantage of this method. In electron microscopy, osmium fixationis widely utilised.

Materials, after osmium fixation, require bleaching to remove the black precipitate produced by fats, usually, with hydrogen peroxide when the slide is brought down to water prior to staining. The slide is transferred from 80 per cent ethanol to a jar containing H_2O_2 and 80% ethanol in equal proportions and kept from 1 to 12 h. For bleached tissues, pre-mordanting in 1% chromic acid solution is necessary.

Platinum chloride solution in water is often applied in place of osmium tetroxide, especially in the somatic tissue of plants. It does not have the same capacity as osmium tetroxide of preserving the life-like structures of the cell, but its capacity for penetration is decidedly greater. It is compatible with formalin and can be used as a substitute for chromic acid in chromic-formalin mixed fixing fluids, but its application is rather limited. Fixation in platinum chloride requires bleaching of the fixed tissue as well.

Chromic acid is a strong precipitant of protein but a very weak precipitant of nuclein. Protein undergoes denaturation and precipitation by the primary action of chromic acid, and the secondary action results in hardening. Chromic acid penetrates the tissues slowly and the hardening induced by this acid makes the tissue resistant to hardening by ethanol in subsequent processing.

Materials fixed in this acid require thorough washing in water, at least overnight, otherwise the deposition of chromic crystals not only hinders staining but also hampers the observation of chromosomes. Because of its slight hardening action it is difficult to use this fluid as a fixative for squash preparations, unless softened by some strong acid. Basic dyes adhere closely to tissue fixed in chromic acid.

In general, chromic acid is considered an essential ingredient of several fixing mixtures. It imparts a better consistency to the tissue and aids staining better than osmium tetroxide.

Potassium dichromate is not as soluble is water as chromic acid. Over pH 4.6, $K_2Cr_2O_7$ can maintain the structure of chromosomes, whereas, with less acidity, only the cytoplasmic structures are preserved. Any of the ions, anions or cations, obtained from $K_2Cr_2O_7$, can react with proteins, depending on the acidity of the fixing fluid.

Potassium dichromate, on the whole, is a good fixative for lipids. It has a rapid rate of penetration and shrinkage is not very marked. It does not harden the tissue significantly. After dichromate fixation cellular constituents respond well to acid dyes, and the response of chromatin to basic dyes can be maintained if the fixation is performed at an acidic level. Because of its rapid rate of penetration, it is often preferred to chromic acid. It is widely used for the fixation of both plant and animal chromosomes, and can be washed away with dilute solution of chloral hydrate in water.

Mercuric chloride is compatible with the majority of the fixing fluids. Slightly acidified mercuric chloride precipitates protein very strongly. Its rapid penetration and strong protein reaction can be advantageously employed in mixtures with compatible fluids. Chromosomes respond well to most of the dyes after mercuric fixation. A serious limitation of mercuric fixation is that a needle-like precipitate of mercurous chloride is often formed in the tissue following this fixation.

Table 3. Some common fixing mixtures

3.1. With ethanol

Name	*Components*	*Proportion*	*Modification*	*Treatment period*	*Application*
Carnoy'sfluid and its modifications	Glacial acetic acid absolute ethanol	1 3	1:2, 1:1, with chloroform 1:6:3 and 1:1:1 add corrosive sublimate to saturation washout with 70 to 955 ethanol; mordant with ferric ammonium sulphate (1-3%) for 3-12 h after fixation	15 min to 24 h	Effective for all materials; with variations in period and concentrations.
	A combination of abs ethanol (1), Sublimate soln (2) and 1-5% glacial acetic acid is a common fixtive for protozoa				
Chromo-nitric acid	aq. nitric acid absolute ethanol aq. chromic acid	10% 4 parts or 20% 3 parts 3 or 4 parts 0.5% ro 1% 3 parts		4-5 h, followed by washing in 70 and 100% ethanol for 2 - 3 days after fixation in dilute mixture or 20-30 min followed by washing for 1 h after fixation in strong mixture	limited to embryonic nuclei
Propionic acid modification instead of acetic acid	(a) propionic acid:ethanol (95%) (1:3)			24-36 h, washed out with 70% ethanol	perithecia of Ascomycetes
	(b) Propionic acid:ethanol (95%) (1:1) + ferric hydroxide (0.4g)			if needed add a few drops of carmine.	plants with small chromosomes
	(c) Propionic acid: chloro form:abs. ethanol = 1:1:2 other proportions are 1:4:2			12-24h	avian chromosomes plant chromosomes
Iron acetate modification	(a) acetic acid:abs. ethanol (1:3) + small amount iron acetate			12h followed by keeping for 15 min in & at. iron acetate solution in 45% acetic acid: 45% acetic acid: 1% for malde hyde sol - (3:5:2); vinse in 45% acetic acid.	authers with small chromosomes

Table 3 *(contd)*

Iron acetate modification (contd)	(b) ethanol (95%) : acetic - carmine soln with added iron acetate (3:1) a flake of rusted iron	12-24h followed by washing abd storage in ethanol (95%) with iron flake for 5-10 days	flower-buds, mainly of Cucurbitaceae
Ethanol - ether mixture	ethanol (1005) : pure ether (1:1)	3min and air dry	fresh liquefied semen
Newcomer's fluid	Isopropyl alcohol : propionic acid : petroleum ether : acetone : dioxane (6:3:1:1:1)	4h - 24h	both plants and animals very stable fixative; can be used in combination with pretreatment chemicals
Lactic acid modification	glacial acetic acid : abs. ethanol : lactic acid (1:6:1)		perithecia of *Venturia*
3.2. With formalin			
Navashin's fluid	Solution A:Chromic anhydride:glacial acetic acid:dist. water (1.5g:10ml:90ml)	24h, followed by washing in running water for 3h	both block and squach preparations of plant material
	Solution B:40% aq. formaldehyde:dist. Water (40:60) mix in equal proportions just before use.	Modifications include altering the relative amounts; replacing acetic acid by propionic acid or preparing all ingredients freshly before use	
Bouin's fixatives	sat.aq. picric acid:40% aq. formaldehyde:glacial acetic acid:urea:chromium trioxide (75ml:15 - 25ml: 5-10ml: 1 - 2g: 1 - 1.5g)	1h 30min - 12h followed by repeated washing in 70% ethanol or ethanol gradesz and anilone oil	mainly animal chromosomes

Table 3.2 *(cont)*

Levitsky's	aq. chromic acid (1%): aq. formaldehyde (10%) different proportions (3:2,4:1,1:1,1:2,1:3)	12 - 24h in cold or room temperature followed by washing for 3h higher concentrations	root-tips, brain and ganglion teleost chromosomes
Mixtures with acetic/propionic acid and ethanol/methanol	acetic acid:aq. formaldehyde (40%): ethanol (95% or 100%) ferent proportions (1:12:30ml+water) (1:6:14); (1:10:85) cetic acid can be replaced by propionic acid and ethanol by methanol	1 - 2h	plant, animal and human tissues
Battaglia's 5111 mixture	ethanol (95%):chloroform: acetic acid: aq. formaldehyde (40%) = 5:1:1:1	5 to 10min	both plant and animal tissues
3.3: With osmic acid			
Flemming's fixatives	aq.chromic acid soln. (1-2%):acetic acid (5 - 100%) : aq.osmic acid soln. (2%), mixed in different proportions and diluted with dist. water; mixed just before use.	1h to overnight in cold, followed by washing for 1h to overnight in running water	both plant and animal tissues
Several modifications are devised with different proportions of the basic components and additions of NaCL/urea/maltose			
Champy's fluids	aq. osmic acid soln, (2%):aq.chromic acid soln. (2%) : aq. potassium dichromate soln. (2-7.5%) in different proportions (22:20:16), (6:5:15), (10:20); (3:6:4), mix just before use and diluted with dist. water	6 - 24h, followed by washing in running water for 24h	mainly animals
La Cour's fixatives	aq.chromic acid soln. (2%) : aq, potassium dichromate soln. (2%) : aq.osmic acid soln. (2%) : acetic acid (10%) : aq. saponin soln (1%) : dist water in proportions of (100:100:60:30:20:210); (100:100:32:12:10:90); (100:100:120:60:10:50) mixed just before use.24h in cold followed by washing in running water for 3 - 12h		mainly plants

Lanthanum salts can also precipitate nucleic acid. In order to secure a good preservation of chromatin, lanthanum salts are added either in the fixative or after treatment with a recognised fixing fluid.

Uranium nitrate may replace potassium dichromate but it is of extremely limited application. It is preferable to use it as an additional component of the fixing fluid rather than as a substituted metallic compound.

Iridium chloride if acidified with dilute acetic acid, has been found to be efective in the fixation of *Triton* sp. However, it is not effective elsewhere, due to its low capacity of nucleo-protein precipitation.

Fixing mixtures

Of the different types of mixtures employed for chromosome studies two categories, at least, can be formulated. In one of these, both metallic and non-metallic fluids have been included, whereas the other is constituted purely of non-metallic fixing fluids. The majority of the fixatives, the principal ones of which are listed below, fall under the first category.

STAINING

The structure and behaviour of chromosomes can be studied only after they are made visible under the microscope. In order to maintain normal activity at the time of observation it is best to mount the dividing cells in the body fluid of the organism and to observe chromosome movements under a phase contrast microscope. The evaporation of the fluid can be prevented by paraffin oil, which has the added advantage of being oxygen-solvent and non- toxic to the tissue.

Staining, as performed on chromosomes, can be classified as *vital* and *non-vital*. For vital staining, non-toxic dyes are applied to the living tissue so that the latter can be studied without being killed. If isolated cells such as blood, bone marrow, etc. are stained, the staining is called supravital. Of the different vital stains so far applied to the study of chromosomes, methylene blue only has been found to be effective in demonstrating cell division in tissue culture. It is basic dye of the thiazine group, $C_{16}H_{18}N_3SCl$, and is soluble in water.

In non-vital staining, the coloration of the chromosomes in the killed tissue is caused by certain chemical agents which are insoluble in the chromosome substance. The principal kinds of dyes that are used to stain chromosomes are synthetic organic dyes, derived from coal tar. The colour of a dye is due to certain chemical configuration, known as *chromophore*; similarly, the tissue must retain the colour which is due to certain chemical configuration in the dye molecule itself, known as the *auxochrome*.

The best example of a chromophoric group is the quinonoid ring. The coloration due to quinonoid arrangement is exemplified by the conversion of colourless hydroquinone to yellow quinone. Auxochromes, responsible for the adherence of the dye to the tissue, are mostly—NH_2 and —OH groups which convert the non-dyeing coloured substance into a form which undergoes electrolytic dissociation in water and is capable of forming salts with acids or bases.

The dyes are generally termed basic or acidic on the basis of their chemical nature and behaviour. In an *acidic* or *anionic* dye, the balance of the charge on the dye ion is negative. In a *basic* or *cationic* dye, however, the dye ion charge is positive. Most of the acid dyes are prepared as metallic salts and are generally neutral or slightly alkaline in reaction, but they react with, and stain substances with, a basic reaction. A basic dye, on the other hand, is manufactured as a salt of mineral or aliphatic organic acids, and it stains substances which are acidic. Several of the dyes used in the study of the chromosomes are amphoteric, such as orcein; they behave both as acids and bases. The majority of the chromosome dyes are triphenyl methane or aniline derivatives, though other dyes have also been found to stain chromosomes.

In order to stain the chromosomes at specific loci, the general procedure is to over-stain it, followed by the removal of the excess stain—a process called *differentiation*. Staining can also be performed progressively by gradually increasing the intensity of the colour. The process of differentiaton allows the stain to adhere to specific sites of the chromosomes. Similarly, acidic dyes may stain chromatin by proper differentiation, where quite likely the dye reacts with the protein moiety of the chromosome. For chromosome staining, basic dyes are applied, chromatin being strongly acidic. Acid dyes colour the cytoplasm which is predominantly basic. The terms basophilic and acidophilic are based on the affinity for basic or acidic dyes, chromosomes being basophilic and cytoplasm acidophilic.

Though chromosome staining is the product of a chemical reaction, yet the intensity of the stain fades with age. The fading may be attributed to the effect of ultraviolet light through continued exposure to daylight, to progressive acidity of the mounting medium or to retention of contamination of elements.

The adherence of a dye to the tissue may also be accelerated through the process of mordanting. Certain metallic hydroxides form compounds with the dye which attach the dye to the tissue and are called the 'lake' for the particular dye. The *mordant* is the term applied to the salt used.

In cases where the 'lake' formed is insoluble in water, the general practice is to dip the tissue in the mordant first, followed by staining. In carmine staining, iron salts are often used as mordants and are added to the stain itself. The process of combination, between the dye and the mordant, is known as *chelation*.

In cytological studies, whenever the dye or the mordant is used separately, the purpose of the mordant is either to modify the isoelectric point of the tissue or to form a chemical link between the stain and the chromosome. In principle, it changes the surface conditions of the fixed chromosomes. Several types of mordants are used in cytology. In cases where mordants are used prior to staining, they evidently modify the chromosome surface in such a manner that the dye adheres strongly to the chromosomes.

Post-mordanting not only helps to retain the stain for a prolonged period, but also clears the cytoplasm. This effect is due to the acidity of the mordant, such as iodine in ethanol, which, being higher than the cytoplasm, removes the stain from its surface. Chromatin, on the other hand, having a stronger acidity, retains the stain. The purpose of postmordanting is more or less to bleach out the undesirable elements. An oxidising post-mordant oxidises the dye, present at certain sites, to a colourless substance.

The term 'mordant' should preferably be restricted to those agents which are applied before staining and which form a complex with the dye or the tissue. Agents, when applied after staining, act more as differentiating chemicals than as mordants.

Fuchsin

Of all the different staining methods employed for the study of chromosomes, the Feulgen reaction is considered to be the most effective with regard to chromosome staining. In 1924, Feulgen and Rossenbeck devised a method based on the Schiff's reaction for aldehydes which stains the nucleic acid of the chromosomes specifically and, as such, has been effectively employed for the visualization of chromosomes.

Feulgen solution or, more precisely, fuchsin sulphurous acid is prepared from the dye, basic fuchsin, which belongs to the triphenyl methane series. The commercially obtained '*basic fuchsin*' is a mixture of three compounds, namely *p*-rosaniline chloride (Michrome No. 722), basic magenta (rosaniline chloride, Michrome No. 623), and new magenta (new fuchsin, Michrome No. 624). The molecular weights are 328.815 for *p*-rosaniline chloride ($C_{19}H_{18}N_3Cl$), 227. 841 for basic magenta ($C_{20}H_{20}N_3Cl$) and 365.893 for new magenta ($C_{22}H_{24}N_3Cl$). All three compounds are characterised by quinonoid arangements within the molecule. Basic fuchsin is easily soluble in water and alcohols, and for Feulgen reagent, 0.5% solution is prepared in boiling distilled water.

The principle of preparing the Feulgen or Schiff's reagent is to treat the basic fuchsin solution with sulphurous acid, the product obtained being colourless fuchsin sulphurous acid. This reagent is the Schiff's reagent, utilized by Feulgen and Rossenbeck (1924) for the demonstration of the DNA component of chromosomes. The procedure involves the preparation of a basic fuchsin solution in warm water, followed by cooling at a particular temperature and the subsequent addition of hydrochloric acid and potassium metabisulphite, needed for the liberation of SO_2, prior to storage in a sealed container in a cool, dark place.

Dissolve 0.5 g of basic fuchsin gradually in 100 ml boiling distilled water. Cool at 58°C. Filter, cool the filtrate down to 26°C. Add it to 10 ml N HCl and 0.5g potassium metabisulphite. Close the mouth of the container with a stopper, seal with paraffin, wrap the container in black paper and store in a cool dark chamber. After 24 h, take out the container. If the solution is transparent and straw-coloured, it is ready for use. If otherwise coloured, add to it 0.5 g of charcoal powder, shake thoroughly and keep overnight in cold temperature (4°C). Filter and use.

Alternatively, after dissolving the dye, bubble a stream of SO_2, through the solution. Filter and store. The addition of activated charcoal, removes the yellowish impurities and, as such, a transparent colourless reagent can be obtained. The colour of Schiff's reagent varies, depending upon the type of dye used, the hydrogen ion concentration and SO_2 content, but without activated charcoal, the solution should be straw-coloured. The reagent is unstable and loses SO_2 on continued exposure to air and becomes coloured again. The basic fuchsin solution, or more precisely, *p*-rosaniline chloride solution, undergoes conversion to leucosulphinic acid, which is colourless, by the addition of sulphurous acid across the quinonoid nucleus of the dye. Sulphurous acid is obtained through the action of HCl on potassium metabisulphite. The excess of SO_2 undergoes reaction with leucosulphinic acid to produce bi-N-aminosulphinic acid, popularly known as Schiff's reagent.

For staining, the procedure involves hydrolysis of the fixed tissue in normal HCl at 56-60°C, for a period varying from 4 to 20 min before immersing the material in Schiff's reagent. The colour develops within a short time and the chromosomes take up a magenta colour and can be observed after mounting in 45% acetic acid. If the test is performed strictly according to the recommended procedure, the chromosomes appear to be specifically coloured against a clear cytoplasmic background. Prolonged keeping in Schiff's reagent is undesirable, as further hydrolysis may take place. A rinse in sulphite solution or SO_2 water is often helpful to remove excess of colour, if any, in the cytoplasm.

The chemical basis of the reaction, includes two principal steps:

(1) By hydrolysis with normal HCl, the purine-containing fraction of deoxyribonucleic acid (DNA) is separated from the sugar, unmasking the sldehyde groups of the latter.
(2) The reactive aldehyde groups then enter into combination with fuchsin sulphurous acid to yield the typical magenta colour. Feulgen reaction is, therefore, based essentially on the Schiff's reaction for aldehydes. After removal of the base, carbon atom 1 of the furanose sugar is so arranged as to form a potential aldehyde, capable of reacting with fuchsin sulphurous acid. The ribose sugar, with an—OH in place of—H at carbon 2, is not hydrolyzed by normal HCl and so does not react with fuchsin sulphurous acid. In the

pyrimidine-sugar linkage, on dissociation, the aldehyde groups are not free to react, unlike the open and reactive aldehydes obtained after breakdown of the purine-sugar linkage. Since different factors are involved in the development of the colour it is always necessary to keep a strict control over temperature and duration of hydrolysis and the type of fixation. With a check through the control, the specificity of the test is unquestionable. Certain other basic dyes of the triphenyl methane series are able to replace basic fuchsin, to some extent, like dahlia violet, magenta roth, methyl violet, brilliant green, malachite green and light green as observed in our laboratory. The methods of preparation and staining are similar to that followed for leucobasic fuchsin. A direct correlation is observed between the colour of the different dyes and the active groups present in them.

Dye	MI No.	No. of batches	Active Group	Colour of nuclei
Basic fuchsin	421	3	$—NH_2$	Mauve
Dahlia violet	105	2	$—C_2H_5$	Mauvish violet
Magenta roth	624	1	$—NH_2$	Magenta
Methyl violet 6B	180	2	$—CH_3$	Violet
Crystal violet	103	2	$—CH_3$	Deep violet
Brilliant green	406	1	$—C_2H_5$	Blue-green
Malachite green	315	1	$—CH_3$	Green
Light green		1	$—CH_3$	Yellow-green
(Golechha, 1968)				

Carmine is one of the most widely used dyes for chromosome staining. It is prepared from the ground-up dried bodies of *cochineal*, the dried females of *Coccus cacti*, a tropical American Homoptera living on the plant *Opuntia coccinellifera*. It is a mixture of substances, the composition of which often varies on the basis of the method of manufacture. The active principle of carmine to which its staining property is due, is carminic acid. Carminic acid (Michrome No. 214) belongs to the anthraquinone group and has the formula $C_{22} H_{20} O_{13}$, the molecular weight being 492.38. The chromophoric property is attributed to its quinonoid linkage and auxochromes are also present. It is soluble in water in all proportions, and is a dibasic acid and claimed to be nearly insoluble at its isoelectric point, pH 4-4.5 (Baker, 1950). If it is dissolved on the acid side of its isoelectric point it acquires a positive charge, behaves like a basic dye and stains chromatin, but if dissolved in alkaline solution it can behave as an acid dye.

In chromosome studies, carmine is used in solution with 45% acetic acid, and the stain thus prepared is known as acetic- carmine. This solution serves the double purpose of fixation and staining, as acetic acid is a good fixative for chromatin and is a rapidly penetrating fluid. In the original schedule of Belling, 1% solution of the dye is prepared in hot 45% acetic acid, certain authors prefer even 5% solution. Belling suggested the addition of ferric hydroxide in acetic-carmine during its preparation to allow the formation of a lake needed for the intensification of colour.

The common procedure of using acetic-carmine as a stain is to squash the tissue in a drop of the dye solution. In the case of bulk compact tissues, such as root tips, leaf tips, etc., materials can be treated in hot acetic-carmine and hydrochloric acid mixture which serves the double purpose of softening and staining. The use of 2% iron alum solution for a few min prior to staining may serve the purpose of mordanting and thus help in the intensification of colour. While squashing, the best way of adding iron is to tease the tissue in a drop of carmine with the help of a scalpel, or penetration can be aided by slight warming. Being present in the form of an acetic solution, carmine is not a suitable stain for sectioned materials. In certain cases acetic acid is substituted by propionic acid.

Occasionally, both plant and animal tissues which present difficulties in Feulgen staining are mounted in acetic-carmine after Schiff's reaction. In such cases, hydrolysis in normal HCl as well as treatment with fuchsin sulphurous acid clears the cytoplasm allowing specific coloration of the chromosomes. The application of carmine as a chromosome stain is widespread. Starting from the lower groups of plants like algae, it can be applied to all other advanced groups, including all animal and human tissues. Even in the study of special chromosomes like the salivary gland chromosomes of *Drosophila,* its effect is remarkable.

Use of acetic-carmine, acetic-orcein, acetic-lacmoid

Materials required

For 2 per cent solution:

Carmine, orcein or lacmoid — 2 g.

Glacial acetic acid — 45 ml.

Dist. water — 55 ml.

For 1 per cent solution:

The same, except for 1 g of the dye.

Add distilled water to glacial acetic acid to form 45% acetic acid solution. Heat the solution in a conical flask to boiling. Add the dye slowly to the boiling solution, stirring with a glass rod.Boil gently till the dye dissolves. Cool down to room temperature. Filter and store in a bottle with a glass stopper. Keep the mouth of the flask covered with cotton wool while the solution is being heated. Store acetic-orcein as 2.2 g. dissolved in 100 ml glacial acetic acid. Dilute as needed.

Procedure

Acetic-carmine: use 1 % solution directly for staining and squashing. Acetic-orcein or acetic-lacmoid: use 1% solution directly for staining. Alternatively, heat the tissue for a few seconds in a mixture of 2% solution and normal hydrochloric acid (9:1) and then squash in 1% solution. The preparation and application of propionic-carmine and propionic orcein are similar to those of acetic-carmine and acetic-orcein except that propionic acid is used instead of acetic acid. Lactic-propionic-orcein, prepared by dissolving 2 g natural orcein in 100 ml lactic and propionic acid mixture (1:1) and diluted to 45% with water, is very effective for p.m.c.s. For mitotic preparations, maceration in 1 N HCl at 60°C for 5 min between fixation and staining is necessary. Lacto-aceto-orcein, suitable for salivary gland chromosomes may be prepared by dissolving separately natural orcein in concentrated lactic acid and glacial acetic acid with boiling to obtain saturated solutions and then mixing the two filtered solutions with distilled water in the proportion 1:1:1.

Orcein

Orcein was first employed as a chromosome stain by La Cour in 1941. The dye has a molecular weight of 500.488, the formula being $C_{28}H_{24}N_2O_7$ (Michrome No. 375). It is a deep purple-coloured dye, obtained from the action of hydrogen peroxide and ammonia on the colourless parent substance *orcinol.* Orcinol is 3,5-dihydroxytoluene, having a molecular weight 160.166 and the formula $C_7H_8O_2$. It is available both in natural and synthetic forms. In nature, it is obtained from the two species of lichens, *Rocella tinctoria* and *Lecanora parella.*

Orcein is soluble in water as well as in ethanol. Under certain conditions, it can behave as an amphoteric dye. In the study of chromosomes it is used in the form of acetic-orcein, that is, 1% solution in 45% acetic acid. It can be used in the same way as acetic-carmine and has the added advantage that no iron mordanting is necessary. In our experience the intensity of the stain, specially for meiotic materials, is not as good as carmine, though it is effective where carmine staining fails. It has been found to be a very effective stain for salivary gland chromosomes as well as the chromosomes of mosses. For the study of root tip and leaf tip chromosomes, the use of a stronger hot solution of acetic-orcein and normal HCl, mixed in a specific proportion, is necessary for softening the tissue before mounting in a dilute solution of acetic-orcein (Tjio and Levan, 1950; Sharma and Sharma, 1957). In various species of fungi, especially Ascomycetes, hydrolysis in normal HCl for a few min at 60°C, after fixation and prior to staining, has been found to be very effective. Acetic-orcein can be substituted by propionic-orcein which has been found to be useful in studying the chromosomes of *Heteroptera.* It is found to be specially useful in the study of somatic chromosomes though applied frequently for meiotic chromosomes as well. It has, however, to be applied with extreme caution, since overheating in orcein-HCl mixture has been found to induce chromosome breakage (Sharma and Roy, 1955). Therefore, at least for the study of chemical effects on chromosomes, orcein staining should preferably not be applied as its effect may often mask the effect of chemicals.

Lacmoid

Lacmoid (otherwise known as resorcin blue) is a blue acidic dye of the oxazine series. Its empirical formula is $C_{12}H_6NO_3Na$,the molecular weight being 235.173. Similar to carmine, it can be used as an acid-base indicator and, when dissolved in acetic acid, it behaves as a basic dye. Unlike carmine, it is fairly soluble both in water and alcohols. Darlington and La Cour used it in place of carmine, and acetic-lacmoid solution has been found to be very effective for the chromosomes of root tips, embryo sacs and pollen grains. Koller utilized this method for the study of the chromosomes of tumours. For comparatively compact tissues of plants, like root tips, similar to orcein, heating in acetic-lacmoid-HCl mixture is needed prior to squashing for dissolution of the pectic salts of the middle lamella. Cedarwood oil and euparal are recommended as mounting media. However, acetic-lacmoid as a stain cannot be universally applied like acetic-carmine. It has a comparatively limited application and may be tried on those materials where other stains have failed.

Chlorazol black

A solution ofchlorazol black E in ethanol has been applied as an auxilary stain for chromosomes, along with acetic-carmine. This dye was applied after fixation prior to acetic-carmine staining and proved effective for species of Rosaceae. It has been found to be effective for the study of root tip chromosomes of plants.

Chlorazol black E, is an acid dye of the trisazo group and has molecular weight of 781.738(Michrome No. 92). Its formula is $C_{34}H_{25}N_9O_7S_2Na_2$. It is highly soluble in water and sparingly in ethanol. Being an acidic dye, the basis of its stainability with chromosomes is not clear, but it is probable that it stains the protein component. Chlorazol black is possibly effective in materials where protein components of the chromosomes are high.

Crystal violet

The discovery of crystal violet as a stain for chromosomes is attributed to Newton, who used gentian violet(Michrome No. 417) which is a mixture of crystal violet and tetra and penta-methyl p- rosaniline chlorides. Crystal violet itself is hexamethyl p- rosaniline chloride. Gentian violet is a basic dye belonging to the triphenyl methane series. Crystal violet (Michrome No. 103), which is supposed to be one of the most adequate stains for chromosomes, is a bluish violet dye. The dye is closely allied to basic fuchsin from which it can be derived by the replacement of the six hydrogen atoms of three amino groups by six methyl groups. It is soluble in both water and alcohols. In chromosome studies, aqueous 1% solution is used. In Newton's crystal-violet-iodine technique, after the application of the stain to the sections or smears, the excess dye is first washed off in water. Then the slides are processed through iodine and potassium iodide in ethanol mordant, followed by dehydration in ethanol; differentiation in clove oil and and cleaning in xylol before final mounting in balsam. The use of iodine as a mordant, after crystal violet staining, is based on the Gram effect on bacteria. During dehydration, crystal violet can easily be washed off in ethanol, but de-staining can be prevented if iodine is added to bacteria. The same principle holds good for chromosome staining. To obtain a proper coloration of the chromosomes in difficult materials, iodine mordanting, which is normally carried out for a few *s* should be further reduced, but in no case should this step be omitted, as acidic components of the cytoplasm also take up the colour which is removed by iodine in ethanol. Differentiation in clove oil is an essential step in the crystal violet technique. Due to the rapid passage through ethanol, dehydration remains incomplete and it is finally completed in a slowly differentiating fluid, clove oil.

Clove oil is an essential oil, yielded by the flower buds of *Eugenia caryophyllata.* It consists principally of eugenol, a guaiacol derivative. Differentiation in clove oil completes the dehydration, clears the cytoplasmic background and imparts a crisp colour to the chromosomes. Final clearing in xylol $C_6H_4(CH_3)$, or dimethyl benzene is an essential step, as clove oil must be completely removed before final mounting. The retention of this oil ultimately leads to the fading of colour. For materials that are difficult to stain, slides can be mordanted in 1% chromic acid and washed prior to staining. In order to secure complete cytoplasmic clearing in materials having a heavy cytoplasmic content, the slides can be further mordanted in chromic acid in between the different ethanol grades, after mordanting in iodine. Crystal violet is widely used as chromosome stain for plants, animals and lower organisms. It is most effective on pollen mother cell smears or for mitotic and meiotic studies from sectioned tissues. Unfortunately it cannot be applied effectively on tissues to be squashed after staining.

Preparation

Dissolve 1g in 100ml water with constant stirring and boiling. Filter. Allow it to mature for a week before use.

Haematoxylin

This natural colouring substance is obtained from the heartwood of *Haematoxylin campechianum.* The dyeing property of haematoxylin is attributed to its oxidation product, haematein (Michrome No. 360). Haematoxylin has the molecular formula, $C_{16}H_{14}O_6$ the molecular weight being 302.272. The molecular formula of haematein is $C_{16}H_{12}O_6$. The presence of quinonoid arrangement in haematein and its absence in haematoxylin is clear. The process of oxidation, which is otherwise known as ripening, may take several weeks spontaneously, but may be hastened by the use of oxidizing agents such as sodium iodate, hydrogen peroxide, chloral hydrate, potassium permanganate, etc. Slow atmospheric oxidation is, however, preferred to the use of oxidizing agents, as too much oxidation may make haematein unfit for the purpose of staining. In view of the necessity of oxidation in the preparation of haematein from haematoxylin, the aqueous solution of haematoxylin is prepared and allowed to ripen for several weeks. Without the use of a mordant, haematoxylin solution is entirely ineffective in staining chromosomes. Commonly used mordants are potassium aluminium sulphate, iron, alum and ammonium alum. They form lakes which become positively charged and behave as basic dyes. For chromosome studies, potasium aluminium sulphate and iron alum are widely used, the latter being more effective. The potash alum lake of haematoxylin is used for progressive staining, whereas iron alum is utilized in regressive staining . Progressive staining implies gradual addition of the stain till the maximum colour is obtained, whereas regressive staining involves overstaining the material and subsequently washing off the excess stain.

Haematein, after ferric mordanting, has a strong tendency to accumulate around densely stained materials, For this reason, it has most often been used in chromosome studies. On the basis of Heidenhain's schedule, the sections or smears in water are first mordanted in a strong solution of iron alum 4% followed by washing in water and staining in haematoxylin. Differentiation is carried out in dilute solution of iron alum or picric acid to wash off the excess stain from the cytoplasm. In properly controlled differentiated preparations, chromosomes appear intensely black. After washing once more in water, the tissue in dehydrated through ethanol, cleared in xylol and mounted in balsam.

So far as plant chromosomes are concerned, haematoxylin staining is not very effective, due to the heavy cytoplasmic content. Animal materials (such as testes, smears of grasshopper and insects, very thin sections, etc.) can be stained with haematoxylin, and being devoid of any strong acid or clove oil in the staining schedule, the chromosomal stain, once obtained, does not fade in permanent slides.

Preparation

Dissolve ammonium alum in distilled water to prepare a saturated solution. Dissolve haematoxylin in absolute ethanol. Add the latter solution slowly to the former. Expose to air and light for one week. Filter, add 25 ml glycerol and 25 ml methanol. Allow to stand, exposed to air, until the colour darkens. Filter. Store in a tightly closed container. Allow the solution to ripen for a month before use.

Wittmann's acetic-iron-haematoxylin (1962) schedule:

Prepare 6 ml of a mixture of HCl and ethanol (1:1). Treat plant material fixed in acetic-ethanol (1:3) in the mixture for 10 min, then in Carnoy's fluid (6:3:1) for

10-20 min, squash in a drop of stain containing 4% haematoxylin and 1% iron alum in 45% acetic acid and heat gently. A further adaptation for materials not requiring hydrolysis, like leucocytes, ascites cells, etc. contains :

Stock solution: Haematoxylin 4g; Iron alum 1g; 45% acetic acid 100 ml. Ripen for 1-7 days.

Stain (working solution): Stock solution, 5 ml:Chloral hydrate 2g.

Toluidine Blue (Michrome No. 641)

Toluidine blue, a basic dye of the thiazine series, is bluish violet in colour. Its molecular weight is 305.825, the formula being $C_{15}H_{16}N_3SCl$. Robinson and Bacsich suggested the preparation of a lake of toluidine blue with mercuric chloride or potassium iodide. The dried dye which is already mordanted, yields a very intense colour, staining both types of nucleic acids, and can be used selectively for DNA if normal hydrochloric acid hydrolysis is performed for a very short period prior to staining. Pelc (1956) utilized toluidine blue for staining through film in autoradiographic procedure, applied in an aqueous solution which is soluble both in water and alcohols. In view of its restricted application, it is not recommended as a general stain for chromatin.

Giemsa (Michrome No. 144)

Giemsa is not a single dye but a mixture of several dyes, namely methylene blue and its oxidation products, the azures as well as eosin Y. The quality of the stain varies with regard to the proportion of the dyes used. The combination stain, Giemsa, is generally prepared by dissolving the powdered mixture in glycerin and methanol, and in staining, chromatin is stained red and cytoplasm blue.

The importance of Giemsa in chromosome staining has increased substantially after the advent of the banding pattern techniques in the early 1970s.

Preparation of stock solution: 3.8 g of powdered Giemsa (R 66-Gurr) is added to 250 ml glycerine or alternatively 1.0 g to 66 ml glycerine. The mixture is maintained at 55 to 60°C for $1\frac{1}{2}$ to 2 h. Then an equal quantity of methyl alcohol is added.

Preparation of phosphate buffer, pH 6.4

Solution A: 11.336 g/100 ml dist. water or 56.68 g/100 ml dist. water.

Solution B: 8.662 g/100 ml dist. water or 43.31 g/100 ml dist. water.

5 ml of each of the solutions A and B are mixed and made up to 1.0 litre with distilled water. The pH is adjusted to 6.4 with 0.1 N HCl.

Preparation of stain

For mammalian materials, the Giemsa stain is prepared by adding 2 ml of the stock to 2 ml of the phosphate buffer (pH 6.4) and making up to 50 ml with distilled water.

Fluorochromes

Dyes for fluorescent staining have achieved increased prominence after the evolution of the methods for fluorescent banding of chromosomes based on the principle of staining with fluorochromes binding to a particular component of the nucleus, followed by observation under ultraviolet light. Several fluorochromes bind specially to DNA or RNA, depending on the nature of the helix.

Acridine Orange (AO, Gurr)

This has been used successfully in staining mammalian chromosomes as a 1:1000 solution in ethanol; the optimum period is 10 min. It is washed off in phosphate buffer at pH 7.6. It fluoresces green in combination with double stranded and red with single stranded nucleic acids. The methods of fixation affect the secondary nature of deoxyribonucleoprotein complex and interfere with AO- binding capacity of chromosomes.

Quinacrine Dihydrochloride ($C_{23}H_{30}Cl\ N_3O.\ O.\ 2\ HCl.\ 2H_2O$) (Winthrop or Sigma)

This dye is employed widely. For mammalian chromosomes, a 0.5% solution in water is suitable. The period of staining ranges from 4 to 6 min, followed by washing for 2 min in distilled water and differentiation in McIlvaine's phosphate buffer solution at pH 5.5. Commercial antimalarial atabrine powder (atabrine dihydrochloride or atabrine hydrochloride) can be used in routine studies instead of chemically pure quinacrine dihydrochloride as 100 mg to 20 ml glass distilled water.

Quinacrine Mustard (QM)

This chemical, initially synthesized by E.J. Modest at Boston, gives highly specific banding patterns, particularly of human chromosomes. It is regarded to bind DNA both through the alkylating group reacting primarily with the guanine content of DNA (Caspersson *et al.*, 1968) and by intercalation of the quinacrine group in the double helix of DNA. The amount of fluorescence exhibited by different segments of a chromosome stained with QM is controlled by the quantity of DNA located on them and by the qualitative differences in QM-binding capacity of the DNA in different segments. The pattern is determined by irregularities in DNA distribution, reflecting the chromosome pattern and also by a superimposed pattern due to particularly strong QM-binding in particular locations. The accessibility of DNA to QM is determined by steric relations between DNA and chromosomal proteins. In the human karyotype, the pattern of QM- binding in approximately 98% of the length of total metaphase chromosomes shows very constant and reproducible banding patterns, capable of identifying the segments.

Weisblum and de Haseth (1972), however, suggest that strong fluorescence with QM reflects the presence of DNA with high A-T content against Caspersson's contention that it indicates local differences in G-C content. Other workers are also of the opinion that since other fluorochromes, like quinacrine dihydrochloride (Q) and ethidium bromide, which lack the alkylating group, can induce similar bands as QM, it is unlikely that the selective binding of QM to the N_7 atom of guanine is responsible for specific banding patterns.

Proflavine and Acriflavine

The fluorescence pattern produced by these chemicals are less clear and in some cases, quite different from the QM ones.

Ethidium Bromide (EB, 2,7-diamino-10-ethyl-9-phenyl anthridinium bromide) reacts specifically with both DNA and RNA by intercalation, to form relatively stable complexes with markedly increased fluorescence. Some plant chromosomes fixed in acetic- ethanol (1:3) and stained with EB give a reverse pattern to QM while other plants stain uniformly. It yields reasonably consistent results with plant chromosomes but is less satisfactory with animal material. (0.005%) at pH 6.8, colours bright orange with native DNA and dull orange with denatured DNA.

Fluorescein-Tagged Reagents

These have been employed in antinucleoside antibody binding methods for the banding of human chromosomes. Base-specific antinucleoside antibodies react with specific nucleoside bases in single stranded DNA *in vitro*. They attach to fixed chromosomes on being treated with denaturing agents like NaOH. A banding pattern may be obtained with antibodies which react specifically with only one of the nucleoside bases in DNA. The anti-adenosine (anti-A) antibodies are prepared by immunizing rabbits to adenosine monophosphate conjugated to bovine serum albumin. Antibodies to rabbit gammaglobulin are induced in sheep and fluorescein-tagged.

Hoechst 33258

This is used as a stock solution of 50 µg/ml in water

Quinacrine Conjugates

Quinacrine derivatives of polylysine stain chromosomes in a banded fluorescence pattern similar to QM. A 2×10^{-6} M solution of the QM conjugate of poly-L-lysine ($\bar{n} = 24$, lysine/dye = 3) is used to stain leucocyte preparations, previously treated with McIlvaine's citrate-phosphate buffer (pH 4.1). Slides are washed and mounted in citrate-phosphate buffer (pH 7.0).

DNA-Binding Nucleoside Specific Antibiotics

DNA-binding guanine-specific antibiotics, chromomycin A3 (CMA) and mithramycin (MM) have been used as chromosome fluorescent dyes, as also the A-T specific fluorochrome 4′-6- diamidino-e-phenylindole (DAPI). Non-fluorescent dyes may be used as counterstain—methyl green with CMA and actinomycin D (AMD) with DAPI.

Some Miscellaneous Double Stains

Gallocyanin and Other Stains

Tissues warmed in gallocyanin solution for 2-4 min can be counterstained in Biebrich scarlet, phloxine or eosin Y.

Safranin O and Aniline Blue

Root tips are stained 15 min in 1% aqueous safranin O and rinsed in distilled water. They are then stained in 1.0% antiline blue W.S in 95% ethanol for 2 min.

Carbol Fuchsin and Methylene Blue

Seminal fluid can be stained in a mixture of carbol fuchsin and rectified spirit (1:1), followed by a rinse in water and staining for 2 min in 1.3 aqueous methylene blue solution.

Orange G and Aniline Blue

These have been used for both mitotic and meiotic chromosomes. Sections fixed in chromic-formalin (1:1) are rinsed in potassium citrate buffer, stained in a mixture containing 2 g orange G and 0.5 g aniline blue dissolved in 100 ml potassium citrate buffer for up to 3 min, washed in the buffer, dehydrated and mounted.

Ruthenium Red and Orange G after Fuchsin Staining

Stem tips, after 30 min hydrolysis, are stained in fuchsin solution for 24 h, rinsed, stained in aqueous ruthenium red solution for 30 min, dehydrated, stained for $1\frac{1}{2}$ min in orange G in absolute ethanol and clove oil, run through clove oil and xylol and mounted in balsam. Chromosomes take up deep purple stain and resting nuclei less intense stain.

MOUNTING MEDIA

Temporary mounts are made mostly of *squashes* and *smears*. The medium for mounting is either the stain itself or its solvent. In the former case, the processes of staining and mounting are done simultaneously, the tissue then being lifted on to the slide containing a drop of the stain and squashed. In the latter, the tissue is stained and squashed on a slide in a solvent of the stain. In both cases, the coverslip is ringed with paraffin wax and then observed.

Permanent mounts involve mounting the tissue, after suitable processing, so that the proparations can be kept for a long period, often for several years, without appreciable distortion of the structure or intensity of stain.

Permanent Mounts of Sections from Paraffin Blocks

The different steps in the process usually depend on the medium in which the stain is dissolved. In general, the entire process is based on first dehydrating the tissue, then impregnating it with the solvent of the mounting medium and finally mounting with the medium chosen.

The most commonly used dehydrating agent for paraffin sections is ethanol, though acetone and various other alcohols are also used. As the tissue has already been embedded in paraffin and therefore has attained a permanent shape, dehydration does not have to be done in gradual stages. If the stain is dissolved in water, the section can be transferred directly to absolute ethanol after staining and mordanting. If, however, a counter-stain in a lower grade of ethanol is applied, the tissue has to be passed through the required grade before transfer to absolute ethanol. Usually two or three jars containing absolute ethanol are kept and the slides are kept for 4-5 *s* in each.

The slides can be transferred directly from absolute ethanol to the mounting medium. e.g. to euparal or to a mixture of the solvent of the mounting medium and absolute ethanol in equal proportions, then to the pure solvent, as in the case of preparations stained in Feulgen and mounted in Canada balsam. Alternatively, they may be transferred to a differentiating medium for removing excess of the stain before transfer to the solvent of the mounting medium, for example, tissues stained in crystal violet and mounted in Canada balsam.

The choice of differentiating or clearing medium usually depends upon the stain used. The most suitable differentiating agent for crystal violet or gentian violet is clove oil. It removes superfluous stain from the cytoplasm, thus rendering the stain in the chromosomes brighter, and also completing the dehydration. Usually the slides are kept for 2-5 min in the clearing medium. As soon as the surplus stain is washed off, they are transferred to the pure solvent for the mounting medium. Often two jars of clove oil are kept as, after some time, clove oil gets slightly coloured. The clearing agent may be a solvent for the stain or a mordant, as iron alum solution for haematoxylin stain.

CHAPTER 2

MICROSCOPY, MICROSPECTROPHOTOMETRY, FLOW CYTOMETRY, IMAGE ANALYSIS, CONFOCAL MICROSCOPY

A. ORDINARY LIGHT MICROSCOPY

In light microscopy, the underlying principle is to obtain a real, inverted, and enlarged image of the material by means of the objective lens, followed by the formation of virtual image by means of the eyepiece lens. In the study of chromosomes, where only light microscopes are required, the compound microscope should have at least the following attachments:

(1) Apochromatic objective and oil immersion lenses (× 100) – (1.3–1.4 n.a.)

(2) Sub-stage aplanatic and achromatic consider –1.4 *n.a.*;

(3) Compensating eyepieces (× 10 × 15 × 20);

(4) Fitted mechanical stage.

The tube length for each microscope is fixed for the operation of particular objectives. In British instruments, it is generally 160mm. Details of the instruments in microscopes are available in several textbooks; the designs are based on two original types of instrument, named after their inventors, the Huygens and Ramsdens.

Special lenses are used principally to eliminate two types of aberration: spherical and chromatic. Spherical aberration is inherent in lenses with spherical surfaces. Here the rays passing through the periphery of the lens focus at a different point to those passing through the entire or close to the axis. In *aplanatic* lenses, which are compound and constituted of different kinds of glass, all the rays are brought to a common focus by suitable corrections.

Chromatic aberration implies that ordinary sources of illumination, being composed of light of differnt wavelengths, of different colours, focus at different points because of a variation in the path followed by the rays. Colour fringes sometimes appear. In suitably constructed lenses, these aberrations are eliminated as far as practicable, so that preferred rays are made to focus at a common point. In *achromatic* objectives the chromatic aberration is corrected for two colours, or more precisely, two wavelengths, and the spherical aberration for one colour. In ***semi-apochromat or fluorite lenses***, a higher degree of correction is achieved. In *apochromatic* lenses, the chromatic aberration is corrected for three colours and spherical aberration for two colours.

Different colours undergo different degrees of magnification, which may result in a number of images of different colours. Because of superimposition, difficulties arise in the proper clarification of the object. This inequality of colour magnification is corrected by means of compensating eyepieces, while apochromatic lenses are used as the objectives. In achromatic lenses, this defect is generally corrected in its own combination. The seperation of finer details in an image is dependent not on the magnification of the lenses but on their resolving power. This power, or resolution depends on the wavelength of the illuminating source as well as on the *numerical aperture* of the lens. The numerical aperture is calculated on two factors; the angular aperture of the lens, and the refractive index of the medium through which the light enters. Resolution is always proportional to the numerical aperture, which is $n \sin \mu$, n being the refractive index of the cover slip, and μ the maximum angle to the optical axis formed by any ray passing through the specimen, before the formation of total internal reflection. With an objective of n.a. 1.4 good resolution can be achieved if the intervening minimum distance from the specimen is 0.24×0.001mm. Brightness increases with increase in numerical aperture, but decreases with increase in magnification. Visual magnification is obtained by multiplying the magnification of the objective with that of the eyepiece.

Oil immersion objectives are used for critical work where the scattering of light, due to its passage through media of different refractive indices is to be avoided. If a fluid of refractive index similar to that of glass and Canada balsam is used to bridge the gap between the cover slip and objective lens (and if necessary, between slide and condenser), then a homogeneous medium can be achieved for the path of light, avoiding loss of light as far as possible. Cedarwood oil is generally used for the purpose,its refractive index (1.510) being close to that of glass (1.518) and balsam in xylol (1.524) but several synthetic media are now available. For the source of illumination, Pointolite, ribbon filament, mercury arc, or even a 100 w ordinary lamp may be used. Proper screens should always be chosen for observation, depending on the colour of the light and stain used. Wratten yellow-green filters for violet-stained preparations and blue filters for redstained preparations are suitable.

For photography, a Zeiss (Ikon) 35mm camera may be used. The camera is fitted over the eyepiece. A number of cameras with microphotographic attachments, like Zeiss, Leitz, Olympus etc. are available. They are designed as a 35mm or plate camera without the lenses, as the real image is formed over the eyepiece. These cameras are always fitted with adaptors for fitting in the microscope tube. Elaborate, and built-in camera attachment can also be obtained with several microscopes, with provision for timing and exposure. Both slow- and high- speed films are satisfactory, depending on the requirements, the former requiring a longer exposure period. Rapid process panchromatic plates, Kodak, Gevaert or Ilford, are most suitable . The plate size is generally $3\frac{1}{4} \times 4\frac{1}{4}$ inches but 9 × 12 cm plates may also be obtained.

B. PHASE AND INTERFERENCE MICROSCOPY

In principle, these two types of microscopy are identical in the sense that the purpose is to bring about visible change in intensity from an undetectable phase change. Interference systems, being more plastic, allow variable phase changes. Variable phase contrast system can also be planned, although for quantitative measurement it is of little use.

The phase change is represented as $\phi = (n_p - n_m)t$, where n_p and n_m are the refractive indices of the object and the immersion medium repectively and t is the thickness of the object. The formula indicates that with increase in the value of t, there will be a decrease in the detectable difference between the refractive indices. The advantages of phase microscopy are:

(1) Simple and easily adjustable arrangement.

(2) Low cost of the apparatus.

(3) Insensitivity to slight variations in slide and coverslip.

(4) Internal details are often better resolved through zone of action effect.

The *principal limitation* of the phase system is that it is not possible to carry out quantitative measurements conveniently and the presence of a halo prevents proper resolution to some extent. This limitation is inherent in the very principle of phase contrast microscopy. The direct light is allowed to fall on a conjugate area (annulus) here, while the diffracted light is separated and falls on the entire phase plate. Consequently, the conjugate area also receives some diffracted light, which is responsible for the formation of a halo, around the object .

In addition to the attachments of an ordinary microscope, the phase contrast microscope is fitted with: (a) a sub-stage annular diaphragm to produce a narrow cone of light for illuminating the object, and (b) a diffraction plate on the rear focal plane of the objectives, where deviated and undeviated light rays are separated after emerging from the object. A layer of phase retarding material is present on that portion of the diffraction plate which is covered by either of the two rays, and it helps in changing the relative phase of the two rays. Two types of phase contrast microscopes are generally available, namely (a) the instruments based on the principle of negative phase contrast, and (b) those depending on positive phase contrast. In the positive phase contrast system, the slightly retarding object details appear brighter against a lighter background, thus resembling visually stained preparations: while in the negative system, the object is lighter than the background. For routine use in cytology, the former is usually preferred due to the excellent contrast of living cells in aqueous media.

For measuring the refractive index and solid concentration of cell structures, the method of immersion refractometry has been applied in phase contrast microscopy. This method is based on the principle that, if an object is immersed in a medium having a refractive index equal to its own, it affords the least contrast to the viewer. In addition, in order to have a choice of media for watching, a large number of fluids (isotonic solutions with different refractive indices) should be used. In the case of living materials, however, the choice of the immersion medium must be restricted to compounds which show no toxicity and which lack the ability to penetrate the cell or deform the cellular structure. The cell is initially observed in physiological saline or body fluid by positive phase contrast. Later the observation is carried out in isotonic protein solution. If the refractive index of the medium is more than that of the cytoplasm, the latter presents a bright appearance, and *vice versa*. After several trials with different strengths of the medium an optimum stage can be obtained, where the two will be nearly identical. The concentration of the solid can be worked out in g/100 ml of protoplasm, on the basis of the formula

$$C = (n_p - n_s)/(\alpha \,.\, n_s)$$

where n_s, is the refractive index of the solvent, which is either water or dilute salt solution. For protoplasm the value of α is 0.0018. In terms of wet weight, the gramme concentration of solid per 100 g of protoplasm is $\frac{C}{1 + C/400}$. Following immersion refractometry, the change in refractive index during cell division has been observed. There is a decrease in cytoplasmic concentration, which reaches minimum density during diplotene or early metaphase. This method allows a study of the changes in the refractive index of the cellular constituents during cell division.

The convenience of securing quantitative results is the principal reason for the development of interference microscopy. The basis is empirical in determining the mass of single and different tissue elements. The actual optical path difference, or phase change, imparts a quantitative aspect to the measuremente by interference microscopy. The light is split into two beams by a beamsplitting mirror, one beam being transmitted through the object and the other passing some distance to the side of it. The interference is produced by the two beams combining at the semi- reflecting mirror.

The advantages of interference microscopy are (i) Phase changes of the material can be measured. (ii) Bright colour effect can be secured. (iii) The contrast can be varied, allowing a proper contrast to be selected with respect to the object, to secure the intracellular details. The system is elastic and variable phase contrast can be obtained. (iv) 'Halo' and 'Zone' effects are absent. (v) Mass per unit area of the cell can be conveniently measured. The flatness of the image obtained and glare are serious disadvantages of the interference system.

Microscope interferometry, in conjunction with immersion refractometry, permits the determination of the refractive index, the solid concentration, the thickness of structures and the dry mass. With the aid of microscope interferometry, the changes in the amount of protein and DNA in the nuclei have been measured.

C. FLUORESCENCE MICROSCOPY

The utilization of the fluorescence shown by some of the cell constituents, as well as of some special dyes, against ultraviolet light, forms the basic principle of fluorescence microscopy. In this system, intracellular constituents are detected either through their own property of autofluorescence, or by secondary fluorescence due to the adherence of labelled or fluorescent dye.

To detect an intracellular constituent on the basis of its own fluorescence, short wavelength ultraviolet light transmitted from a quartz condenser, fitted with a microscope is required. Several compounds can be identified on the basis of their own fluorescence, provided an understanding is obtained as to the exact wavelength which excites this, as well as the wavelength of the excited substance.

The general practice in the majority of laboratories is to utilize the property of secondary fluorescence obtained by flurochrome preparations, but such dyes become effective only in long blue or near ultraviolet light. These agents are either acidic, basic or amphoteric in nature and can be used in accordance with their application in cytological practice; but a special technique is necessary for the detection of secondary fluorescence of specific compounds within the cell. The method involves the observation of frozen sections or squashes against ultraviolet light and through a standard microscope fitted with adequate condensing lenses and filters. A special advantage of this technique is that unfixed materials can be studied, thus eliminating the artefacts which often arise due to fixation.

Conjugated planar dye molecules, in solution, form complexes having properties different from those in monodispersed form. The metachromatin dye polymer complexes allow observation of fluorescence and absorption spectra. One of the planar dye molecules is acridine orange, dissociating near pH 7. This dye, as a monomer at pH 6.0, has fluorescence and absorption peaks at 535 and 490 nm, and as a polymer the peaks are at 660 and 455 nm respectively. It is difficult to distinguish spectroscopically between acridine orange-DNA and acridine orange-RNA complexes. However, in fixed preparations, it may be possible to distinguish DNA and RNA on the basis of metachromasia of acridine orange, a red colour specifying RNA, and the yellowish-green colour indicating DNA. But under conditions *in vivo*, no such differentiation could be recorded. On the basis of the assumption that amino groups in the 3.6 positions of acridine react with phosphoric groups of nucleic acids, Lerman suggested internucleotide binding of planar dye molecules. Dyepolymer complexes and their optical properties have also been studied.

Several azo dyes become fluorescent after combining with tissues. Aromatic compounds in the dye control its fluorescent property. In other classes of dyes, ring closure, coplanarity of chromophores, accumulation of ring system and low dye concentrations exert an inhibiting influence. There are also other factors, which control the fluorescence of triphenyl methane, anthraquinone and quinone-imine dyes.

Quantitative measurements can be carried out by fluorescence microspectroscopy. Long exposure time for photographs should be avoided as the intensity may fade with continuous exciting irradiation of the fluorochrome-stained cell, in certain cases.

In preparing tissue sections meant for observation under the fluorescent microscope, chemical fixatives are avoided as far as practicable, freeze-dried or chilled preparations of unfixed materials being preferable. In cases where natural fluorescence is absent, the choice of proper and specific dyes, with fluorescent properties, is the most critical step in the procedure, as the dye imparts fluorescence to its specific substrate which can be detected *in situ*.

Frozen sections of unfixed tissue can be cut in a microtome and maintained in a refrigerated cabinet at –20°C. Freeze- drying, during which the temperature of the tissue should be maintained at about –40°C by means of a slush of diethyl oxalate, can also be employed.

The most adequate light source, emitting ultraviolet rays, is the carbon arc, preferably with a direct current and fitted with an electromagnetic field. Among other sources, high pressure mercury vapour arcs may be mentioned. Suitable condensing lenses of single bispheric type are generally fitted in front of the light source so that the black lens of the dark-field condenser gets the image of the horizontal carbon crater. In all fluorescence microscopes, barrier filters for cutting off undesirable wavelengths and excitor filters for excitation of fluorescence are used. The latter may be of wide and narrow band types. Barrier filters are chosen in appropriate combination with exciter filters. For photography, 35 mm films, with exposure time necessary for fast films, are usually employed; the negatives are generally thin and strong contrast printing is desirable.

The advantages of fluorescence microscopy were realised in attempts to detect specific antigens. In principle, the method involves the forced production of antibodies by the injection of antigens labelled with fluorescin isocyanate into the tissue. The conjugation of the gamma globulin of the antibody with fluorescein isocyanate in the antigen, makes visible the site of the antibodies within the tissue. For the study of the effect of chemicals on the nucleus and cytoplasm, fluorescence microscopy is often recommended. In the study of cellular nucleoproteins and nucleic acids, diaminoacridine dyes, such as acriflavine, acridine orange and acridine yellow have been employed because of their affinities. When applied to living cells, they react principally with nuclear DNA and RNA.

With acridine orange, RNA and DNA can be differentially stained, as the former appears red and the latter looks green under a fluorescence microscope.

Schedule for the Study of Chromosome Fluorescence Following Treatment with Anti-DNA Serum

Use as substrate, intact white cells from mixed short-term unsynchronised primary cultures, washed in globulin-free serum for preliminary experiments. Smear the cells on a slide, applying spirally with a glass rod. Fix the slide in cold acetone for 5 *s*. Wash in phosphate-buffered saline (pH 7.2, 0.15 M —'PBS') for 5 min at room temperature, and stain.

Prepare osmoticaly isolated interphase and metaphase nuclei of lymphocytes from partially synchronised phytohaemagglutinin- stimulated cultures at 45-50 h. Add required amount of distilled water to dilute cultures to 100 mM, determined for each batch of medium with the help of an advanced instrument freezing point osmometer. After 20 min, centrifuge at 500 r/min for 5 min and decant supernatant. Add 2 ml of pre-cooled acetic acid-methanol (1:3) mixture to the tube drop by drop, while gently agitating the cells in an iced water-bath. Keep overnight at 4°C. Change fixative to acetic acid-methanol mixture (1:2). Suspend by pipetting. Centrifuge and remove fixative, keeping only 0.2 ml. Add 0.2 ml of 50% acetic acid and resuspend the cells. Prepare air- dry smears by spreading a drop on a slide at 55°C and drying for 30 *s*. Wash and stain.

To one set of unstained slides, add deoxyribonuclease solution (0.1 mg/ml) at 37°C, in PBS containing 0.05 mg $MgCl_2$/ml. Treat controls in enzymefree solutions, wash all slides in PBS and stain.

Conjugate 'reticular' anti-DNA serum procured from microbiological laboratories at a fluor : protein ratio of 1 : 100. Remove the free dye by DEAE sephadex chromatography, using saline (buffered at pH 7.0 by 0.0175 M phosphate) as eluant. Concentrate to the original volume by dialysis against 25% pyrrolidone and store at 4°C with 1 : 10 000 merthiolate as preservative.

Treat slides, in petri dishes lined with damp filter paper, with serum for 1 h at 37°C. Wash in 500 ml PBS for 15 min with gentle stirring. In preliminary experiments, a two-layer technique with unlabelled LE serum and conjugated horse antihuman globulin serum (Sylvana, Millburn, B.J.) may be followed.

Mount cover slips with 10% PBS in glycerol. Illuminate by a Sylvania L50 mercury vapour lamp attached to a microscope with a dark-field condenser. Observe through apochromatic objectives, using UG2 (4 mm) exciter and No. 47 barrier filters. Photograph through BG12 exciter and No 47 barrier on Agfachrome 35 mm film with 3 and 6 min exposures.

From anti-DNA sera tests on osmotically isolated, and acetic- methanol fixed nuclei, obtained in cultures partially synchronised by overnight treatment at room temperature after 8 h initial incubation at 37°C, discrete attachment of the antibody is seen in interphase and metaphase nuclei. In the brightly stained metaphase chromosomes, the more frequent attachments are telomeric, or near secondary constriction. Pre- heating to 80°C for 5 min intensifies these patterns; treatment with deoxyribonuclease reduces them and brief pre- heating in an open flame destroys them (after Razavi 1968).

Methods have been developed for identifying Y chromosomes in buccal smears of human interphase nuclei as well as for differentiating specific sites in human lymphocyte chromosome preparations, using quinacrine (0.05% quinacrine mustard in deionised water for sperm, or 0.5% aqueous or 1% solution in absolute ethanol). Quinacrine mustard (250-300 μg/ml in glass distilled water) was also used for human chromosomes. In all these methods, except for sperm, the material after treatment with colchicine or colcemid, is fixed in acetic- methanol or ethanol, washed, rinsed in buffer (pH 4.1-5.5), stained for a few minutes and mounted in distilled water, buffer or buffered glycerol. For photography, Kodak-Tri X-Pan film with exposure time of 10-30 *s* has been recommended. The same principle is used to identify Y chromosomes in blood smears.

D. MICROSPECTROPHOTOMETRY

Under Ultraviolet Light

Ultraviolet light, in place of visible light, has the unique advantage of clasifying unstained living cells, due to the strong ultraviolet absorption by nucleoprotein (Caspersson, 1950). Moreover, it aids the quantitative estimation of the cell nucleoprotein owing to the characteristic absorption of purine and pyrimidine components of nucleic acid at 265 nm. A linear relationship between absorption and section thickness and the concentration of DNA has been clearly demonstrated in nuclei. Ultraviolet absorption spectra of cytological objects generally show absorption between 20-40 nm; nucleic acids at 260 nm, and proteins free from nucleic acids at about 280 nm.

The absorption of cellular components is principally attributed to covalent unsaturated groups such as C=O, C=N and N=O in organic compunds, i.e. purines and pyrimidines, indole groups of tryptophane, benzene and imidazole rings of tyrosine and histidine, respectively.

The principal difference between an ordinary light microscope and an ultraviolet microscope lies in the fact that in the latter, transparent fused quartz lenses are used in place of optical glasses, which are opaque to shorter ultraviolet wavelengths. The source of ultraviolet rays is generally the mercury vapour lamp. The slides and cover slips are made of quartz and in order to secure a monochromatic beam, a quartz monochromator is fixed between the source and the microscope. The photographic image is obtained by using a photographic plate and a photoelectric cell. Focusing is generally performed in visible light or on a fluorescent screen. Computation of the images can also be obtained. As illuminator, the commonly accepted method is to have the source through the exit slit of the monochromator. The aperture of the condenser is generally kept at about 0.3 or less. Discharge lamps, such as hydrogen and deuterium lamps and low pressure mercury lamps, with good achromatic objectives, have replaced the original rotating electrode resonant metallic arc. Such lamps give a wide band at 254 nm and can be used with suitable monochromators or interference filters.

Most equipments take advantage of Xenon compact arcs, which combine intensity with output through the ultraviolet range at 260 nm. In the monochromators meant for selecting the particular wavelengths, a band of energy is emitted whose wavelength distribution is controlled by the dispersion of the elements showing diffraction and refraction and the size of the slits. Several microspectrophotometers are equipped with grating monochromators where the change in wavelength is done by grating rotation. A number of interference filters have been developed, to serve as protective filters and as wavelength selectors. Photography may be adopted for the integration of absorption of objects and the negatives may further be scanned through densitometry. Photoelectric recording is utilized for a study of the series of absorption spectra needed for each wavelength. The most convenient method is to allow the light to pass through a selected area for final recording in a photomultiplier tube. The intensities may be recorded at different wavelengths and the background intensity measured by removing the object and allowing the light to pass through the empty space.

In principle, to measure changes in the quantity of nucleic acid and protein, the microscope is generally used as spectrophotometer. The monochromatic beam of ultraviolet light may be split into two beams, in the split beam device, one falling directly on the photoelectric cell (the blank) and the other passing through the ultraviolet microscope to another photoelectric cell. The sample to be measured is placed in the path of the beam passing through the microscope. The light passing through the material is reduced in intensity; this is calculated by counting the difference in the photoelectric current yielded by the two beams, as indicated by a galvanometer. The data can also be electronically computed and televised.

The measurement of absorption is based principally on Lambert- Beer's law which may be stated as follows:

$$I_X = I_O \times 10^{-kcd}$$

where I_X is the changed intensity of the beam of I_O; I_O the incident intensity, after the ray passes d : d the thickness (in cm) of c; c the concentration (in g/100 ml) of the absorbing molecules; and K the extinction coefficient. As in cytomicrospectrophotometry, relative amounts are obtained, the constant K is ignored, since absolute values are not necessary. The value of K, when necessary, can be worked out from biochemical data.

The values of I_X and I_O can be obtained without changing the position of the material. With the aid of these values, the percentage of transmission (T) can be worked out (I_x/I_o) and the presence of the components per arbitrary units, showing ultraviolet absorption, can be computed. A limitation of ultraviolet microscopy is that the ray may have some deleterious effect on the absorbing material, but in the above method the period of exposure to ultraviolet is very much reduced.

The fixing fluids for objects meant for spectrophotometric analysis should not affect the light scattering properties of the specimen and must not contain compounds which undergo deposition under ultraviolet rays. For the study of DNA, ethanol-acetone mixture or acid or *neutral* formalin fixation are suitable. In the case of RNA, however, most of the aqueous fixatives cause extraction of soluble RNA. Freeze substitution method with ethanol-potassium acetate has been shown to retain soluble RNA. Nucleoproteins can best be studied in cultured cells either by ethanol-freezing substitution or by chemical fixation with ethanol-acetone (1 : 1) at 4°C for 24 h.

For a correct assessment of the absorbance data, it is desirable to follow the extraction of specific cellular components simultaneously. Such procedures, in addition to aiding identification of the absorption of particular chemical constituents, may also serve as controls. Further, the most significant use of extraction is to secure a ***blank***, so that non-specific light loss and light scatter can be corrected. These purposes are served through digestion with proteases or nucleases. In mounting the specimens in ultraviolet microspectrophotometric work, media like pure glycerine, glycerine-water mixture, 45% zinc chloride and paraffin oil possess the essential prerequisites for checking the non-specific light loss.

During the application of the ultraviolet rays, caution is recommended, since continuous exposure for even 10 min at 257 nm in a Köhler microscope may result in the loss of absorption capacity by the chromosomes treated with acidic fixatives. Techniques for the ultraviolet microscopy of cells in culture have been developed to study conditions and changes *in vivo*, for which perfusion chambers with quartz cover slip windows are generally used. Modifications of Eagle's medium have been used for cell culture, which have to be replaced later by a salt solution transparent to ultraviolet rays. In the scanning procedure, improvements have been devised in which large suspended cell populations are quickly analysed with the rapid cell spectrophotometer.

Photometric observations through ultraviolet rays have been extended to chromosomes, or chromosome segments, extracted out of the cell with the micromanipulator. Micrurgical extraction of polytene chromosome has been dealt with in the chapter on micrurgical techniques. Extraction of nucleic acid through nucleases and the analysis of a drop of the extract has been made through microspectrophotometry and methods have been developed for the determination of base constituents in nucleic acid, extracted from incised segments of chromosomes. The extract is applied to a treated cellular fibre, through which the discharge of electric current results in separation of different ultraviolet-absorbing bands. The photomicrographs of these bands, following densitometric analysis, yield the quantitative values of the substances.

For the identification and quantitation of constituents in the chemical make-up of chromosomes, reliance on microspectrophotometry is continuously increasing.

Under Visible Light

Microdensitometry or cytophotometry in visible light is based on the principle that between 400-700 nm, light is partially absorbed by the matter due to interaction with outer constellation of electrons. This interaction mainly depends on the physical and chemical nature of the matter and the wavelength of the light. The process follows Bougner-Beer law which holds that the absorbance is dependent on the concentration of the matter and the pathlength. It is expressed as logarithms of the reciprocal of transmitted light after absorption. It is in fact the optical density (OD) or extinction (E) i.e. $\log\frac{I_o}{I_s}$ = OD = E = A. As the absorbance is dependent on the concentration of the absorbing matter, the technique has been adopted for quantitative estimation of cell *vis-a-vis* chromosome constituents. Absorption measurements are carried out on materials stained with Feulgen solution, methyl green pyronin, azure B, Millon dye and such other compounds capable of staining specific cell constituents.

For the fixation of materials, Carnoy's fluid or 10% neutral formalin is widely used to measure chromosomal DNA.

Staining may be carried out in Feulgen solution or in pyronin- methyl green mixture or in any other specific dye for binding with nucleic acids or proteins.

Microscope and accessories

The appliances needed for visible light photometric or densitometric work are quite simple. For example, in Leitz MPV, they include, in principle, a strong tungsten light source; monochromatic unit, fitted with condensing lens; a microscope (Aristophot) having condenser with low numerical aperture; phototube, measuring variable field diaphragm, a photomultiplier, a power supply unit and galvanometer. Alternate arrangements are available for observing specimens in ordinary light (without monochromator) as well as for photography. Similarly, through suitable accessories, the data can be recorded by a recorder or oscilloscope, instead of galvanometer. In Reichert Zetopan photometer, interference filters are used between tube and photocell over the microscope, instead of in monochromator. The guidelines for use are available with each model.

Method of analysis

As in ultraviolet microspectrophotometry, in taking measurements, the data required are the intensity of the background light (I_0) which is taken from a blank area on the slide, and the reduced intensity of light after absorption by the objective (I_s), the *transmission* being calculated as I_s/I_0. The transmission indicates the fraction of light that remains after loss by absorption. In a completely transparent object, this value is I and there is a logarithmic decrease in transmission with increase in absorbing molecule, due to concentration or thickness. Therefore extinction or absorbance is:

$$E = \log_{10}\frac{1}{T} = \log_{10}\frac{I_0}{I_s}$$

The above law is fundamental for all cytophotometric calculation. The law, developed for dilute solutions, holds equally well for all cytophotometric, i.e, densitometric observations if reactions are carried out under strictly controlled conditions (Berlyn and Miksche, 1976). From the optical density or absorbance data, the mass of material can be measured on the basis of the fact that what is obtained in cytophotometry is transmittance (T). In conversion to mass (M), the following equation is adopted: $M = \frac{A \log 1/T}{K}$ where A is the area (πr^2, of which π may be omitted) and K, the extinction coefficient. The K is a function of the wave length and in relative mass determination, k being constant, this value is not needed in calculation. This method holds good for homogeneous samples such as interphase. But in densitometry, the spatial heterogeneity of the specimen presents distribution error which is quite common to nonuniform microscope materials such as the distribution of chromosomes. More precisely, it is due to unequal distribution or heterogeneity of DNA in the nucleus. In such cases, the total absorbance of a projected area is dependent on the total amount of such material present in the projected area. As such it should be the sum total of the absorbance of individual units within the area. To some extent, the distribution error is corrected by measuring the transmission at two different wavelengths(Patau, 1952), as outlined below:

The prerequisite of two-wavelength technique of absorption analysis is the uniform illumination of the area with a monochromatic source and absence of light scatter through the use of proper mounting medium. In Feulgen- stained sections, it is desirable to carry out the analysis on the same slide, since acid hydrolysis- an essential feature of Feulgen staining-affects the reaction to a significnt extent. The method of analysis is as follows (Pollister, Swift and Rasch, 1969):

If the two wavelengths selected are λ_1 and λ_2 the extinction (E_1) at the former should be half the extinction (E_2) at the latter, so that $E_2 = 2E_1$. As mentioned above, $E_1 = \log\frac{I_0}{I_s}$ at λ_1, and $E_2 = \log\frac{I_0}{I_s}$ at λ_2. Choice of the proper wavelengths is important. The total amount of absorbing material (M) in a measured area (A) is, $M = KAL_1D$ where K is constant; ($K = \frac{1}{e_1}$, e being the extinction coefficient in λ_1, K may be disregarded for relative determination in cell microspectrophotometry). L_i is respective light loss and D is the correction factor for distributional error. Area (A) is measured as πr^2 where π may be omitted. From transmission T_1 and T_2, at λ_1 and λ_2 the degree of light loss may be worked out as follows:

$L_1 = 1 - T_1$ and $L_2 = 1 - T_2$

With the ratio L_2/L_1 at hand, the value of D can be worked out from Table 9.2 given by Garcia (1962), with detailed principles. The following table taken from Pollister, Swift and Rasch (1969) gives the value of D corresponding to each L_2/L_1 ratio:

The best method to avoid distributional error is to take integrated aborbance. As the mean of the logarithms is not equivalent to the logarithms of the mean it is always desirable to scan absorbance from each individual point and to integrate the data.

Table 1 — Values of D for different values of L_2/L_1

L_2/L_1	0.00	0.01	0.02	0.03	0.04	0.05	0.06	0.07	0.08	0.09	
1.0	—	4.033	3.461	3.134	2.907	2.734	2.595	2.479	2.380	2.294	1.0
1.1	2.218	2.150	2.089	2.033	1.982	1.935	1.892	1.851	1.813	1.777	1.1
1.2	1.744	1.712	1.683	1.655	1.628	1.602	1.578	1.555	1.533	1.511	1.2
1.3	1.491	1.471	1.453	1.435	1.418	1.400	1.384	1.368	1.353	1.339	1.3
1.4	1.324	1.310	1.297	1.284	1.271	1.259	1.247	1.235	1.224	1.213	1.4
1.5	1.202	1.191	1.181	1.171	1.162	1.152	1.143	1.133	1.124	1.116	1.5
1.6	1.107	1.098	1.091	1.083	1.075	1.067	1.059	1.053	1.045	1.038	1.6
1.7	1.031	1.024	1.017	1.011	1.004	0.998	0.991	0.985	0.979	0.973	1.7
1.8	0.968	0.962	0.956	0.950	0.945	0.940	0.934	0.928	0.923	0.918	1.8
1.9	0.914	0.909	0.903	0.899	0.894	0.890	0.884	0.880	0.876	0.871	1.9
2.0	0.867										

Different types of microdensitometers are available, designed with the same objective but they differ with regard to their mechanism of operation. The difference mainly lies in the system of scanning. They can be classified under two categories: (a) in which the object is stationary and the scanning is done along the whole length of the object and in the other, (b) the object is moved with a motor-driven stage and the measuring area is fixed. The moving object densitometers are manufactured by Zeiss, Leitz, etc. while in the instruments of Vickers, Barr and Stroud, the specimens to be scanned remain stationary. In the Barr and Stroud type, the image of the object is enlarged and scanned by a mechanical device, whereas in the Vickers model the fixed object is scanned by moving the reduced image of an illuminated aperture with the aid of a pair of oscillating mirrors. In more sophisticated televised models, a television camera receives the image which is electronically scanned. The Vickers models are quite flexible in several respects and have the advantage of utilization for dry mass scanning through micro-interferometry with suitable adjustments. Gradual refinements in models have also resulted in high resolution scanning densitometry from photographic negatives of individual chromosomes.

Use in fluorometry

The principle of microscope photometry can be advantageously employed for fluorescence quantitation as well. The energy of fluorescent light being much lower than that of ordinary transmitted light, it is always desirable to use incident illumination(through suitable accessories), the illuminating lens being the objective itself. A xenon arc lamp may serve as the source of light. In order to secure large optical flux, it is desirable to use a fluorite or apochromat objective with a high numerical aperture.Automatic switchover device from low power transmitted light to fluorescence light for measurement is available.The illuminated field should be large but not exceed the area to be measured as illumination of other objects may cause change in fluorescence of the specimen to be measured. The immersion medium should also be free from any fluorescence effect.

For fluorescence analysis of chromosomes, it is necessary to get an integrated value of the intensity of the whole fluorescence spectrum. The transmission is obtained also from the barrier filter and the photocathode is so chosen as to be responsive to the entire spectrum. If necessary, spectral distribution of fluorescence can be measured by attaching a continuous spectrum monochromatizing device, and a suitable photocathode with a wide range of response.

Methods are also available for high speed quantitative karyotyping by flow microfluorometry. The technique involves isolation of metaphase chromosomes from cells, staining with a DNA specific fluorochrome and measuring for stain content at the rate of 10^5/min in a flow microfluorometer. The results tally well with scanning cytophotometry.

E. FLOW CYTOMETRY

Flow cytometry is an elegant and rapid method for sorting and quantifying chromosomal data at the microscopic level. It normally involves, at the chromosomal level, the isolation or preparation of chromosome suspensions, staining with DNA specific fluorescent dye and finally analysis through flow cytometer, such as Facstar plus Cytometer or Slit Scan Flow Cytometer (SSFCM). As approximately 10^6 chromosomes can be analysed per second, statistically large number of chromosomes can be processed within a few minutes. In SSFCM, the chromosomes are flown though a 1.3 µm thick laser beam (width - $1/e^2$ points) produced by an argon ion laser emitting 1.0 W at 488 µm. Fluorescence intensity detected along its length is detected photometrically and digitized every 20 *ns* by means of a waveform recorder. The time varying fluorescence intensity is recorded as a measure of the distribution of the fluorescent dye along the length of the chromosomes. The profile analysis also provides data on chromosome aberrations since normal chromosomes produce bimodal profiles whereas dicentrics yield trimodel profiles (Lucas *et al*, 1991).

Protocol

Use cultured human lymphoblastoid cells maintained in suspension at 37°C in upright T-75 flasks in 50ml of minimum essential L-medium containing 20% fetal calf serum. Collect cells in mitosis after suitable interval following treatment with 0.1 μg/ml colcemid. Spin down the cells and resuspend in 0.4 ml hypotonic KCl (0.075 M), and incubate at 20°C for 10 min. Add 0.2 ml isolation buffer (0.075 M KCl, 0.05 mg/ml propidium iodide and 1% Triton x -100). Incubate the cells for additional 3 min at 20°C. Release the chromosome from cells by syringing twice through 22 mm gauge needle, spin down, and resuspend in 0.5 KCl to enlarge the chromosomes. Analyse through a Slit Scan Flow Cytometer (SSFCM).

In plant system, such as in *Triticum monococcum*, chromosomes may be isolated from the cell lines following enzymic digestion of the cell wall and polyamine isolation buffer. The suspension can be stained with Hoechst 33258 and chromomycin A_3 and analysed on a Facstar plus Cytometer (Leitch et al,1991).

F. CHROMOSOME IMAGE ANALYSIS

The refinement of methodologies has been responsible for automatic karyotyping and microelectronic analysis of image of human chromosomes. The plant system, because of the tough cell wall and high degree of variation, presents difficulties in this respect. Lately methods have been developed for Chromosome Image Analysis in plants which permit quantification of density profile in chromosome architecture, thus bringing out even cryptic differences. The analysis of image through digital means is involved. The essential requirements are well macerated air dried preparations with good prometaphase chromosome spreads, proper staining and underexposed, overexposed and normally exposed photographs for analysis. The contour lines are extracted from the overexposed photographs and then superimposed on the underexposed ones. The density profile of each chromatid is measured, and the condensation pattern is studied under three different exposure conditions. Image analysis compensates for the unclear staining among the plates as well as between chromosomes within one plate.

Equipment and protocol

The image analysis system involves a fluoresence microscope with autoscanning stage and autofocussing unit. A photometer and a filter system are also attached. A high resolution TV camera is directly mounted on the top of the microscope. Images on the photographic paper or film can be taken through a zoom lens and a close up lens. Image analysis units are selected as the image processing component of the system. For output of data and images, a printer and a color image recorder are chosen. A videotape recorder can also be selected as additional input (Zeiss Oberkochen- Photomicroscope III).

Pretreat roots with .05% colchicine at 18°C for 2 h. Fix in methanol-acetic acid 3 : 1 for one h to overnight. Macerate on a slide in a moist chamber using a mixture (0.3% pectolyase Y-23, 2% cellulase Onazuka RS, 1% macerozyme-R-200, 1 mM EDTA -pH 4.2) for 15 min at 37°. Keep the tips in 0.2% Triton X-100 for 15 min. Break the samples into five pieces with forceps in fixative. Air dry. Stain in Wright's stain (5% phosphate buffer, pH 6.8).

Alternatively: wash in distilled for 15 min and macerate at 37° for 50 min in enzyme mixture (4% cellulase Onazuka RS, Yakult Honshu Co Tokyo, 1% pectolyase Y-23, Seishin Pharmaceutical Co. Tokyo, pH 4.2). After thorough washing, add a few drops of fixative, chop up the pieces, air dry, stain in 3% Giemsa (pH 6.8) 40 min and dehydrate in 100% methanol for more than 10 min.

Photograph images of good prometaphase chromosomes and analyse by the CHIAS (Fukui, 1986). The density profile along the midrib of each chromatid (CP) is measured. Compare the CP of pretreated and not pretreated chromosomes.

G. CONFOCAL MICROSCOPY

Confocal Scannning Optical Microscopy (CSOM) is one of the outstanding advances through which three dimensional images can be reconstructed. It permits non-invasive high resolution cellular images through in focus three dimensional system, virtually free from out-of-focus limitation. A set of optical sections at different focal level series gives ultimately a three dimensional image which can be processed for digital display as computer reconstruction, a stereo pair or animation sequences.

The alternative to video scanning is laser scanning. In this system, a focussed laser source of 0.1 - 0.5 micrometer is passed across the subject, building up an image. The laser scanning microscopy is comparatively slower but resolution is higher than dynamic video imaging. Confocal microscopy, combined with laser scanning and immunofluorescent staining, has opened up new dimensions of research. Confocal images are optical sections through X - Y plane of the object. The light from above and below the plane-in-focus is prevented from forming images through a pinhole diaphragm. As such the contrast and resolution are very high. At present, Zeiss, Leica and other manufacturers are supplying laser scanning confocals.

CHAPTER 3

STUDY OF BANDING PATTERNS OF CHROMOSOMES

Differential banding patterns of chromosomes, usually observed at specific regions on particular levels, were initially developed for the analysis of human chromosome segments. These bands are made visible through low and high intensity regions under the fluorescence microscope or as differentially stained areas under the light microscope. The methods were then extended first to different animals and later to plant chromosomes.

The protocol for molecular hybridization, that is, denaturation at the cytological level, if followed by renaturation and staining with different dyes, particularly Giemsa, gives intensely positive reaction at similar segments of chromosomes which otherwise show repetitive DNA. Obviously such treatment is capable of revealing repetitive segments in chromosomes. This banding, following denaturation-renaturation and Giemsa staining, is termed *C*-banding.

Earlier, Caspersson and his colleagues had recorded differential fluorescence of different chromosome segments following staining with various fluorochromes and observation under the ultraviolet microscope and had successfully employed it in formulating a banding pattern analysis of the human karyotype. Such bands were referred to as *Q*-bands. Other fluorochromes produce banding patterns as well, eg. Hoechst 33258 similar to *Q* and ethidium bromide the reverse.

Saline treatment, followed by Giemsa staining, gave a set of patterns, similar in relative staining intensities to the *Q*-bands. Such bands-known as *G*-bands, did not require any denaturation-renaturation. Since then *G*-banding has been obtained following treatments in a variety of chemicals and even mere heating.

The *R* banding or reverse banding, where the pattern is opposite to that of *Q* and *G* bands, was obtained following controlled heating. Acid extraction has been employed for *N*-banding for the nucleolar-organising region. A *CT*-banding procedure produces bands in the telomeric and centromeric regions alone. Orcein - staining after incubation in XSSC results in *O*-banding in plants which is similar in general to the *G*-band pattern.

Various causes have been ascribed to the occurrence of the chromosome bands. Of them, four factors appear to be of particular importance, namely the occurrence of repetitive DNA, and differences in the base composition of DNA, in the protein components and in the degree of packing of DNA or DNP-complex.

The different banding patterns have been employed in locating marker segments of chromosomes, particularly those with repetitive DNA sequences, in plant and animal systems. They were, with the exception of *O* and *Hy*-banding, developed initially for human chromosomes and have been successful in establishing standard human karyotypes and in ascribing certain syndromes to changes in bands on specific chromosomes. The techniques were later extended first to other mammalian chromosomes and then to other animal systems, with modifications. The chemical principle underlying *Q*-banding has been well worked out. With quinacrine mustard (QM), the fluorescent amino-acridine nucleus becomes intercalated within the double helix of DNA. The basic nitrogen atoms form ionic bonds with DNA phosphate and the alkylating side group binds covalently with guanine from DNA. Quinacrine dihydrochloride or atebrin does not form a covalent link due to the absence of alkylating side group. Thus the primary *Q* binding with both compounds has been suggested to be through intercalation of the acridine nucleus in the double helix.

FLUORESCENT BANDING OR *Q*-BANDING

Quinacrine mustard binds to chromatin by intercalation of the three planar rings with the large group at position 9 lying in the small groove of DNA. Most pale staining regions are caused by a decreased binding of *Q*, predominantly due to non-histone proteins. DNA-base composition influences the fluorescence by correlation with G-C bonding at lower *Q*:DNA ratios and by conversion of dyes bound near G-C bases into energy sinks at higher ratios. Chromosomal proteins possibly have a much less pronounced effect.

With Quinacrine Mustard (QM) or Quinacrine Dihydrochloride (Q)

A: Schedule for mammalian chromosomes: The "Caspersson method"

Transfer air-dried chromosome preparations (fixed in acetic acid-methanol, 1:3) to absolute ethanol and bring down to water. Soak in McIlvaine's or Sörensen's phosphate buffer (pH 4.1 to 7.0) and stain in QM (50 μg/ml) solution for 20-30 min at 20°C. Wash in three changes and mount in buffer of same pH. Observe under phase contrast to mark out suitable metaphase plates, transfer to ultraviolet microscope and photograph immediately since the fluorescence fades progressively under ultraviolet light.

B: Q-banding using quinacrine dihydrochloride

Immerse airdried slides in distilled water adjusted to pH 4.5 (with HCl) for 10 sec. Stain in 0.5% quinacrine dihydrochloride solution in distilled water, pH 4.5, for 15 min. Rinse in three changes of acidulated water (pH 4.5) for a total period of 10 min. Mount in water (pH 4.5), cover with a coverslip and seal. Observed under a fluorescence microscope as given in A, above.

Modifications

(1) The ethanol grades for progressive hydration of the slides and also the staining solution may be prepared by using the buffer instead of distilled water.

(2) In an alternative schedule keep air-dried slides for 5 min in 0.005% QM solution in glass distilled water. Wash in distilled water for 25 *s* three or four times, mount and observe in distilled water.
(3) Vinblastine sulphate (0.0075-0.015 μg/ml) for 1 h may be used as a pre-treating chemical instead of colcemid for malignant cells. Stain in 0.5% atabrine (Q) solution for 5 to 10 min, rinse (2 min) in three changes of isotonic salt solution and mount in the same solution.
(4) To analyse the human Y chromosome stain air-dried slides in a solution of 5 mg QM in 100 ml phosphate buffer (pH 7.0) for 30 min, rinse and mount in the same medium.
(5) Stain HeLa cell preparations in 2% Q solution containing 0.1% thymol, wash in running water, differentiate and mount in McIlvaine's buffer. This method is used extensively in various primates, rhinoceros, amphibian embryos and in assessing human newborn for chromosomal disorders.

Schedule for plant chromosomes

Pre-treat root tips in aqueous 0.05% colchicine, fix in acetic- ethanol (1:3) overnight, squash directly in 45% acetic acid and stain in 0.5% aqueous quinacrine dihydrochloride for 10-15 min. Rinse in water and mount in distilled water. Alternative stains are H33258 (0.02%) and ethidium bromide.

With Hoechst 33258

Hoechst 33258 (2-[2-(4-hydroxyphenyl)-6-benzimidazolyl]-6- (1-methyl-4-piperazyl)-benzimidazole), shows enhanced fluorescence with both AT and GC-rich DNA. It can be used as a probe for identifying all types of AT-rich regions in chromosomes, including those which are not demostrable with quinacrine. Its interaction with DNA and chromatin is characterized by changes in absorption and circular dichroism measurements. It may be used as a 2×10^{-6} M solution in 0.1 XSSC in mammals and has been employed in the study of spontaneous and induced chromosome aberrations and in mouse-human heterokaryon analysis as an alternative to autoradiography.

Schedule 1: Hoechst fluorescence by heating in plants

Fix root tips, after pre-treatment in colchicine (4 h), in acetic-ethanol (1:3). Treat 1-day old slides in barium hydroxide/saline mixture and stain in H33258 (40 μg/ml) in McIlvaine's buffer, pH 4.1, for 10 *s*. Rinse and mount slides in same buffer and seal. Heat the sealed slides for about 6 *s* on a hot plate (120 °C) and cool rapidly by placing them, cover slip down, between two slabs of dry ice. A combination with Giemsa banding has been employed in rye.

Schedule 2: In Drosophila chromosomes

Dissect out ganglia in Ringer's, incubate 45 min at 24°C in Schneider's cell culture medium with 0.05 μg/ml colcemid. Keep in 1% trisodium citrate for 10 min, 0.5% trisodium citrate for 5 min, fix in three changes of acetic-methanol (1:3) for

30 min. Warm to 50°C in a drop of 60% acetic acid. Add a drop of (1:3) fixative to warm slide.

Rinse slides for 10 min in PBS, stain 10 min with a 1/100 dilution of stock 33258 Hoechst in PBS, rinse 10 min in PBS, rinse 4 min in water, drain and store for less than 12 h. Mount in McIlvaine's buffer pH 4.0 and observe in fluorescence microscope.

Reverse Fluorescent Banding with Chromomycin and DAPI

Two DNA binding guanine specific antibiotics, chromomycin A_3 (CMA) and closely related mithramycin (MM) are used as fluorescent dyes. Metaphase chromosomes in roots can be sequentially stained with CMA or MM and the DNA-binding AT- specific fluorochrome 4′-6-diamidino-2-phenylindole (DAPI). Non-fluorescent counterstain may be used-methyl green with CMA and actinomycin D (AMD) with DAPI. In general, C-bands which are bright with CMA and MM, are pale with DAPI and *vice versa*, Human metaphase chromosomes show a small longitudinal differentiation in CMA fluorescence, which is reverse that of AMD/DAPI double staining banding but of a lower contrast. CMA- banding appears to resemble *R*-bands. The anthracycline antibiotics daunomycin and adriamycin produce bands on human chromosomes. The fact that both AT and GC specific fluorochromes are available and that sequential fluorescent staining of chromosomes (Holmquist, 1975) and counterstaining with non-flurescent DNA ligands is feasible, means that suitable combinations may facilitate analysis of molecular mechanism.

Schedule

For plants treat root tips with colchicine (0.05%, 3-6 h), fix in acetic- ethanol (1:3, overnight), squash in 45% acetic acid and air-dry after removing coverslip by dry ice. Air-dried human or mammalian chromosome preparations are used.

For CMA staining

Pre-incubate slides in McIlvaine citric acid-Na_2HPO_4 buffer (pH 6.9-7.0), containing 10 mM $MgCl_2$ for 10-15 min. Stain slides in buffer containing 10 mM $MgCl_2$ and 0.12 mg/ml CMA ('reinst', Serva, Heidelberg) or 0.11 mg/ml MM ('rein', Serva) for 5-10 min, wash and mount in McIlvaine's buffer (pH 6.9-7.0) and seal with rubber solution.

For counterstaining, pre-incubate slides in McIlvaine's buffer (pH 4.9) for 10-25 min, stain for 5-15 min in a buffered solution (pH 4.9) of 0.1% methyl green GA (Chroma, Stuttgart), from which methyl violet has been removed with chloroform extraction. Rinse in buffer (pH 4.9) and in neutral buffer containing 10 mM $MgCl_2$, stain in CMA as described before.

For sequential staining with CMA and DAPI

Observe and photograph CMA or MM fluorescence as for *Q*-banding. Remove coverslip by dry ice or rinsing in buffer and then acetic-methanol and methanol successively. Remove chromosome-bound CMA or MM extracting with pyridine for 2-3 days. Stain with DAPI solution (1 μg/ml in McIlvaine's buffer, pH 7.0).

For AMD+DAPI double staining

Dissolve AMD ('rein', Serva) in small amount of methanol and make up to 0.25 mg/ml with McIlvaine's buffer (pH 6.9-7.0). Pre- incubate human chromosome slides in buffer for 5 min and stain with AMD for 15-20 min, rinse with McIlvaine buffer (pH 6.9-7.0). Stain in DAPI at concentrations between 0.1 to 0.4 μg/ml (McIlvaine's buffer, pH 7.0) for 5-10 min. CMA and MM fluorescences fade rapidly and can be slightly stabilized by ageing the preparation and the solution. DAPI fluorescence is observed with excitor filters UG 1, BG 3 and TK 300 (dichroic mirror of Leitz Ploem epi-illuminator) and barrier filters K-400 of the epi-illuminator and K-470, K-490 or K-510.

For sequential fluorochrome staining with CMA and Hoechst

Immerse slides in freshly prepared 0.02% Hoechst 33258 (Sigma) in ethanol for 3 min; drain and transfer to ethanol for 2 min, air-dry. Place about 30-50μl of a 0.5 mg/ml solution of chromomycin A3 (Sigma) in McIlvaine's buffer (pH7.0) plus 5mM $MgCl_2$ over each preparation and cover with a coverslip. Incubate the slides for 1h in a moist box at room temperature in dark. Rinse off coverslips with distilled water and briefly airdry. Mount each slide in a small droplet of 50% Citifluor PBS (Agar Scientific) in McIlvaine's buffer plus $MgCl_2$, pH 7.0.

Observe chromosome fluorescence in a Zeiss Axiophot microscope with mercury vapour lamp. Use Zeiss filter set 1 (excitation 365 nm; emission 395 - 297 nm) for *Hoechst fluorescence* and Zeiss filter set 6 (excitation 436/8nm; emission 460-470 nm) for *chromomycin staining*.

Bands that are bright with Hoechst will be dim with CMA and *vice versa*. All positive G bands could be differentiated into two types with these two stains, indicating the presence of AT-rich or GC-rich heterochromatin.

For Distamycin A/DAPI staining in human male meiosis

Incubate meiotic chromosome preparations from seminiferous tubules (human) in distamycin A solution (10-50μg/ml) for 10 min and then in DAPI solution (0.3μg/ml) for 10 min. Rinse in McIlvaine's buffer. Mount in saturated saccharose solution. Observe in Zeiss fluorescence microscope with HBO 50 W/AC lamp and 2FL exciting filter. Always prepare DA solution in distilled water freshly before use since it is not stable. DAPI stock solution (1:50) in distilled water or McIlvaine's buffer can be stored at 4 °C for several months.

DAPI is AT-specific fluorescent dye while DA is AT- specific non-flourescent counterstain. DA/DAPI staining has been used widely in mammalian systems and non-mammalian vertebrate species using peripheral blood cultures to obtain mitotic plates.

For Distamycin (DA)/DAPI staining in human mitotic cells

Incubate mitotic chromosome preparations from human leucocyte cultures in DA (50μg/ml) in McIlvaine's buffer, pH 7.0, for 20 min at room temperature. Wash in buffer and stain with DAPI (1 μg/ml in McIlvaine's buffer, pH 7.0) for 20 min. Rinse in buffer, mount in mounting medium with glycerol and McIlvaine's buffer (1:1). Observe with BP 365/FT 395/LP 397 (Zeiss) filter combination.

D-bands with antibiotics

Of the anthracycline group of antibiotics, daunomycin (Cerubidine-HCl) and adriamycin (Doxorubicin-HCl) give well defined and reproducible orange-red flourescent banding patterns on human chromosomes at concentrations of 0.5 mg and 0.2 mg respectively per ml in 0.1 M sodium phosphate buffer (pH 4.3) and the staining period 15 min. The schedule and pattern are similar to Q-bands. The G-C specific DNA binding antibiotic-olivomycin produces reverse fluorescence banding patterns in human, bovine and mouse chromosomes.

Ethidium Bromide as Counterstain for Quinacrine

Fix cellular smears, in case of plants in 95% ethanol. Use air- dried preparations for animal materials. Keep in 95 and 70% ethanol for 3 min each; distilled water, 3 min; citric acid phosphate buffer (0.01 M, pH 5.6), 3 min staining solution (50μg/ml QM in citric-phosphate buffer, 0.01 M, pH 5.6, containing 0.5% zinc sulphate), 10 min; two changes of 6 min each in pH 5.6 buffer; 5% zinc sulphate solution, 5 min; two changes in distilled water of 4 min each; ethidium bromide solution (2μg/ml EB in 7.0 pH phosphate buffer, 0.01 M), 5 min; 7.0 pH phosphate buffer (0.01 M), two changes of 2 min each; 7.6 pH buffer (0.01 M) two changes of 4 min each, mount in same buffer and seal. With this counterstain, cytoplasm fluoresces pale green, nuclei pale orange and interphase fluorescent bodies appear as bright yellow spots, as also the brightly fluorescing bands in metaphase chromosomes.

The mechanism of the counterstaining by EB is not yet clear. Presumably the brilliant areas contain AT-rich DNA binding firmly with QM, and EB removes, replaces or obscures the quinacrine in weakly stained areas, so that densely stained chromatin, by contrast, seems more prominent.

Banding with Acridine Orange (AO)

Acridine orange reverse banding (RFA) is more useful than *Q*-banding for characterizing variations in certain acrocentric human chromosomes but does not give any results consistent with *Q*-banding. Terminal bands or *T*- bands were obtained by controlled denaturation of chromosomes at 87°C using Giemsa and AO in diluted buffer.

Other chemicals observed to give quinacrine-like fluorescence include alcoholic extracts of the alkaloids from fresh roots of eight genera of Papaveraceae and Fumariaceae as also from *Chelidonicum majus, Macleaya cordata and*

Glacium flavum. Of limited use are sarcolysinoacridin, berberine sulphate, and 2,7-di-*t*-butyl proflavine DBP.

GIEMSA OR *G*-BANDING

Giemsa is a complex mixture of thiazine dyes and eosin. Of them, methylene blue and Azure A, B and C alone give good banding. Thionin, with no methyl groups, gives poor banding while eosin has no effect.

The surface topography of the human chromosomes at each stage of the *ASG* banding, observed with electron microscopy, shows that chromosomes collapse with 2 XSSC treatment at 60°C for 1 h and, on subsequent staining with Giemsa, fibrillar appearance and transverse ridging of chromatids are observed. The banded patterns under light microscope may be due to the ridges absorbing more dye than other parts of the chromosome. With trypsin and QM, the collapse occurs before staining and with Q, it occurs during washing after staining. The collapse is pH- dependent. The band and interband regions may reflect the concentration or compaction and arrangement of nucleoprotein fibres (Bahr, 1973; Takayama, 1976). On the other hand, Khachaturov *et al.* (1975) suggest that differential staining due to Q, QM and G is due to their binding with various acid mucopolysaccharides. Strong banding is favoured by presence in stain of high concentrations of methylene violet and its immediate homologues. Chromatography shows that only small amounts of these dyes occur in dry Leishman and Giemsa powders. However, additional active dye may be generated during preparing solution from dry powder. Heated Giemsa solution produces more consistent bands on mammalian chromosomes.

According to Matsui and Sasaki (1975) during *G*-banding, macromolecules like DNA and proteins are lost, leading to an uneven distribution of chromatin. The amount of macromolecules residual after *G*-banding approximates 2 morphological bands. Non-histone proteins of relatively larger molecular sizes are removed by banding methods. The residual proteins in the *G* bands are relatively small molecules containing a large number of *S-S* bands and possibly contain more stabilized DNA. Thus the *G*-positive bands represent relatively thermostable chromatin consisting of smaller non-histone protein molecules.

G-banding in Plants

Schedule 1a for root tip: Grow roots at 10°C; pre-treat in 0.05% colchicine in the dark for 4 h at room temperature, fix in acetic-ethanol (1:3) for 48 h and store in 70% ethanol at 4°C.

Hydrolyze root tips for 8 min in 1 N HCl at room temperature, squash in 45% acetic acid, remove cover slips by dry ice method, dip in absolute ethanol, air-dry and store in desiccator for 5 days. Treat with aqueous 0.064 M $Ba(OH)_2 8H_2O$ for 50 min at room temperature, wash in two changes of deionized water for 5 min, incubate in 2 XSSC (pH 7.0) at 60°C for 40 min, rinse in two changes of deionized water and air-dry. Stain in fresh Giemsa stock solution, diluted 50 times with M/15 Sörensen phosphate buffer, pH 6.8 for 5-8 min at room temperature, rinse in deionized water and mount in Permount.

Schedule 1b: Pre-treat root tips with colchicine (0.1% for 1 to 2 h or 0.05% for 2 to $3\frac{1}{2}$ h), fix in acetic-ethanol (1:3) for 3 h, store overnight in 90% ethanol; squash in 45% acetic acid on slides 'subbed' with Haupt's adhesive. Remove cover slips with dry ice, rinse in ethanol, air-dry and incubate in hot $Ba(OH)_2$, varying between 15 min at 50°C in 5 % $Ba(OH)_2$ to 20 min at 60°C in 6.5 %. Rinse in distilled water, incubate in 2 XSSC at 65°C for 1 to 2 h, rinse, stain in Giemsa (Gurr's R66 improved stock solution diluted with 50 times M/15 Sörensen buffer pH 6.9) for 3 to 24 h. Wash in distilled water, air-dry and mount in DPX. Modifications of the BSG techniques have been employed with varying degrees of success in several angiosperms.

Schedule 2 For flower-buds: Fix buds for 24 h in Pienar's fixative (6:3:2, methanol, chloroform, propionic acid). store at 4°C for 1 week in 90% ethanol. Squash in 45% acetic acid, remove cover slip with dry ice and air-dry. Immerse slides in 45% acetic acid for 20 min at 60°C, wash for 15 min in tap water. Rinse finally in distilled water.

Place in fresh saturated aqueous $Ba(OH)_2$ for 5 min at room temperature, wash in tap water for 1 h. Incubate in 2 XSSC at 60°C for 1 h, rinse in distilled water, stain in 2% Giemsa (GT Gurr's R66 improved stock solution diluted 50 times with M/18 Sörenson phosphate buffer pH 6.9) for about 1 h. Rinse rapidly in distilled water, air-dry, rinse in Euparal essence and mount in euparal.

Schedule 3 Kinetochore staining in meiosis: Immerse air-dried slides for 5 min in 2 XSSC (pH 7.0) at 90°C, 1 h in 2 XSSC at 60°C, stain in Giemsa.

Or keep air-dried slides for 10 min in 90-94°C potassium phosphate buffer (pH 6.8, 0.12 M); 15 *s* in ice-cold phosphate buffer quinch; 1 to 24 h in 60°C phosphate buffer; 15 *s* in ice-cold phosphate buffer quinch and stain in Giemsa. Kinetochore takes up stain. These methods were tried in *Triticale*, wheat-rye hybrids and *Secale*.

Giemsa G-banding in Animals

Is possibly related to variation in the DNA-protein (non-histone) interaction between large blocks of the chromosome which may be coordinately organized in some way, probably as units of replicon organization or of transcription.

A. Trypsin G banding

Use airdried preparations after hypotonic swelling and fixation in acetic acid - methanol (1:3). Age in dry conditions at room temperature for 7 - 10 days. Alternatively, immerse slides in 10% solution of hydrogen peroxide for 3 min, rinse in physiological saline (0.9% sodium chloride in distilled water) and air dry. Moniter changes in subsequent stages using a 40X phase contrast objective.

Prepare stock trypsin solution (Difco Bacto-trypsin, cat. no. 0153- 60-2) by dissolving the contents of one vial in 10 ml distilled water. Keep at 4°C. Prepare working solution freshly before use by diluting one ml of stock with 14 ml of physiological saline. Immerse slide with chromosome preparation in working solution for 5*s*. Wash immediately with physiological saline and observe under phase contrast. Banding is shown by a dark outline of the chromatids. Stain the slide in one part of Leishman stain (BDH) added to two parts of phosphate buffer, pH 6.8 for 3 - 8 min. Rinse immediately in buffered water. Blot dry.

For destaining, rinse in acetic acid - methanol (1:3) for a few seconds and repeat trypsin treatment for fresh banding. Alternatively, stain in 10% Giemsa in phospate buffer, pH 6.8, for 4-8 min, rinse in phosphate buffer and distilled water and airdry. The time of trypsin digestion may vary but the optimum is usually below 10 *s*.

For *G-banding in male human meiosis* immerse fragments from testes in 0.44 g% hypotonic KCl solution for 8-10 h. Treat in acetic-methanol (1:3) for 12-18 h and macerate. Resuspend in 5 ml 45% acetic acid, centrifuge at 800 rev/min for 5 min, air-dry on pre-cooled slides. Keep slides at room temperature for 15 days, keep for 2 h in 2 XSSC at 40°C. Rinse and cool, treat with 0.25% trypsin solution for 10 *s* at 18°C. Rinse and stain in Giemsa.

For *G*-banding in *Heteropeza pygmaea*, remove embryos in 0.9% sodium citrate and keep for 5 min. Transfer to a subbed slide. Cover with 45% acetic acid. Squash under a coverslip. Remove with liquid N_2, rinse in absolute ethanol, air-dry. Incubate slides in 2.5% trypsin solution in Ca- and Mg-free Hanks BSS pH 7.2, for 3 to 5 min at 37°C. Wash in 70 and 100% ethanol and air-dry. Stain in Giemsa diluted 1:20 (v/v) with 0.1 M phosphate buffer, pH 7.0.

In another method treat one part of slide with trypsin solution (0.025 to 0.05% in Ca- and Mg-free BSS) or with trypsin-versene (one part each of 0.025 to 0.05% trypsin and 0.02% EDTA at pH 7.0) for 10-15 min at 25-30°C. Keep other part as control. Rinse in two changes each of 70 and 100% ethanol, air-dry, stain for 1-2 min in Giemsa, rinse twice in distilled water, air-dry and mount.

B. Acetic acid - Saline - Giemsa (ASG) banding

Use air-dried preparations of chromosomes fixed in acetic acid- methanol (1:3).

Incubate the slides for 1 h in 2XSSC (0.3M NaCl, 0.03M trisodium citrate, pH 7.0) at 60°C. The period may be increased if needed upto 18 h.

Rinse the slides for a few seconds in distilled water. Stain in Giemsa (Gurr's R66 in phosphate buffer, pH6.8), either 5% for 10 min or 2% for 1.5 h at room temperature. Rinse in distilled water and airdry.

Alternatively, blot, soak in xylol and mount in DPX.

C. Alkali treatment

Immerse air-dried slides in 0.07 N NaOH at 20°C for 90 *s*, wash in 70% ethanol, transfer to 96 and 100% ethanol and dry; incubate for 24 h in Sörensen's buffer at pH 6.8 at 59°C and stain with buffered Giemsa solution (pH 7.0) for 20 min. The concentrations of NaOH may be varied between 0.002 and 0.007 and periods of incubation in buffer between 12 to 18 h. 2 XSSC may be used instead of Sörensen's buffer (Schnedl, 1973).

Both ASG and alkali denaturation techniques are applicable to human chromosomes at prophase and from fibroblast and amniotic fluid cells when combined with H33258 technique. Minor modifications have been used for spider mites, carnivores, hamsters and dogs.

D. Protein denaturation and oxidation treatments

Immerse slides in a mixture of 3 vol urea (8 M) and 1 vol Sörensen's buffer (pH 6.8) at 37°C at 10 min, wash in water and stain with Giemsa diluted 50 to 70 times with Sörensen's buffer for 5 to 20 min at room temperature. Other protein denaturing substances used are: two strongly anionic detergents (sodium lauryl sulphate and sodium deoxycholate, both 1.25%); a nonionic detergent, Nonidet P-40, 5% (Shell Canada Ltd); a commercial detergent for cleaning glassware, '7X', 1% and mixtures of urea and detergents dissolved or diluted in deionized water and applied to air-dried slides for 10-20 min. The patterns are similar to those obtained with proteolytic enzymes.

Incubate air-dried slides for 20-40 min in 10 mM potassium permanganate solution prepared in 33 mM phosphate buffer (pH 7.0) and containing 5 mM magnesium sulpate in an ice-bath. Wash in running water for 30 *s*, soak in absolute methanol for 5 min and air-dry. Stain with Giemsa for 5 min and seal with Enkitt containing 1% butyl hydroxytoluene as anti-oxidant. This staining after oxidation may be associated with sulphydryl or disulphides in the chromosome. Peracetic acid and cupric sulphite reagent give similar patterns. A combination of trypsin-urea banding yields reproducible results in birds and invertebrates.

Modifications of G-banding

Reagents were studied for producing bands in Chinese hamster cells after Giemsa staining. Many salts induced bands in alkaline pH. Strong bases were more potent. Bands were mainly obtained by staining with Giemsa diluted either with NaCl solution or phosphate buffer (PB-Giemsa). Treatment for 10 min in collagenase (50 µg/ml) solution in Ca- and Mg-free Hanks' solution (pH 6.5) of mitotic slides from leucocyte cultures, followed by Giemsa staining, gives *G*-bands.

Fresh air-dried preparations of leucocytes may be immediately banded by either of the following methods:

(1) Immerse in 4% formalin for 5 min, rinse in running tap water for 5 min and stain in 1% Giemsa solution for 30 to 40 min: *or*

(2) Keep in absolute methanol for 5 to 10 min, air-dry and stain for 30 min with 1/80 Giemsa solution in 0.13 M phosphate buffer, pH 6.7, *or*,

(3) Dehydrate slide for 1 to 2 h in vacuum. Stain for 20 min in 1/100 dilute Giemsa solution.

Stains other than Giemsa may produce *G*-banding after the usual banding schedule. For Alcian blue staining incubate the slide for 3 h at 37°C in a solution containing 0.9% NaCl, 0.05 M sodium EDTA, 0.075 M phosphate buffer, pH 6.7 and 10 USP units/ml sodium heparin. Wash in running water for 5 min and stain for 30 to 60 min in Alcian blue 8GX (Allied Chemical: 1% in 3% acetic acid at pH 1.07).

In the study of meiosis in mammalian male, air-dried preparations, after three days, may be stained for 30 *s* in carbol fuchsin, followed by BSG banding schedule. For the study of lampbrush chromosomes, the dissected ovaries of *Triturus marmoratus* may be fixed in ethanol vapour and stained in 4%. Giemsa in 0.01 M phosphate buffer at pH 7.0 for 20 min, followed by washing in distilled water and air drying prior to banding.

Methylene blue results in preferential staining pattern of *Vicia* chromosomes similar to *G*-staining. It binds to nucleic acids and interacts weakly mainly as outside binding with chromatin.

Human meiotic and mitotic chromosomes may be stained with N-N′ diethyl pseudoisocyanine (PIC) stain. With Budr or denaturation before staining in PIC, numerous banding techniques can be adopted. Chromosomes can be stained with PIC either directly or after esterification and oxidation. The bands are visible in ultraviolet light by fluorescence and in ordinary light by intensity of colour. The most suitable wavelength is 579 nm.

Metrizamide is another medium for banding the chromosomes according to the buoyant density. Whole mount electron microscopy of Chinese hamster chromosomes, banded by exposure to actinomycin D duringG_2, indicates that differential condensation of chromatin plays a role in banding by this technique.

Feulgen-bands

Feulgen staining has been employed in lieu of Giemsa in producing *G*-bands.

For animal material, treat the banded slide, before staining, with N HCl at room temperature for 20 min and keep in Feulgen solution for 40 min in dark. Wash in two changes each of tap and distilled water, dehydrate through successive grades of 70, 80 and 95% ethanol and xylol and mount in Cargille oil R.I. 1.556. The banding, though slight, is distinct and may be enhanced by contrast photography. It has been used in family and linkage studies

For plants, store pre-treated fixed root tips in 70% ethanol at 4°C for at least four days. Hydrolyse in 10% HCl at 60°C, wash thoroughly, stain in leucobasic fuchsin solution, wash and stain in aceto-carmine for 5 min. Squash in a drop of aceto- carmine and heat over a steambath for 1 to 2 min. This process has, however, limited application.

C-BANDING

C-banding techniques were discovered as a byproduct of the *in situ* RNA/DNA hybridization procedure. The methods were initially devised for mammalian chromosomes but have been found to be effective in other systems as well.

Giemsa C-Banding in Animals

Possibly results from a specific DNA - protein interaction rather than as a function of DNA itself, as originally postulated.

A. C-banding after barium hydroxide

Use air-dried chromosome preparations after ageing for 3 to 4 days. Store in dry dust-free chamber either at room temperature or 4°C for complete drying and evaporation of fixative. Incubate the slide in 0.2N HCl for 1h at room temperature to remove cell debris and protein. This step is optional. Rinse in distilled water.

Keep the slide in freshly prepared solution of barium hydroxide octahydrate at 50°C for 5 to 15 min. The solution is saturated. Remove the scum on surface before immersing the slide. Rinse the slide thoroughly in several changes of distilled water. Incubate the slide in 2XSSC (0.3M NaCl:0.03M trisodium citrate, pH 7.0) for 1h at 60°C. Rinse in distilled water. Stain in 2% Giemsa in phosphate buffer, pH 6.8 for $1\frac{1}{2}$ hours (for giemsa staining, see details in chapter 2). Rinse in distilled water, air dry and mount or keep dry.

B. C-banding after sodium hydroxide

Incubate airdried metaphase chromosome preparations in 0.2M HCl at room temperature for 15 min. Rinse in distilled water. Incubate in 0.07M NaOH for 2 min. Keep in 70% ethanol for 5-10 min and then three changes of 95% ethanol for a total of 5-10 min and air dry. Incubate the slide in 2XSSC at 60-65°C overnight, preferably in a covered petri dish.

Rinse the slide in three 5 min changes in 2XSSC, 70% ethanol and 95% ethanol consecutively. Air-dry. Stain in 2% Giemsa in phosphate buffer for 5-30 min, pH 6.8. Mount.

Modifications:

In a modified version, after treatment with 0.2N HCl, treat with pancreatic ribonuclease (100µg/ml in 2 XSSC) at 37°C in a moist chamber for 60 min. Rinse slides again in 2 XSSC, 70 and 95% ethanol and air-dry. Treat with 0.07 N NaOH for 2 min, rinse immediately again in several changes of 70 and 95% ethanol to remove NaOH completely. Incubate in 2 XSSC and 6 XSSC at 65°C overnight and rinse successively in 70 and 95% ethanol several times. Stain for 15 to 30 min in Giemsa solution (5 ml stock: 1.5 ml methanol; 50 ml distilled water; 1.5 ml 0.1 M citric acid, pH adjusted with 0.2 M $NaHPO_4$ to 6.8 to 7.2). Rinse with distilled water, air-dry and mount in Permount. This method has been used for newt chromosomes.

In a schedule for a shorter duration, treat air-or heat-dried slides with 0.07 N NaOH for 1 min; rinse in 70 and 95% ethanol; air-dry; incubate in 2 XSSC at 60°C for 16 h; rinse and air- dry as before; stain in Giemsa for 1 h at pH 6.8 (1 ml stock Giemsa in 9 ml buffer); rinse in distiled water and air-dry; rinse in xylol and mount in neutral medium.

Alternatively, treat air-dried slides for 30 *s* in 0.07 N NaOH in 0.112 M NaCl (2.8 g NaOH and 6.2 NaCl in 1l distilled water, pH 12.0) at room temperature; rinse thrice in 12 XSSC (pH 7.0) for 5 to 10 min each; incubate in 12 XSSC at 65°C for 60 to 72 h; pass through 3 changes each of 70 and 95% ethanol, keeping 3 min in each; air-dry; stain in buffered Giemsa solution for 5 min, rinse briefly in distilled water, air-dry and mount in Permount.

Alternatively, NaOH can be used at 0°C or room temperature for 0 to 180 *s*. Incubation may be carried out in 2 X, 6 X, 12 X and 24 XSSC for intervals between 15 min and 136 h. Periods of Giemsa staining may vary from 5 to 90 min with or without pre- treatment of chromosomes. Treatment in dibasic sodium phosphate (pH 11.8 to 12.0) for 10 to 30 min followed by 5 to 90 min in Hanks solution and Giemsa staining gives *C*-bands.

Giemsa C-banding in Plants

C-banding should be considered as a technique that stains all constitutive heterochromatin whereas N-banding and modified C- banding are specialised staining techniques.

C-banding

A. Pre-treat root tips in 0.05% colchicine for 4 h, fix in acetic-ethanol for 12-15 h, hydrolyse in 1 N HCl at 60°C for 25 *s*, squash in 45% acetic acid with albuminized cover slips. Separate cover slips in absolute ethanol or by CO_2 freezing; dry at 60°C; denature in $Ba(OH)_2$ for 5 min at room temperature. Rinse in running distilled water, dry, incubate in 2 XSSC for 1 h at 60°C, rinse and dry again. Stain in 0.5% Giemsa at pH 6.8 for 5-15 min, wash in distilled water, dry, dip in xylol and mount in canada balsam or DPX. Modifications involve changes in time of incubation and temperature.

B. Hydrolyze pretreated and fixed root-tips in 45% acetic acid for 25 min, incubate for 5 min in 5% barium hydroxide at 18°C. Stain in 2% Giemsa (Gurr's R66) in Difco buffer, pH 6.8 for 5-10 min. Rapidly airdry slides by shaking. Mount in Euparal (Kenton 1978).

C. N-banding

Treat freshly excised root-tips with ice-cold water for 22 to 24 h. Fix in glacial acetic acid - 99% ethanol (1:3) upto 3 days in room temperature or even upto one month in cold. Stain root-tips in 1% acetocarmine solution for 1-2 h in room temperature and squash in 45% acetic acid. Remove the coverslip by freezing. Treat the preparation with 45% hot acetic acid at 55 to 60°C for 5 - 10 min. Air dry overnight. Incubate the preparation in hot phosphate buffer (1M NaH_2PO_4) at 94±1°C for 2 min, rinse briefly in tap water and airdry. Next stain preparation in Banco Giemsa stain (1 drop per ml Sörenson's phosphate buffer) for 30-40 min at room temperature, rinse briefly in tap water and airdry. This method has been used in Gramineae, specially wheat (Gerlach 1977, Gill *et al* 1991).

D. Modified C-banding (MC) combines parts of the protocols of N- banding and C-banding procedures.

Treat freshly excised roottips in cold water for 24 h and then fix in glacial acetic acid-ethanol (99%) 1:3 for one day. Stain in 1% acetocarmine solution for 1-2 h at room temperature and squash in 45% acetic acid. Remove coverslip by freezing, treat with 45% acetic acid at 50°C for 5-10 min and airdry overnight. Next treat preparations in a saturated barium hydroxide solution at 50°C for 2.5 min, rinse in tap water, immerse in 2 XSSC at 50°C for 3-10 min. Rinse in tap water. Stain wet slides in Wright or Leishman Banco Giemsa staining solution (one drop per ml Sörenson's phosphate buffer) for about 20 min, rinse in water and airdry.

Alternatively, treat freshly excised roottips with 0.05% colchicine for 3 h. Fix in glacial acetic acid-ethanol (99%) 1:3 for upto 3 days at room temperature. Keep in 45% acetic acid for 2-3 min. Squash. Remove coverslip by freezing. Keep in 99% ethanol overnight and air dry for a few min. Incubate the preparations in 0.2M HCl at 60°C in a water bath for 2 min. Wash in distilled water, incubate in saturated barium hydroxide solution at room temperature for 7 min, rinse in distilled water and incubate in 2 XSSC for 1 h at 60°C. Transfer slides directly to 1 to 5% Giemsa staining solution (Fisher) in phosphate buffer for upto 30 min. Wash and airdry. Place airdried slides in xyleme, mount in Permount.

E. Combined C and N banding for plants

Preparation of slides: Excise fresh root-tips from germinating seedlings. Pretreat with distilled water at 0°C for 16 h. Fix in freshly prepared acetic acid-methanol (1:3) and keep at - 20°C for about a week. Wash the roots with distilled water and then macerate with an enzymatic mixture (4% Onozuka RS; 1% pectolyase Y-23; 75mM KCl, 7.5mM Na_2EDTA, pH 4.0) at 37°C for 45-60 min. Rinse in distilled water. Place roottip in a few drops of fixative on a glass slide, tap carefully with the tip of a pair of forceps and airdry.

For C banding incubate the air-dried slides after ageing for a week in 0.2N HCl at 55°C for 3 min, in 5% barium hydroxide solution at 25°C for 5 min and then in 2 XSSC at 55°C for 20 min. Rinse in distilled water between each step.

For N-banding: incubate air-dried aged slides in 1M sodium dihydrogen phosphate solution at 90°C for 3 min. Rinse in distilled water. When staining for either C or N-bands stain the treated slides with 8% Wright solution (Merck) diluted with $\frac{1}{30M}$ phosphate buffer (Na_2HPO_4 or KH_2PO_4, pH 6.8) for 1 h at room temperature. Rinse and airdry.

For combined staining: photograph after staining for either C and N band and then subject to the procedure for the other banding without destaining.

The two patterns are in general similar. However, in barley, every centromeric site showed an N-band positive and C-band negative heterochromatin at the centromeric site. It is suggested that the differential alteration of chromosome structure rather than the differential extraction of chromatin, may be responsible for such N+C- heterochromatin.

Alternatively, pretreat freshly cut roottips with 0.05% colchicine solution containing either 10 ppm actinomycin D or ethidium bromide, for 2 h at 25°C. Transfer to Ohnuki's hypotonic solution (55 mM KCl, 55 mM $NaNO_3$, 55 mM CH_3COONa, 10:5:2) for 30 min to 1 h at 25°C. Fix in acetic acid-methanol (1:3) for 1 to 4 days at - 20°C in a freezer. For actinomycin D-treated roottips, the fixation can be reduced to 2 h at 25°C. Smear or squash roottips either with a tweezer in fresh fixative or macerate enzymatically. For enzymatic maceration, wash fixed roottips for about 10 min and macerate in an enzymatic mixture (pH 4.2) containing 2% cellulase RS and 2% macerozyme R-200, both of Yakult Honsha Co. Ltd. Tokyo for 20 to 60 min at 37°C in a 1.5 ml Eppendorf tube. Rinse with water 2 or 3 times. Pick up macerated roottip onto a glass slide using a Pasteur pipette, add fresh fixative and cut into small pieces. Squash, air dry for 2 days in an incubator at 37°C. Stain samples prepared by actinomycin pretreatment in 10% Wright solution diluted with 1/15M phosphate buffer (pH 6.8) for 10 min at 25°C. Wash and air dry.

Fix enzymatically macerated samples again in a 2% glutaraldehyde solution diluted with the phosphate buffer for 10 min at 25°C and wash. Immerse post-fixed slides either in 2% trypsin (Merck Art. 8367) dissolved in PBS (pH 7.2) for 10 min at 25°C, or in 0.02% SDS dissolved in tris-HCl buffer (20 mM, pH 8.0) for 2-5 min at 25°C. Briefly wash and airdry. Stain in 5% Wright solution in $\frac{1}{30M}$ phosphate buffer (pH 6.8) for 5 min.

Three types of G bands are obtained: by actinomycin, trypsin and SDS. Pretreatment with colchicine or colchicine and ethidium bromide aid in the accumulation of prometaphases in which the fine patterns can be observed. Image analysis can be done from enlarged microphotographs through the digitally manipulted chromosome image analysing system CHIAS (Fukui 1986).

CT-BANDING

CT-banding is a further amplification of *C*-banding techniques. In human material, treat air-dried preparations with $Ba(OH)_2$ solution at 60°C, incubate in 2 XSSC at 60°C and stain in cationic dye 'Stains All'. The bands are of *C*- and *R*-types, located mainly at telomeric regions.

Cd-BANDING

Cd-banding reveals two identical dots (centromeric dots *Cd*) at the centromeric region, one for each chromatid. Store mammalian airdried preparations for one week at room temperature. Incubate in Earle's BSS medium (pH 8.5 to 9.0) at 85°C for 45 min. Stain in 4% Giemsa in 1/300 M phosphate buffer (pH 6.5).

Combined method

In general, more than one banding pattern is utilized in assessing relationships between taxa. For example, combined *C* and *G* techniques have been effective in checking homology in bat genera, *Q*-and *G*-banding in primates and *C*-banding and autoradiography in cell line of kangaroo rat. In the dinoflagellates, no *Q*, *G* or *C* bands could be recorded, indicating their primitive nature. A comparison of *G,Q* and *EM* banding patterns has been made in Indian muntjac and of *Q,G* and *C*-banding in *Miaster* sp. (Diptera). Mouse chromosomes show an almost exact reversal of Giemsa 11 banding pattern for human chromosomes. In man-mouse hybrid, this method, associated with *Q*-banding, serves to identify human chromosomes against mouse chromosome background.

O-BANDING OR ORCEIN BANDING

It was developed primarily for use in plant chromosomes (Sharma, 1975, 1977). The technique principally involves an elimination of denaturation and consists of pre-treatment of the tissue, fixation treatment in a strong concentration of XSSC, washing, staining in acid-orcein and mounting in 45% acetic acid. Orcein- positive bands appear in different segments of chromosomes, including the centromeric and intercalary ones. The mechanism of reaction possibly involves the DNA-protein linkage, since orcein is an amphoteric dye, capable of staining both DNA and protein. The gradual removal of non-histone protein through SSC treatment is principally responsible for *O*-banding at the sites of stronger DNA-protein linkage. As the removal of protein by SSC application is gradual, mild treatment results in intercalary bands comparable to *G*-banding whereas strong prolonged treatment ultimately shows only *C*-bands where the linkage is strong due to highly homogeneous repeats. The method therefore allows localization of major and minor reiterated sequences in chromosomes.

Fix pre-treated tissue in acetic acid-ethanol (1:2) for 2-12 h. Treat in 45% acetic acid for 5 min. Wash in water. Treat in a mixture of 1 M sodium chloride and 0.1 M sodium citrate (1:1) at 27-28°C for 2-3 h. Wash in water. Warm for a few seconds at 90°C in a mixture of 2% aceto-orcein solution and N HCl (9:1). Keep in the mixture at 27-28°C for 1 h. Squash under a cover slip in 45% acetic acid and seal.

HKG BANDING TECHNIQUE FOR PLANTS

Pretreat germinated seeds directly with 0.05% 8-hydroxyquinoline solution for 2h 30min at 32°C. Dissect out the embryos, wash with distilled water and fix in three changes of chilled acetic acid-methanol (1:3), keeping 15min in each change. Store at -20°C for one to several days. Wash in water thoroughly. Dissect out root-tips, mince in tiny fragments in distilled water. Centrifuge at 150g for 2min. Resuspend pellet in 2ml freshly prepared enzyme (0.2ml flaxzyme NOVO and 1.6ml distilled water). Incubate at 35°C for 2h 30min. Remove supernatant after centrifuging for one min at 150g. Gently resuspend pellet in 2ml distilled water. Filter through nylon mesh into another centrifuge tube and make up volume to 5ml with distilled water, Centrifuge for one min at 150g. Wash pellet with cold acetic acid-methanol (1:3) and resuspend in 0.5ml of same fixative. Drop 4 or 5 drops of cell suspension on a slide covered with 45% acetic acid, air dry, place slide on a hot plate at 50°C for 10min. Age at 35°C for one to several days.

Treat two-weeks old slides with 5% barium hydroxide solution at 56°C with continuous agitation for 10-15s, wash in two changes of 70% ethanol, then absolute ethanol and then transfer to acetic acid-methanol (1:8). Dry on a hot plate for a few minutes, stain with 3% Giemsa (Merck) in phosphate buffer, pH 6.8 for 8-10min. Wash twice in distilled water and air dry. Observe for *C-banding*. ***For HKG banding*** take one to 5 day old slides. Hydrolyse at 60°C in N.HCl for 4-6min, wash four times in distilled water for a total of 10min. Briefly immerse the slides in 0.9%NaCl, dip 10 times in 70% ethanol, dry on a hot plate at 50°C for a few *s*. Immerse in 0.06N KOH solution for 8-12 *s* in room temperature with continuous agitation. Wash in two changes of 70% ethanol, absolute ethanol and transfer to acetic-acid-methanol (1:8). Dry on a hot plate and stain with Giemsa as described earlier for C-banding. Dark distinct bands are produced which are different from C nands. The method has been applied in both wheat and maize Shang *et al* 1988, De Carvalho and Saraiva (1993).

REVERSE BANDING OR *R*-BANDING

R-banding in animals

The pattern is the exact reverse of G-banding, indicating a similarity of the mechanism involved. The pattern is produced by incubation at high temperature or a suitable pH.

A. R-banding by high temperature

Incubate air-dried chromosome preparations in 20mM phosphate buffer, pH 6.5 for 10 min at 87°C. Rinse with tap water. Stain in 2% Giemsa solution, in phosphate buffer, pH 6.8 for 10 to 30 min.

B. R-banding by low pH treatment

Incubate air-dried preparations in 1M sodium dihydrogen phosphate (pH 4.0 to 4.5) at 88°C for 10 min. Rinse in water. Stain in 5% solution of Giemsa in distilled water for 10 min.

C. R-banding with acridine orange

Allow air-dried chromosome preparations to age at room temperature for 7 to 10 days. Incubate slides in fresh phosphate buffer at 85°C for 20 to 25 min. Stain slides in 0.01% acridine orange dissolved in the same phosphate buffer for 4 to 6 min. Rinse and mount in the same buffer. Observe under fluorescence microscope.

D. Telomeric R-banding Telomeres become strongly stained and the rest are faintly banded.

R banding

Allow air-dried slides to age for a few days. Incubate slides in phosphate buffered saline (PBS) with pH adjusted to 5.1 for 20-60 min at 87°C. Rinse in PBS. Stain in 3% Giemsa in phosphate buffer, pH 6.8 at 87°C. Leave for 5-30 min. Rinse.

NOR-BANDING

NOR-banding has been employed principally in the localisation of nucleolar organising regions.

A. Giemsa Staining

For human chromosomes incubate air-dried slides in 5% aqueous trichloracetic acid for 30 min at 85-90°C; rinse in water; reincubate for 30-45 min at 60°C in 0.1 N HCl; rinse in tap water and stain for 60 min with phosphate buffered Giemsa, pH 7.0 Deoxyribonuclease (100 μg/ml for 60 min at 37°C and pH 6.6) and pancreatic ribonuclease (pH 7.0) can be used instead of TCA.

B. Giemsa Staining

For animal and plant chromosomes incubate air-dried slides at 96± 1°C for 15 min in 1 M NaH_2PO_4 solution (pH 4.2 ± 0.2 adjusted with 1 N NaOH). Rinse in distilled water, stain in Giemsa (diluted 1 in 25 in 1/15 M phosphate buffer, pH 7.0), rinse in tap water and air-dry.

C. Silver Staining Methods

For Animal and Human Chromosomes: This technique specifically stains the nucleolar organising regions. The specificity is particularly high in mammalian material under controlled conditions. It is suggested that the initial incubation with silver nitrate impregnates the nucleolus and the NOR. Subsequent incubation in ammoniacal silver solution and a formalin developer results in the deposition of more metallic silver at the sites selectively impregnated.

a) Ag-AS staining

50% silver nitrate solution: Dissolve 10 g silver nitrate (Analar) in distilled water to a total volume of 20 ml. Filter and store in dark at room temperature.

Ammoniacal silver (AS) solution: Dissolve 8g silver nitrate in 10 ml distilled water, and add 15 ml conc. ammonia solution (35%w/w NH_3, pH 12 - 13) slowly with stirring. Filter and store in dark. It should be made up atleast two days before use and can be stored upto two weeks.

3% formalin: Dissolve 8g paraformaldehyde in 100 ml distilled water, bring the pH to 10 with 10 M NaOH, heat until dissolved to give 20% stock formalin solution. Filter and neutralise with conc. HCl. Store at 4°C. From this stock, dilute an aliquot to 3% final concentration with distilled water. Adjust pH first to 7.0 with sodium acetate crystals and then to pH 5.5 with formic acid.

Protocol: Age airdried chromosome preparations for at least two days. Place 3-4 drops of 50% silver nitrate on the slide and cover with a coverslip. Incubate under a photoflood lamp (Philips 240 V, 500W) to heat the slide between 60 - 70°C. After the silver nitrate starts to crystallize around the edge of the coverslip in 10 to 15 min, remove the lamp. Allow the slide to cool slightly and wash off coverslip and $AgNO_3$ under running tap water. Air dry the slide.

Mix quickly two drops of AS solution with two drops of 3% formalin on a coverslip and lower the pretreated slide, preparation side down, on the coverslip. Moniter staining first under phase contrast and then under bright light. Chromosomes turn pale yellow and nucleoli and NORs black in 30 to 60 *s*. Rinse slides in distilled water, dehydrate in ethanol series, soak in xylene and mount in Permount. If needed the slides can be counterstained with Giemsa (2% Gurr's R66 in phosphate buffer pH 6.8) to differentiate the chromosomes.

b) Simplified method with limited application

Place 3 or 4 drops of 50% silver nitrate on previously aged chromosome preparations and lower a coverslip over it. Place in a moist chamber prepared by soaking filter paper in covered Petri dish with distilled water. Incubate moist chamber with slide at 37 to 65°C for 4 to 18 h, depending on the temperature. Moniter staining periodically under phase contrast. If needed, chromosomes can be counterstained by Giemsa as given in the more detailed procedure.

This method has been applied to several mammalian cell lines in culture and stains selectively the same chromosome areas (NORs) displaying heavy labelling by *in situ* RNA/DNA hybridization for 18s and 28s ribosomal cistrons. The available cytochemical data suggest that the Ag-As reaction stains the chromosomal proteins at the NOR rather than the rDNA itself.

HY-BANDING

HY-banding was initially applied to somatic chromosomes of some members of the Liliflorae. Root tips, fixed in acetic- ethanol (1:3), on treatment with 0.1 or 0.2 N HCl between 60 and 80°C and staining with aceto-carmine, gave banded chromosomes. Heterochromatic regions stain differentially, mentioned as Hy^+ and Hy^- bands but do not always coincide with *G*-bands.

RESTRICTION ENDONUCLEASE BANDING

Two categories of chromosome bands are observed,

a) Related to DNA base composition in certain areas, by simply staining fixed chromosomes with quinacrine or Hoechst 33258 or chromomycin A_3.

b) Depending on the different classes of chromosomal proteins, in their interaction with each other and/or with DNA by physical or chemical mistreatment of chromosomes followed by staining in dyes like Giemsa. Restriction endonucleases selectively digest fixed chromosomes and subsequent staining with Giemsa or Ethidium bromide gives differential staining.

Digestion by restriction endonucleases gives chromosome band patterns that may be directly related to the DNA sequences in mammalian chromosomes. *AluI*/digestion gives a C-band-like pattern in mouse chromosomes localized in the centromeric regions, which are highly enriched with SAT DNA sequences without *AluI* recognition site (Lica and Hamkalo 1983). *HaeIII*, which recognises the base sequence GGCC, preferentially digests the G- enriched inter-G-band regions of human chromosomes, inducing a G- band like pattern (Miller *et al* 1983). *Hind III* and *EcoRI* recognise the base sequences AAGCTT and GAATTC respectively, giving G-bands (Mezzanotte *et al* 1983). Each of the isoschizomeric pairs of restriction enzymes, *MboI-Sau3A* and *EcoRII-BstNI* produced different band patterns in human and mouse chromosomes but the same band pattern in grasshopper (Gosalvez *et al* 1989), indicating that organisation of chromatin may influence the activities of these enzymes.

RE Banding in Human Chromosomes

General protocol

Prepare working solution of the restriction endonuclease by dissolving the enzyme in the assay buffer suggested by the manufacturer to give a final concentration of units/ml according to enzyme (see Table). Keep human metaphase preparations from peripheral blood for 1 to 5 days at room temperature or 4°C before treatment for ageing.

Place 30 µl of the working solution on each slide and cover with a coverslip. Incubate slides in a moist chamber at 37°C for 6 to 16 h. After incubation, rinse slides in several changes of distilled water. Shorter periods are used in some cases (see Table).

For Giemsa staining, stain the slides with 4% Giemsa (Merck) in Sörensen's buffer (pH 6.8) for 5 min.

For fluorescence staining, destain preparations in ethanol or ethanol-water (3:1). Incubate in ribonuclease (100 µg/ml in phosphate-buffered-saline, PBS) at 37 °C for 1 h. Stain with the intercalating dye Propidium iodide (PI, 5µg/ml in PBS). RNase is essential to improve quality of staining, since PI binds to double - stranded DNA, RNA or DNA-RNA hybrids. If the slide has not been stained in Giemsa, destaining in ethanol prior to fluorescence staining is not needed.

Alternatively, stain in ethidium bromide .

For double digestion procedure: stain chromosomes previously incubated with *EcoRI* with Giemsa, photograph, destain in 70% ethanol for 15 min, incubate overnight in *AluI* (10 units/ml), restain with either Giemsa or Ethidium bromide and photograph again. Observe with Leitz Dialux 20EB microscope with BP 516-560 and LP 580.

Short treatment followed by Giemsa staining gave G-band like and longer treatment C-band type patterns. With propidium iodide staining, short treatment gave uniform fluorescence but C bands were seen on longer treatment. It was suggested that C-banding is the result of a uniform DNA removal in non-centromeric regions taking place after a critical time point while the initial G-like banding is produced by changes in DNA-protein interactions.

RE Banding for Drosophila (Mezzanotte *et al* 1986)

Age metaphase chromosome preparations for 3 to 10 days, incubate in incubation mixture containing 50 units of AluI or Eco RI (Boehringer) dissolved in 100 µl of incubation buffer indicated by the manufacturer, by placing drops of the mixture on the slide and using a coverslip to spread the enzyme solution evenly. Incubate for the required period (between a few min to 16 h) and then halt enzyme treatment by washing the slides at 4°C in 5mM EDTA solution, pH 7.0. Stain with ethidium bromide — a DNA-specific dye.

RE Banding in Mammalian Chromosomes in vitro

Cell culture: Grow mouse LM cells as monolayers in 5:3:2 Hanks balanced salt solution (BSS): medium 199: horse serum. Synchronize log phase cultures with excess thymidine (final concentration, 10 mM) for 24 h. Remove medium, wash cultures with BSS and reculture cells in fresh medium. After 8 h, add colcemid (0.075 µg/ml) and incubate overnight. Collect accumulated mitotic cells from monolayer by selective detachment.

Banding: Treat aliquots of mitotic cells with 0.075M KCl for 10 min at 37°C and three changes of glacial acetic acid-methanol (1:3) for 20 min each. G-band with trypsin or C-band with $Ba(HO)_2$/SSC for comparison (optional).

Isolation of chromosomes (modified from Wray 1975): Wash mitotic cells in Hanks BSS, centrifuge at 480 g for 5 min, resuspend in Hanks BSS: distilled water (1:1) for 20 min on ice and centrifuge as above. Wash in cold PMH (0.10 Pipes buffer, 1mM $MgCl_2$, 1.0 M hexylene glycol, pH 6.5) by centrifuging at 1000 g for 5 min, resuspend in 10 ml PMH and incubate at 37°C for 10 min. Syringe through a 22 gauge needle to liberate the chromosomes and moniter under phase contrast. Pellet unbroken cells by centrifugation at 300 g for 10 min, recentrifuge the supernatant at 1500 g for 30 min to pellet the chromosomes. Resuspend chromosomes in a small amount of PMH and store at 4°C.

Endonuclease digestion: Resuspend stored chromosomes by centrifuging at 1500 g for 30 min and then resuspending at a concentration of 300 µg/ml DNA in the enzyme digestion buffer recommended by the manufacturer. Incubate at 37°C. Remove suspension and add EDTA for a final concentration of 10 mM. Cool on ice, keep overnight in cold to allow chromatin extraction and centrifuge at 1500 g for 30 min. Store suspension with extracted chromatin. Observe under phase contrast. For light microscopy, transfer extract onto a glass coverslip after centrifuging at 1500 g for 15 min. Fix in acetic-methanol (1:1) and stain with Giemsa.

RE Bands in Plants

Allium cepa, Hordeum vulgare, Vicia faba

Pretreat roottips suitably (in 0.05% colchicine or 0.025% in a saturated solution of alphabromonaphthalene) for 3h. Fix in acetic acid-ethanol (1:3). Digest roottips at 37°C in 1% cellulase for 90min and 2% pectinase for 2 h; diluted in 0.01M sodium citrate buffer, pH 4.4 to 4.8. Squash roottips in 45% acetic acid. Remove coverslip with dry ice. Wash in 96% ethanol and airdry. Slides can be stored in glycerol for several months.

Before enzyme treatment, wash out glycerol by 2XSSC for 10 min. Air dry. Equilibrate the airdried slides in the incubation buffer for 1 h. Put 20μl drops of buffer and buffer-containing- enzyme respectively on the slides. Incubate in moist chamber at 37°C to 45°C (according to enzyme) for 2 to 24h. Remove the drops by pipettes. Briefly rinse slides in distilled water and stain in 4% Giemsa(Merck) for 2-5 min, airdry and mount in entellan(Merck).

RE banding in *Vicia faba* gave almost the same patterns as conventional Giemsa banding but the method is still rather variable(Schubert 1990).

RE Bands in Plants (Barley)

Germinate barley seeds in a Petri dish at 25°C. Excise primary roots when they are 1-1.5 cm long and dip in 0°C chilled distilled water for 14-20h. Fix in acetic acid-methanol (1:3) overnight. Wash thoroughly. Macerate with the enzyme cocktail containing 2% cellulase, 1.5% macerozyme R200, 0.3% Pectolyase Y-23; 1mM EDTA, pH 4.2 at 37°C for 20 min on the glass slide (Kamisugi and Fukui 1990). Rinse and tap the root tips briefly by a fine forceps with the fresh fixative. Airdry the slides.

Enzymes used : MboII, GAAGA8/7; RsaI, GT↓ AC; HAE III, GG↓CC;
Hinfl, G↓ANTC; and DraII, PuG↓GNCCPy

Working solution for each : 200 units in 100μl of incubation buffer with suitable salt concentration. (NaCl concentration for L buffer OmM; M buffer 50 mM; H buffer 100mM)

Drop 20μl of the working solution on chromosome spreads on the slides; apply a clean coverslip. Incubate at 37° C for 2 to 6 h in a humid chamber. Rinse, stain with a 2% Giemsa solution (1/15M phosphate buffer, pH 6.8) for 15-30 min. Specific bands were seen with the different enzymes, while buffer solutions could alter these bands.

RESTRICTION ENDONUCLEASE/NICK TRANSLATION

In fixed Mouse Chromosomes

Prepare metaphase spreads from mouse cell cultures. Digest chromosomes by spreading 50µl of enzyme solution, containing 50 units of the enzyme dissolved in buffer recommended by the manufacturer, on the slide and cover with a 22X 40 mm coverslip to ensure reaction with all the material. Keep for digestion in a humid box for periods required (5 or 20 min or 1h) at 37°C. Remove the coverslip, wash in distilled water.

For Giemsa staining, keep slides for 30 min in 2% Giemsa (GurrR66, BDH) in phosphate buffer, pH 6.8. Rinse with distilled water, air and mount in DPX.

For ethidium bromide staining, keep slides for 1h in 0.04% ethidium bromide (BDH) solution in phosphate saline. Rinse thoroughly in distilled water, dry mount in 1% dithionite in phosphate-buffered saline and seal with rubber solution.

For nick translation, incubate in a mixture containing 5 units of DNA polymerase I (endonuclease free) and 10µM each of dATP, dCTP, dGTP and biotin-16dUTP in 50µl of the appropriate buffer recommended by the manufacturer. Cover with a coverslip to prevent evaporation and ensure reaction with all the material. Keep at room temperature for 45 min. Wash for 5 min in 5% trichloracetic acid at 4°C to remove all unincorporated nucleosides. Rinse in Buffer 1(stock solution : 121 Tris base, 58.4 NaCl, 4.07g $MgCl_2.6H_2O$ made up to 900 ml with distilled water; add 65ml conc HCl to bring pH to 7.5. Autoclave, cool, add 5ml of Triton X-100. Dilute stock solution ten times before use). Block chromosome preparation by incubation for 20 min at 42°C in Buffer 2(2% w/v bovine serum albumin Sigma in Buffer 1) and then for 10 min at room temperature in Buffer 1.

Visualize the incorporated biotin-16-dUTP by binding with the streptavidin-horse radish peroxidase conjugate (SA-HRP, Bethesda Res Lab). Use 1:250 solution of SA-HRP in Buffer 2 and water(4µl SA-HRP : 96 µl Buffer 2 : 900 µl double distilled water). Add 100µl of the diluted solution to each slide. Cover by a 22X40 mm coverslip, incubate at 37°C for 30 min in a humid box (paper towel soaked with PBS). Remove the coverslip in a beaker of PBS, wash the slides twice in PBS (5 min each). Wash again for 2 min in PBS-TX (0.1 % Triton X100 in PBS) and keep in PBS.

For developing, use 5ml of 0.5mg/ml 3-3′-diaminobenzidine (DAB, Sigma) in PBS to which 50 µl of 1 vol hydrogen peroxide is added immediately before use. Add 400 µl of this solution per slide and incubate in dark at room temperature for 45 to 60 min. Wash slides thoroughly in distilled water, rinse twice in PBS and finally in distilled water. Dry thoroughly and mount in DPX.

Only AvaII and Sau96I were observed to attack readily the mouse major satellite in fixed chromosomes. Chromatin conformation is seen to be an important factor in determining patterns of digestion of chromosomes by restriction endonucleases (De la Torre *et al* 1991).

In the RE/NT method, sites of attack by the restriction endonucleases on the chromosomal DNA are labelled directly by nick translation with labelled nucleotides. These patterns on human chromosomes sometimes differed from those produced by RE/Giemsa staining showing that both distribution of recognition sites and accessibility of the chromosomal DNA to the enzymes are important factors(Summer *et al* 1990).

In situ Nick Translation of Human Chromosomes Directed by Restriction Endonucleases

Wash flame or air-dried preparations of human chromosomes from peripheral blood or cell line cultures with 10mM Tris-HCl pH 7.4. Incubate with 50 units per slide of the endonuclease dissolved in the buffer solution recommended by the supplier at room temperature. The period of incubation is 30 min for HpaII and 120 min for MspI and EcoRI. Stop the reaction by rinsing the slides thoroughly in 10mM Tris-HCl, pH 7.4.

Nick translate the chromosome preparations for 10 min at room temperature. Cover slides with 50µl of a mixture containing 50 mM Tris-HCl, pH7.9; 5mM $MgCl_2$; 10mM 2-β-mercapto-ethanol, 50 µl bovine serum albumin; 10 units /ml DNA polymerase (endonuclease-free); 4µm dATP, dCTP and dGTP and 0.3 µm (^{3}H) TTP (96Ci/mmol). Stop the reaction by rinsing the slides in 10mM Tris-HCl, pH 7.4. Stain the chromosomes in Giemsa (4%) or by chromomycin A_3-methyl green technique.

MspI and HpaII are 5'CCGG3' recognition sequences used to map CpG-rich sequences. CCGG sequences are seen to be preferentially located on RE-positive bands(Prantera and Ferraro 1990).

RE-Banding and Electron Microscopy of fixed Mouse Metaphase Chromosomes

Tissue Culture : Grow L-929 mouse cells in monolayers at 37°C in RPMI 1640 medium. Add 33258 Hoechst (40µl/ml) during last 12h of culture to produce undercondensation of centromeres. Add colcemid (0.5µ g/ml) for the last 12h. Detach mitotic cells from monolayers by gently shaking the flasks. Treat with 0.075M KCl hypotonic solution for 10 min at 37°C.

EM of isolated chromosomes: Centrifuge hypotonically treated cells at 1200 rpm for 7 min. Incubate pellet for 30 min in acid isolation buffer (1% citric acid, 1% Triton X, 6mM $MgCl_2$). Resuspend the cells and syringe 7 to 10 times through a 22 gauge needle. Centrifuge isolated chromosomes for 5 min and 2000 rpm through a cushion of isolation buffer onto Formvar - coated grids. Immerse grids in a series of 50, 75% and absolute ethanol, keeping 10 min in each. Transfer to 100% amyl acetate for 10 min. Air dry.

For chromosome digestion, use AluI (AG/CT), HpaII (CC/GG), HinfI (G/ANTC) and HaeIII (GG/CC) in concentrations of 45 units per µl dissolved in the respective buffers supplied by the manufacturer (Amersham). Immerse the grids in a 50 µl drop of working solution for 12 h at 37°C in a moist chamber. Stop the treatment by washing the grids in double distilled water and subsequently in a series of ethanol, 50%, 75% and 100%, keeping 10 min in each. Transfer to 100% amyl acetate, keep for 10 min and airdry.

LM of isolated chromosomes: Digest airdry preparations of mouse chromosomes in the same concentrations of the enzymes as EM. Stain with 3% Giemsa in phosphate buffer, pH 6.8 for 5-10 min.

Results show that REs producing identical effects in LM(AluI and HinfI) produce different effects in EM. All enzymes affect C-bands but Alu I reduces the density of these regions while the others increase the density indicating that conformational changes are involved. The appearance of chromosomes in EM is related to the relative digestion of isolated DNA by RE (Gosalves and Goyanes 1988, Gosalves *et al* 1990).

Antibody Binding Methods for Human Chromosomes

Anti-adenosine antibody binding method:

Base specific antinucleoside antibodies react with specific nucleoside bases in single-stranded DNA *in vitro*. They attach to chromosomes on being treated with denaturating agents like NaOH, giving a banding pattern similar to *Q* and *G* banding.

Prepare anti-adenosine (anti-A) antibodies by immunizing rabbits to adenosine monophosphate conjugated to bovine serum albumin. Antibodies to rabbit gamma globulin are induced in sheep and fluorescin-tagged. Denature air-dried leucocyte preparations by heating for 1 h at 65°C in 95% formamide solution in XSSC, adjusted to pH 7.2 with concentrated HCl. Dilute the previously prepared anti-A antibody to 1:10 with phosphate buffered saline (PBS). Wash slides twice in 70% ethanol, once in 95% ethanol and air-dry. Keep in PBS for 5 min. Layer the anti-A over the slides and maintain at room temperature for 45 min in a moist chamber. Wash in PBS, add the fluorescin-tagged anti-rabbit gammaglobulin (1:50 in PBS) and keep for 45 min in moist chamber. Rinse, mount in PBS and examine under fluorescence microscope.

Antiguanosine, anticytidine and antithymidine, after heating in formamide, give similar bands.

Combined chemical and immunochemical procedure for human chromosomes:

Photo-oxidise air-dried leucocyte preparations in a solution containing 33.4 μM methylene blue in 0.1 M Tris-HCl buffer (pH 8.75) by placing the slides in a coplin jar in 50 ml of the dye solution and bubbling oxygen through it for 10 min.

Seal the jar, place in a glass water bath at 25°C 15 cm away from a 150 W Sylvania flood lamp at an illumination of about 64 600 lx overnight. Rinse the slides in PBS. Layer with anti-C, prepared in rabbits (diluted 1:10 with PBS) and keep at 25 °C for 30 min. Rinse slides in PBS, layer with fluorescin-labelled antibody, prepared in sheep against rabbit immunoglobin G (diluted 1:50 with PBS), incubate at 25 °C for 30 min, rinse and mount in PBS. Alternatively, the slides can be treated directly in QM after photo-oxidation. Observe under fluorescence microscope.

The photo-oxidation in presence of methylene blue destroys guanine residues in DNA. The chromosomes, on reaction with cytosine-specific antibody, reveal a fluorescent banding pattern identical with *R*-banding and reverse to the anti-A antibody binding method and QM and *D*-banding patterns described earlier. This pattern thus appears to reflect the DNA- base composition.

SISTER CHROMATID EXCHANGE

The initiation of banding methodology led to the improvements in differentiating sister chromatids. It was initiated by incorporating bromodeoxyuridine (BrdU) during one or two successive replications and staining with fluorochromes afterwards. Alternatively the slides were stained with Giemsa after treatment with a fluorochrome, exposure to light or storing and heating. This technique permits the identification of two chromatids of each chromosome on the basis of differential staining intensities. The chromosomes of *somatic tissues* show the formation of sister chromatid exchanges (SCEs) by this method.

Another set of methods utilizes BrdU-Giemsa staining to identify late DNA-replicating sites by pale colour or dot formation. Dotted chromosomes may also be produced with sodium phosphate solution supersaturated with $NaHCO_3$. A combination of *G*- banding and autoradiography has been employed also to map sister chromatid exchanges in human chromosomes.

BrdU-dye techniques provide a new approach to study both structural and functional properties of metaphase chromosome bands. Incorporation of BrdU for an entire cycle followed by 33258 Hoechst staining does not give marked bands but its incorporation for only part of one S phase differentiates between early and late replicating chromosome regions, the latter corresponding to the *Q* and non-centromeric *C*-band positive chromosome segments. After two cycles of BrdU incorporation, sister chromatids can be identified. The random assortment of sister chromatids of homologues may be observed after a third cycle. Baseline sister chromatid exchanges (SCEs)- spontaneous or BrdU-induced-are more frequent in

Q + bands. They are sensitive indices of DNA damage by alkylating agents and light.

Fluorescence Plus Giemsa (FPG) Technique

Protocol 1 For animal tissue

Culture chinese hamster ovary (CHO) cells in McCoy's 5 A medium supplemented with 13% fetal calf serum. Treat the exponentially growing cells with BrdU (final concentration 10 µm) for 24 h during which two rounds of replication take place (Perry and Wolff 1974).

Keep the cultures in dark to minimise the number of sister chromatid exchanges caused by the photolysis of BrdU-containing DNA. Add colcemid (final concentration 2 x 10^{-7} M), keep for 2 h and collect the mitotic cells by shaking. Treat the cells for 8 min with 0.075 M KCl to spread the chromosomes and fix in acetic- methanol (1:3). Place cells in fixative on slides and allow to dry. Stain in Hoechst 33258 (0.5 µg /ml deionised water) for 12 min. Rinse briefly in deionised water. Mount the preparations in water and ring coverslip with rubber solution. Sister chromatids exhibit differential fluorescence under ultraviolet light.

Allow the same preparations to age for 24 h. Remove the coverslip. Incubate for 2 h at 60 °C in either 2 XSSC or water. Stain for 30 min in 3% Giemsa solution (Gurr's R66, pH 6.8). The bands are permanent and do not fade.

To fish, inject 0.5 mg BrdU in Hanks' BSS (Ca and Mg free)/g body weight intraperitoneally. After two to ten days inject 0.25 mg colcemid/g fish, keep for 6 h; sacrifice; treat scales, intestine, gill and kidney in 0.4% KCl for 20 min, fix in acetic- ethanol, prepare slides and follow the same method as for CHO. Three-day old chick embryos can be similarly treated after exposure to BrdU (12.5-50 µg) *in ovo* for 26h.

Protocol 2

Cell culture: Culture chinese hamster V79 cells in alpha-modified Eagles' minimum essential medium (Sigma) with 10% fetal bovine serum (M.A. biproducts) and penicillin-streptomycin. Incubate at 37°C in a humidified CO_2 atmosphere and add 0.02M Hepes to the medium (Ikushima *et al* 1990).

SCE analysis: Inoculate cells (4 × 10^5) into 25 cm^2 plastic tissue culture flasks (Falcon). Add 5 µM 5-bromo-2'- deoxyuridine (BrdUrd) to the culture after 24 hours. Add colcemid (0.2 µg/ml, Gibco) to culture 24 h after BrdUrd addition. After 1 h, harvest cells, detach with 0.25% (w/v) trypsin, centrifuge, treat hypotonically with 0.075M KCl for 18 min at room temperature and fix in acetic acid: methanol (1:3).

Stain slides in 0.5 μg/ml Hoechst 33258 for 20 min, illuminate by black light for 30 min after mounting in 2 x SSC. Stain with 2% Giemsa in Sörensen's phosphate buffer (pH 6.8) for 15 min (modified fluorescence-Giemsa technique).

Protocol 3 For Plant Tissue

Expose lateral rots to aqueous solution containing 100μM 5--BrdUrd, 0.1 μM 5-FdUrd and 5 μM (Urd) for 22 h. An alternative solution is 10^{-7} M FdU, 10^{-4} M BrdU and 10^{-6} M Urd. Transfer to aqueous solution containing 100μM thymidine (dThd) and 5 μM Urd for 21 h. Treat in 0.05% colchicine for 3 h, fix overnight in acetic-methanol cold (1:3) in dark at 20°C. Rinse in 0.01 M citric acid-sodium citrate buffer (pH 4.7), incubate for 75 min at 27°C with 0.5% pectinase, dissolved in same buffer. Squash in 45% acetic acid, coat with a mixture of 10:1 gelatin and chrome alum. Remove cover slip by dry ice and bring preparation to water through descending ethanol grades. Incubate in moist chamber for 60 min at 27°C with RNAse (1 mg RNase in 10 ml 0.5 XSSC) 200μl. Cover. Rinse in 0.5 XSSC, stain 20 min in H33258 (1 mg dissolved in 100 ml ethanol; 0.1 ml of this solution added to 200 ml 0.5 XSSC). Rinse and mount in 0.5 XSSC.

Store over distilled water for four days at 4°C, incubate for 60 min at 55°C in 0.5 XSSC. Rinse in 0.017 M phosphate buffer, pH 6.8, stain for 6-7 min in 3% Giemsa (R 66) solution in same buffer. Rinse in phosphate buffer, then water, air dry, pass through xylene and mount in Canada balsam (Kihlman and Kronborg 1975).

Acridine Orange Staining Method

Grow Chinese hamster cell line D-6 in dark in culture medium containing BudR at concentrations ranging from 0.1 to 40 μg/ml for two cell cycles and fix as usual for chromosome preparation. Stain slides with 0.125 mg/ml acridine orange in phosphate buffer, pH 6.0 for 5 min, wash for 15 min and mount in same buffer. Sister chromatids are clearly demarcated, one brightly and the other faintly stained. AO fluorescence may be quenched partially due to BUdR as reported for 33258 Hoechst fluorescence.

4'-6-Diamidino-2 Phenylindole Staining

Grow mouse cells in modified Eagle's MEM supplemented with 10% fetal calf serum. Incubate the cells in dark and grow in a medium containing 10^{-5} M BrdU for 18-20 h or 30-32 h (one or two cycles of DNA replication). Add colcemid (0.1 μg/ml) to culture, harvest after 2 h by trypsinization and prepare usual air dry smears.

Stain with 1μg/ml of 4'-6-diamidino-2 phenylindole (DAPI) in phosphate buffered saline (pH 7.0) for 10 min, rinse in deionised water, mount in 0.2 sodium phosphate (pH 11.00). Observe chromosome fluorescence with Zeiss microscope with HBO 200 W/4 Hg lamp, dark field illumination, a BG 12 exciter filter and a 470 nm barrier filter.

Giemsa Staining

Protocol 1

Expose exponentially growing Chinese hamster ovary cells to BrdUrd for 24 h. Add colcemid (2 x 10^{-7} M) for 2 h. Collect cells by mitotic selection, centrifuge, resuspend pellet in 0.1 M sucrose for 3 min or 0.075 M KCl for 8 min. Recentrifuge, fix in acetic-methanol (1:3), air dry. Immerse slides in 0.5 µg/ml Hoechst 33258 in deionised water for 12 to 15 min, rinse in deionised water and dry. The slides may be mounted in deionised water with a cover-glass or sealed or placed in a moist chamber. Expose slides for 24 h to daylight, remove cover slip, incubate for 2 h at 62.5 C in 2 XSSC or water. Stain in 3 per cent Giemsa (Gurr R66, pH 6.8), rinse in water, dry, pass through xylene and mount in DPX (Wolff and Perry 1974).

Protocol 2

Grow Chinese hamster ovary cells for two generations in CS-F 10 (calf serum) containing 10^{-4} M BrdU in place of thymidine, with 0.1 µg/ml colchicine for last 2 ml of culture. Shake culture bottles and centrifuge suspension at 90 x g for 5 min at 4°C. Resuspend the pellet with approximately 10^{7} cells in 4 ml hypotonic buffer at 37°C, fix in cold acetic-methanol (1:3), air dry (Korenberg and Freedlender 1974).

OR

Grow human peripheral leucocytes for 72 h in F 10 medium with 15% fetal calf serum, PHA, 10^{-4} M BrdU and for the last 3 h, colcemid at 0.08 µg/ml. Centrifuge cell suspension at 70 x g for 5 min, resuspend and incubate at 37°C for 8 min in 0.075 M KCl. Fix the pellet in acetic-methanol (1:3) and air dry.

For staining, heat slides, at least one day old, at 87-89°C for 10 min in 1.0 M NaH_2PO_4, pH 8.0, rinse in distilled water at 23°C, stain with Giemsa for 2 to 10 min (2.5 ml Harleco Giemsa to 50 ml distilled water), rinse, air dry and mount in Permount. Chromatids singly and doubly substituted with BrdU acquire differential Giemsa stain affinities after treatment at 88°C for 10 min in 1.0 M Na-phosphate buffer (pH 8.0).

Combined SCD Staining and *G*-banding

Treat monolayer mammalian cell cultures with BUdR (5 µg/ml) for two cell cycles. For human lymphocyte culture, add BUdR 24 h after initiation and reincubate for an additional 48 h. Grow cultures in dark. Apply a 2 h colcemid treatment (0.05 µg/ml) before harvesting for usual air-dried preparations.

Protocol 1 trypsin treatment

Dilute trypsin solution used in routine cell cultures and Hanks' balanced salt solution (without Ca and Mg) to 20% of its original strength. Treat slides-two days to two months old-with this solution for 1 to 2 min, rinse successively in Hanks' BSS, 70 and 95% ethanol and air dry. Stain in 2% Giemsa solution in 0.01 M phosphate buffer (pH 7.0) for 2 min and rinse in deionised water.

This procedure gives both sister chromatid differentiation (SCD) and *G*-banding. Short treatment produces SCD; a slightly longer one yields *G*-bands on the bifilarly substituted chromatid and longer treatment gives bands on both chromatids. Continued treatment obliterates SCD pattern and only *G*- bands are seen.

Protocol 2 urea treatment

Prepare 8 M urea solution, mix with M 15 Sörensen's buffer (3:1) and keep at 37°C. Treat slides in this solution for 5- 15 *s*, rinse and stain in Giemsa as given in protocol 1. The SCD pattern is formed at first and is replaced by *G*-banding. Alternatively, *G*-banding may be induced first by urea treatment and later SCD induced in the same cells by the earlier technique.

Premature Chromosome Condensation-Sister Chromatid Differentiation (PCC-SCD) Techniques for Chromosome Replication Studies

Evidence for chromosomal replicons as units of sister chromatid exchanges

BrdUrd treatment in vivo:

Implant one paraffin-coated 50mg BrdUrd tablet subcutaneously in the lower right abdominal region of mouse, aneasthesized with Metofane (methoxyflurane: Pitman-Moore). Sacrifice by cervical dislocation after 21-22h. Remove both femurs. Flush out bone marrow and prepare air-dried slides after hypotonic treatment and fixation as described in earlier schedules. Immerse air dried slides in Hoechst 33258 (Sigma Chemicals, 50µg/ml in distilled water) at room temperature for 10min. Rinse slides twice in distilled water, mount with MacIlvaine's buffer, pH 8.0. Place on a 50°C slide warmer and expose to a blacklight (GE-F15T8/BLB) at 2 in for 20min. Stain slides in 8% Gurr's Giemsa stain (Bio/Medical Specialities) containing 20% MacIlvaine's buffer for 6-10min at room temperature.

The observation of sister chromatid differentiation (SCD) requires incorporation of 5-BrdUrd into cells through two consecutive S-phases for *in vitro* studies.

'Reverse' Staining Method

Prepare chromosome slides from cultures continuously labelled with BrdU (5-10 µg/ml final concentration). Incubate in trypsin solution (0.1% in PBS, pH 7.2) for 2 *s* or more, and rinse. Stain in basic fuchsin solution for 4 min. The solution consists of 0.1% NaOH, adjusted to pH 10.2 with the aid of HCl. Rinse, dry and examine. Metaphases treated in this way show one heavily stained and one pale chromatid. The pattern is the reverse of that obtained with Giemsa methods for differential staining of sister chromatids. Use diluted enzyme solutions or staining solutions with reduced pH for preparations which are very sensitive to trypsin. The method has been successfully applied to fresh as well as to 7 month old preparations (Scheres *et al* 1977).

Table : Concentrations of some restriction endonucleases for human chromosomes

Enzyme	Final conc in working solution units per µl buffer	Incubation time (h)	Type of band	Ref
Isoschizomers NdeII-Sau3AI (↓ GATC) and MboI	2U/µl	6 h	NdeII similar to MboI; sau3AI different	Ludena *et al* 1990
PleI (5'-GAGTCN4 ↓N5-3')	0.5U/µl	16 h	Discriminate HinfI target sites Reveal regions within major C bands that include the major sites of SatII DNA	(A or T as central base) Tagarro *et al* 1992 a
TfiI (5'-G↓A (A/T) TC-3')	1U/µl	16hs		
TfiI (GA [A/T] TC)	0.5 - 1.0U/µl	16 h	Do	Do 1992 b

BrdU-33258H Analysis of DNA Replication

Culture peripheral lymphocytes at 37°C in Eagle's MEM with 20% fetal bovine serum, 2 mM L-glutamine and saline extract of red kidney beans (PHA).

To find late-replicating regions in S-phase, grow cells for 44-46 h in medium containing 10^{-4} M 5-BrdU, 4 x 10^{-6} M 5- fluorodeoxyuridine (FdU) and 6 x 10^{-6} M uridine (U). Replace it 5-9 h before harvest by Ham's F-10 medium containing 20% fetal bovine serum and thymidine (dT), 0.6 or 1.2 × 10^{-5} M. In most cases, duration of the terminal thymidine pulse is 5-7 h.

To suppress fluorescence of regions last to complete DNA synthesis, a converse 'B-pulse' protocol may be used, in which nearly three days of growth in control medium is followed by a 5- 7 h pre-harvest pulse of BrdU. FdU and U at same concentration as above. A modification of both protocols is the addition of 10^{-4} M deoxycytidine (dc) together with the BrdU.

For harvesting the cells, centrifuge 2 h after adding colcemid (0.1 µg/ml) and treat successively with 0.075M KCl + acetic-methanol (1:3). Stain slides with 0.5µg/ml 33258H in 0.14 M or 0.40 M NaCl, 0.004 M KCl, 0.01 M phosphate, pH 7.0 and mount in 7.0 or 7.5 pH McIlvaine's buffer for fluorescence microscopy.

Incubate the same slides, not subjected to autoradiography, to 30 min at 65°C in 2 XSSC (pH 7.0) and stain 20-30 min in 20% Gurr's R66 (pH 6.8). For autoradiography dip the slides in Kodak NTB-2 emulsion:water (1:1), expose for two to three weeks and develop with Dektol.

Dot Formation on Chromosomes

Dotted chromosomes are produced in both BrdU and non-BrdU- substituted Chinese hamster cells after treatment with 1.0 M Na- phosphate solution (pH 9.0), with a supersaturated amount of $NaHCO_3$ at 80-95°C. *The temperature needed is always slightly higher than that needed for differential staining of chromatiads.*

SCE

Grow Chinese hamster cells in monolayer for 24 h in McCoy's 5A medium supplemented with 10% fetal calf serum. Add BrdU at a final concentration of 10 µM and incubate in dark at 37°C for 20 h for twice replication in BrdU without photolysis. Two hours before harvesting, add colcemid at a final concentration of 0.03 µg/ml. Collect arrested metaphase cells by shaking culture bottles or by gently pipetting medium over monolayer for 5 min; treat with 0.075M KCl for 10 min at 37°C, fix in acetic-methanol (1:3), air dry. Heat 1.0 M NaH_2PO_4 in distilled water to 90°C, add solid $NaHCO_3$ for supersaturation (pH 9) with $NaHCO_3$ precipitate. Treat slides in this solution for 3-4 min at 30-95°C, rinse in distilled water, stain for 3-5 min in Giemsa (2 ml Fisher Scientific in 40 ml Sörensen buffer at pH 6.8); rinse in two changes of distilled water and air dry.

G-banding and SCD appear prior to dot formation. The late DNA-replicating sites become unifilarly BrdU-substituted and are identified by pale colour or dot formation.

Table : Concentrations of some restriction endonucleases for human chromosomes

Enzyme	Final conc in working solution units per µl buffer	Incubation time (h)	Type of band	Ref
Isoschizomers NdeII-Sau3AI (↓ GATC) and MboI	2U/µl	6 h	NdeII similar to MboI; sau3AI different	Ludena *et al* 1990
PleI (5′-GAGTCN$_4$ ↓N5-3′)	0.5U/µl	16 h	Discriminate HinfI target sites	(A or T as central base)
TfiI (5′-G↓A (A/T) TC-3′)	1U/µl	16hs	Reveal regions within major C bands that include the major sites of SatII DNA	Tagarro *et al* 1992 a
TfiI (GA [A/T] TC)	0.5 - 1.0U/µl	16 h	Do	Do 1992 b
TaqI (5′...T↓CGA...3′)	2.0U/µl	6 h under coverslip	Under controlled conditions, DNA loss is seen in sat II and some sat III DNA domains rich in Taq I sites	Tagarro *et al* 1990

CHAPTER 4

AUTORADIOGRAPHY — LIGHT, FIBRE AND HIGH RESOLUTION

A. LIGHT MICROSCOPE AUTORADIOGRAPHY

The purpose of autoradiography is to locate radioactive material in a specimen with the help of photography. Whereas with the Geiger counter, an electronic wave is measured by an electrical device, in autoradiography, the radioactive material is detected bya photographic process of development. The method is based on the principle that if a photographic emulsion is brought into contact with radioactive material, the ionizing radiation will so convert the emulsion as to show spots at certain points after being developed.

Specimen and film are brought into contact with each other for a certain period of exposure and there is decay of radioactive atoms. The radiation thus emitted affects the emulsion, activating silver halide crystals. The final result is the formation of a latent image which can be developed to denote the location, intensity and distribution of radioactive material. If the radioactive substance is tagged with a metabolic precursor, the distribution of the grains in the autoradiograph will indicate the distribution of the precursors and thus the metabolic path can be determined with accuracy.

The stripping film method was considered, for a long time, to be advantageous for the study of chromosome structure, bringing out a sharp contrast between radioactive spots and background tissue. In this technique, the outline of the schedule is to cover the preparation with a sensitive emulsion with a bottom layer of inert gelatin against a glass plate.

However, application of emulsion in the liquid form has been found to be most suitable for autoradiography. It is now widely used for plant, animal and human tissues and also for their cultures. This method allows the formation of a monolayer of emulsion on the tissue. In the study of high resolution autoradiographs (*see* section on high resolution autoradiography), the liquid emulsion technique has been found to be the most convenient one.

In order to use a radioactive molecule as a tracer, it is necesary to utilize the specific precursors of the molecule, the distribution of which is to be studied within the cell. For this purpose, in the study of chromosome metabolism, labelled uridine is used for the detection of RNA, thymidine for the detection of DNA, and specific amino acids for the detection of proteins. With ^{32}P alone, the localization of DNA is diffficult, as phosphorus is incorporated in different metabolic products of the cell unless such products are extracted or digested. The measurement of radioactivity for quantitative study, however, depends on several factors, such as the amount of radioisotope, its half-life, and the energy of β-rays emitted.

To obtain an accurate autoradiograph, the tissue should be treated with at least a certain minimum concentration of the labelled substance. Approximately 10 grains/100 μm^2 of emulsion is just enough, requiring an expsure of 10/δ/β-particles (where δ denotes the yield of developed grains per particle hitting the film), which were obtained from the decay of double the number of radioactive atoms at the time of exposure. The half-life should also be considered, because for short-lived isotopes, the exposure of two half-lives results in the decay of three-quarters of the atom. In consideration of the various factors, the minimum concentration *c* for short-lived and long lived isotopes has been suggested (Pelc, 1958) as:

For *short-lived isotopes*: (half-life *H* in days, in 5 μm sections and emulsions, and *f* denoting the proportion of labelled to unlabelled tissue)

$$C = \frac{11.5f}{d^{\delta}} \text{ in } \mu{}^{\circ}\text{Ci/ml.}$$

For long-lived isotopes: $C = \frac{12.5f}{d^{\delta}}$ in μCi/ml.

The isotope being long-lived in this case, the decay during the time of exposure (*d*) is negligible.

For fine-grained stripping film, the value of δ can be taken as I. In addition to the factor of minimum concentration, the actual amount of radioactive substance required for feeding the tissue depends on the metabolic role of the metabolite in which it is supposed to be incorporated. For example ^{32}P, when injected into the tissue, mixes with the free phosphate in the organism. To obtain an autoradiograph, therefore, a sufficient quantity of tracer is needed, but if the tracer is labelled with a metabolite present only in small amounts in the tissue, the quantity of tracer needed will be very low. A specific example is seen in mice, where labelling of DNA requires at least 1 mCi of ^{32}P per mouse while ^{14}C-labelled adenine is required in an amount of 25 μCi. Similarly, in leucocyte culture from peripheral blood, application of tritiated thymidine at 1μCi/ml for 5-6 h before harvesting is needed for incorporation in chromosomal DNA.

Technical Steps in the Preparation of Autoradiographs

The different steps followed in autoradiography are: (a) administration of the tracer into the tissue; (b) fixation; (c) paraffin embedding or smearing; (d) staining; (e) application of the photographic emulsion; (f) drying; (g) exposure, and (h) photographic process. Staining can be done before or after the application of the emulsion. For details please *see* section on high resolution autoradiography.

The radioactive tracer can be obtained is specific salt solutions or tagged with metabolic precursors. For instance, ^{35}S is available as H_2SO_4 in dilute HCl solution or as sulphate in isotonic saline solutin. Tagged isotopes are available in the form of tritium (^{3}H) labelled thymidine, ^{14}C-labelled adenine, thymine, uracil, etc., representing nucleic acid bases. Similarly, for proteins ^{35}S-labelled methionine, phenylalanine, etc. are used.

Fixation can be performed either through freeze substitution or through a number of non-metallic fixatives; however, the fixative should not leach the radioactivity during preparation. A method for slow fixation at –8°C for 4 h, followed by washing in ethanol has been devised. As far as practicable, metallic fixatives should be avoided: ethanol, a mixture of ethanol and acetic acid, neutral formalin, and formol-saline can all be employed for fixation, especially where isotopes are applied for incorporation into nucleic acid or protein of chromosomes, but the most reliable method so far found is freezing.

In order to secure a high resolution *in paraffin sections*, the sections should not exceed 5 µm in thickness, should preferably be cut in a freezing microtome, and should be mounted on slides coated with a film of alum—gelatin (0.5% aqueous solution of gelatin and 0.1% chrome alum) or with egg albumen. Absolute drying for at least 48 h is necessary before mounting the sections. Similar slides should be used in *the case of smears* though preparation made with ordinary slides have occasionally yielded good results. In order to soften the tissue for smearing after fixation, the use of pectinase or cellulase is preferable instead of prolonged hydrolysis in N HCl, which often results in the removal of nucleic acids:

Staining of the preparations can be done prior to, or after, the application of the photographic emulsion, but in the former, the protection of the material by a very thin layer of impermeable substance, like celloidin, may be necessary. Experience in this laboratory has shown that uncoated tissue, especially in the study of chromosomes, allowed good resolution. Leucobasic fuchsin, methylgreen-pyronin, toluidine blue and Leishmann-Giemsa are satisfactory stains for autoradiography.

The photographic emulsion can be applied over the section either in the form of a liquid or smear, or as a film pressed on the material, being composed principally of a fine layer of gelatin, containing mumerous silver halide crystals. The formation of radioactive spots is based on the principle of the production of ion pairs due to an electronic event, ultimately manifested in the single gtrain of black silver. Nuclear emulsion of small grains should not be used because of the possibility of their being washed off during washing. The emulsion should be stored at 4°C to prevent melting and should be applied at 25-30°C.

The tissue, after being coated by emulsion, either in liquid form or else as stripped film, must *be dried quickly*, preferably in a strong current of air in a cold chamber. Quick drying is necessary to prevent the formation of air bubbles within the tissue and also the production of artefacts in the presence of excess moisture.

The *period of exposure* in a cool dark chamber depends on the type of isotope used and its concentration, and is the period required for a specified number of ionized particles to heat a unit area. The time needed for any emulsion can be increased to secure a satisfactory image with a smaller dose—from a few days to even one or two months continuous exposure may be necessary in certain cases.

The photographic process to secure autoradiographs consists of the following steps: development of the latent image, fixation, washing and drying. The principle of the first step is to develop the latent image in an aqueous solution of a reducing agent in a dark chamber. The widely used developers are hydroquinone and methyl-*p*-aminophenol, which have the property of reducing the silver halides. The fixation of the intact image involves removal of the silver halide from the emulsion and hardening of the image. In autoradiography the use of an acid-hardener fixing bath becomes necessary for the protection of film or plate. Continuous washing in running water is necessary to remove traces of any excess sodium thiosulphate, and a temperature of 15.5-21.0°C speeds the washing and prevents softening of the emulsion.

Representative Schedules for Light Microscopy Autoradiography

Method of Administration of Isotope

For plants

For the study of somatic chromosomes, grow young seedlings of *Vicia faba* in medium containing 2 μCi of ^{32}P (as orthophosphate)/ml in tap water. Fix young healthy roots at different intervals ranging from two days to one month in acetic-ethanol or chilled 80% ethanol.

For the study of meiosis in plants, 20-75 μCi of ^{32}P in 1 ml of water can be administered on flower buds of *Tradescantia paludosa* or *Rhoeo discolor*, or 4μCi/ml labelled ^{32}P in White's medium, on inflorescence of *Lilium henryii*.

For animals

For the study of nucleic acids in mammalian chromosomes, 40 μCi of ^{32}P/g can be injected intraperitoneally in rodents. Similarly, ^{14}C adenine can be injected in the dose of 0.3 μCi/g and the animal can be killed 24 h after injection. Fixation can be made in acetic-ethanol mixture. For salivary gland chromosomes of *Drosophila*, autoradiographs can be prepared after feeding *D. melanogaster* with 20 μCi/ml of ^{14}C adenine for at least 24 h.

For human material

Human bone marrow cells can be cultured in a medium containing 0.5 μCi/ml of ^{14}C adenine or 1 μCi/ml of ^{32}P. Fixation is performed in methanol. For human peripheral blood culture autoradiograph, preparations can be made following the method of Moorhead and colleagues (1960). Tritiated thymidine (1 μCi/ml) may be added 5-6 h before termination of the culture. Colcemid may also be added 2 h before harvesting the cells. Air-dried preparations can be made after treatment in hypotonic salt solution and staining the chromosomes with 2% acetic-orcein solution or Giemsa. Slides are then covered with liquid emulsion (Ilford K5) for two days' exposure. Developing is done in Amidol solution for 4 min, followed by fixing in 30% sodium thiosulphate solution for 7 min. Slides are dried in air and mounted in DPX.

Liquid emulsion autoradiography for plant, animal and human materials

In order to study materials given labelled amino acids or nucleosides, acetic-ethanol (70-95%) 1:3 fixative is recommended. Stock solution of the emulsion should be kept at 22-24°C. Kodak NTB series is used with different grain sizes. With decrease in grain size, higher resolution and sensitivity are expected. The operations must be performed in a dark room with a covered safe light (15 W bulb—Wratten Series 1).

Melt the emulsion at 42-45°C using a constant temperature water bath. To avoid background effect, it is better to develop a clean dry plate, without any material, as control.

Subbed slide is not necessary if Kodak emulsion is used. Fit two slides with materials back to back and dip in emulsion, kept in a long trough, for 4-5*s*. Drain off the emulsion., After separating the slides, dry in racks against a stream of air and keep in slide boxes, sealed with tape, in a cool dark chamber, for the required period of exposure (2-3 weeks).

After exposure, develop the slides in Kodak Dektol, D-11 ro D-19 developer for 2 min.

Fix in Kodak Acid-Fixer for 2-5 min.

Rinse in running water for 20 min. Give a final rinse in distilled water and dry.

For staining, three alternative stains are recommended:

(1) 0.25% aqueous solution of toluidine blue (pH 6);

(2) methyl green-pyronin;

(3) Giemsa stain.

Wash in 95% ethanol, make air-dry preparations and mount in euparal.

Stripping film autoradiography of peripheral blood culture for human chromosomes (also applicable to bone marrow and fibroblast cells)

For culture: Allow 10 ml of heparinized blood to stand for 2-3 h at 22-24°C. Take 2-4 ml of supernatant in a 2 oz flask and make up to 10 ml by adding the medium (McCoy's or Eagle's medium + 20% fetal calf serum). Add 0.1 ml of phytohaemagglutinin M or P.

For continuous labelling of DNA, add ^{3}H-labelled thymidine (1 μCi/ml of medium) 6 h before termination of culture and add colcemid at a final concentration of 0.03 μg/ml. For short pulse labelling, monolayer cultures are necessary, cells need to be centrifuged and colcemid at a concentration of 0.06 μg/ml is needed.

For squash preparations from leucocyte suspension or fibroblasts, detach cells with 0.2% trypsin and centrifuge in 12 ml conical tubes for 4 min at 500 r/min and pour off the supernatant. Add 10 ml of 1% sodium citrate and leave for 10 min. Re-centrifuge and add 10 ml of 50% acetic acid without disturbing the cell pellet. Decant off the fixative after 20 min and make air-dried smears. If staining is desired, add 2% acetic-orcein and squash. Follow dry ice schedule for permanent preparations, treat in absolute ethanol and store the dried slides. For blood culture, trypsinization is not necessary and the usual schedule of air-dried preparations can be followed without staining.

All operations should be performed in the dark room in the safelight at 20-22°C.

Cut 40 × 40 mm squares of AR 10 film mounted on glass slides and keep the plate for 3 min in 75% ethanol. Transfer to another tray containing absolute ethanol. With forceps, lift a single square of film and float, with emulsion side down, in a tray of distilled water, so that the film automatically spreads out. Bring the glass slides containing the material below the film under water and lift the'slides so that the film adheres neatly on the surface of the slide with the material on it. Dry in a stream of air and store in a slide box, sealed with tape, containing silica gel desiccant for the required period in the cold. Prior to developing, paint the reverse side of the slide with a paint and dry. Develop with Kodak developer D-196 for 5 min. Treat with the acid fixer as usual, wash and dry. Detach paint and film with a blade and follow the usual procedure for mounting. For staining, where this has not been performed before, stain with Giemsa (distilled water, 100 ml; 0.1 M citric acid, 3 ml; 0.2 M Na_2HPO_4, 3 ml; methanol and Giemsa stock solution, 5 ml) for 4—7 min. Rinse in distilled water and dry. Detach paint and film and follow the usual procedure for mounting.

B. FIBER AUTORADIOGRAPHY

The autoradiography of the spread fibers of DNA is otherwise termed fiber autoradiography. It permits a study of replicons and their spacing as well as the duration of the "S" phase. It also enables one to analyse the degree of synchrony of the onset of replication. The duration of the "S" phase in different tissues is a function of development and differentiation. However, the space between the replicons has been shown to be an important factor in controlling "S" phase duration.

Protocol

Use phytohaemagglutinin-stimulated culture in Eagle's medium supplemented with AB serum, in case of cultured cells or during premeiotic spermiogenesis.

At 2 h before termination of cultures, give two pulse treatments of thymidine: 2 h before harvesting, hot pulse (Sp. act. 45 Ci mmol, 100 Ci/ml Radiochem, Amersham) for 1 h and warm pulse (Sp. act. 12.7 Ci/mmol : 20μCi/ml) - 1 h.

Stop ^{3}H TdR uptake by repeated washing in calcium or magnesium-free phosphate buffered saline (PBS). Suspend the cells in PBS highly diluted with water for swelling and spreading.

Take a subbed slide and add 1% SDS on a drop of suspension containing sufficient number of cells. Spread the suspension and air dry. Treat the slides with cold TCA for 5 min. Wash in running water and rinse in ethanol. Coat the slides with equally diluted E_4 emulsion. Keep in dark for 6-10 months. For developing, use Kodak D19 b developer for 10 min at 10°C and photograph autoradiograms. Measure the track length in photographs and measure replicon spacing.

C. HIGH RESOLUTION AUTORADIOGRAPHY

Since the pioneering work of Liquier-Milward (1956) on the combined use of electron microscopy and autoradiography on tumour cells, high resolution autoradiography has proved to be an effective tool in the study of ultrastructure as correlated with function.

High resolution technique has been used on *in situ* RNA/DNA hybrids and its combination with immunofluorescence has allowed thedelimitation of functionally specialized segments of chomosomes. As with autoradiography in general, this aspect of the subject has been much refined through the use of tritiated compounds of high specificity available in forms with specific activity even more than 15 Ci/mM— Schwartz Bio Research Inc. They have a very short range of radiation (0.018 MeV—1 μm), and thus eliminate the difficulty of using high energy β particles emitted by ^{14}C (0.155 MeV/40 μm in water) where the range of radiation exceeds the thickness of emulsion. Moreover, the use of fine-grained nuclear emulsions, with grains as small as 0.1 μm (Ilford Nuclear Research Emulsion K5-

L4) has facilitated the preparation of emulsion of uniform thickness and rendered the technique more convenient. The need for an autoradiographic resolution of less than 1 μm, prompted research on the modification of autoradiographic techniques used in the case of ultrathin sections observed under the electron microscope. This requirement was due to the fact that the size of the grains reaches beyond the limit of resolution of light optics and the thinness of the section does not allow sufficient contrast under light microscopy.

It has, however, been found necessary to observe thicker sections (0.3-0.5 μm) under light microscopy, prior to the sudy of high resolution ultrathin autoradiographs, principally to get an idea about the approximate exposure time needed for ultra thin autoradiographs and for their comparative assessment. Experience has shown that exposure time needed for ultra thin sections is about 10 times more than that required for comparatively thicker sections.

In addition to the need to predict the exposure time for ultra thin sections, light microscopy is required to obtain a demarcated picture of the area to be studied in ultra thin sections. For both these purposes, the desirable thickness of the sections is between 150 and 500 nm.

Fixation and Embedding

In the analysis of chromosome structure, osmium tetroxide solution buffered to pH 7.2-7.4 is often recommended. The addition of divalent cations like calcium (10^{-2M}), in the fixative, checks swelling and helps in maintaining uniformly the packed macromolecular configuration as observed in erythrocyte nuclei. Lafontaine (1968) obtained very satisfactory results with chromosomes of *Vicia faba*, by fixing in 1% unbuffered (pH 6.0-6.4) solution of OsO_4 in double distilled water, to which varying amounts of calcium chloride were added. Freeze-drying methods can also be adopted for dehydration after quick freeze fixation.

For embedding, any of the usual media, such as methacrylate, epoxy resins, araldite, Epon or polyester, Vestopal W, can be employed.

Section Cutting

The slides have to be kept ready before cutting the sections. Clean slides with frosted ends should be dipped in the subbing solution (1 g Kodak purified calfskin gelatin is dissolved in 1litre hot distilled water, cooled, 1 g chromium potassium sulphate is added and the solution stored in the cold). The subbed slides are dried in a dustfree chamber and stored in boxes.

To secure thick sections meant for predicting the exposure time needed for ultra thin sections, as well as for comparative assessment, the block is trimmed so as to obtain a much larger face than that needed for ultra thin sections. As a result, more material, greater ease of operation, and serial ribbons can be secured. A glass knife with a smooth cutting edge is used and is adjusted with a metal or a tape boat. A metal boat is preferred since it presents a larger area. After mounting on the microtome, sections of desired thickness (150-500 nm) can be cut by setting the section indicator and observing the interference colour by adjustments of the water level and illumination (*see* Reid, 1974).

When a ribbon with 2-4 sections is cut, the sections are picked up with a damp, fine-haired, clean nylon brush and transferred to a drop of water placed near the edge of a subbed slide. The sections adhere to the end of the brush where the bristles are narrow rather then in the middle, where it would be difficult to separate the section from the brush. The slides are then dried at 45°C or at 60-80°C.

In order to locate specific regions, the block is trimmed in such a way that ultra thin sections can be obtained. A slightly better quality of glass knife is selected. First a thick section (120 nm) is cut and transferred to the slide by the method given previously, followed by several ultra thin sections (60-100 nm) which are shifted on the boat, prior to cutting another thick section of the original thickness. The latter is also mounted on the slide and both are observed. If the desired region is present in both the thick sections, the intervening thin sections are mounted on the grid, since the presence of the desired zone or material in the ultra thin section has been ensured.

The section may be marked with a diamond pencil on the underside of the slide. It is preferable to observe the very thin sections under the dissecting microscope and to mark the locations. Sections 200-500 nm thick may be even examined under phase contrast microscope without staining and comparatively thinner sections (150-200 nm) may be examined following staining with aniline dyes. To find out the exposure time needed for ultra thin sections, 2-4 days exposure on 0.4 µm section gives an approximate indication.

Coating of Ultra Thin Sections

Section embedded in methacrylate, epon, araldite or Vestopal may be used. The success of the operation principally depends on obtaining fine compact monolayer of silver halide crystals. With increase in the thickness of the layer, the sensitivity increases, at the cost of resolution. Selection of a proper emulsion is one of the most important factors in high resolution autoradiography. Its sensitivity depends on the extent to which it can register and develop the latent images formed by electrons in their path on silver halide and is measured by the number of grains developed per unit distance in the track of particles with minimum ionization. The particle energy and the distance the electron has to traverse through the silver halide, control the formation of the latent image. Normally, ^{32}P, ^{131}I, ^{14}C and ^{35}S have long range ionizing particles. With tritium (^{3}H), nearly all electrons emitted into the upper hemisphere can be developed. During exposure, oxidation of the latent image affects sensitivity. Protection against oxidation becomes essential with smaller crystals and fine-grained development. With tritiated compounds, the problems of sensitivity and resolution are not so severe as the emitted β-particles are heavily scattered within one silver halide crystal of Ilford L4 emulsion. With finer grained emulsions, such as Gevaert 307 or Kodak NTE, (crystal size 50 nm), multilayered crystals add to the sensitivity. More suitable are Ilford L4 (crystal size 120 nm) for electron microscopic and Ilford K5 (180 nm) foe light microscopic autoradiography. A closepacked monolayer of silver halide adds to resolution by preventing the spread of electrons from the source.

For mounting the sections, both the collodion film and the sections must be perfectly smooth. Electroplated Athene-type copper grids are used for coating the collodion with a thin carbon layer. For extra strength, the grids are generally covered with 0.25 or 0.5% parlodion-carbon and dried. A thin film is spread so that resolution is not hampered and at the same time, breakage is avoided during the procedure. Sections are mounted on the grid which is add at one edge by a piece of scotch tape (double coated) to a slide. Several grids can be placed on one slide.

Various methods have been proposed for applying a uniform layer of emulsion. It may be applied, either by dipping the slide in the emulsion or dropping it on the slide (5 ml of distilled water per 1 g of emulsion). A thin layer may be allowed to form on a specially constructed loop before applying on specimen grids. The emulsion may be centrifuged directly on the specimen grids or may be finely layered on agar before application on sections. Salpeter (1966) suggested the preparation of a substrate of uniform property before applying the emulsion and recommended the formation of the emulsion layer on a fine layer of carbon or silicon monoxide, dried on sections mounted on collodionated glass slides.

Thickness of Emulsion Layer

The thickness and uniformity of the emulsion layer should be checked by the interference colour, based on the principle that interference colours of emulsion layers in reflected white light depend on their thickness. This can be viewed even in the dark on density differences using a yellow safe light (Filter AO). The most appropriate method for ascertaining the thickness needed for quantitative work is to use a developer (Devtol) which does not affect silver halide crystals. After developing, and prior to fixation, the slides can be air-dried and viewed in white light for interference colours. It is always desirable to use Dricrite in the storage chamber to keep the slides dry. The time of exposure has to be deduced from the time required for thicker (0.5 μm or so) sections as mentioned above.

Developing

After adequate exposure, developing of the photographs should be carried out in clean and dustfree conditions. The use of developer is meant to reduce the exposed silver bromide crystals, carrying the latent image to metallic silver. It is always necessary, by trial, to find out the optimum period of development which would permit the maximum number of grains to be developed with least background effect. The grids must always be kept in absolute ethanol for 3-4 min before developing. This hardening schedule is an essential step as it checks the sudden swelling caused by aqueous developer and the resultant loss of grains. In all the steps followed for taking electron microscopic autoradiographs, the problem of contamination is a serious one. To check against this limitation, it is always necessary to use a small quantity of fresh, filtered solution for each plate.

Several developers are in vogue and their adequacy depends on the type of emulsion used for coating. *Chemical developers, which reduce silver halide crystals* result in a coil of silver filament of 0.3-04 μ diameter in certain developers like D19, or a long filament as in Microdol X. Since the latter is comparatively easier to interpret in that the initial image may be considered in the middle of the line, it is often preferred. In a procedure used, Athene-type grids are placed on filter paper for drying after being lifted from ethanol with forceps. The dried grids should be floated in an inverted position on the convex surface of the developer and kept for 6 min at 22-24°C. After immediate transference to a watch glass containing distilled water, for a few seconds, the grids should be placed in the fixer with the sections facing upwards. The fixer effectively removes all unexposed silver halide crystals and helps in the later removal of the gelatin. With *physical developers*, the highest possible resolution can be obtained. The principle is the *dissolve completely the silver bromide crystals*, only the latent image with silver ions is kept, with the use of 1.0 M sodium sulphite and 0.1 M 4-phenylene diamine, on silver Mitrate in varying proportions. Development for even 1-2 min at 20°C is sufficient. This developer is, however, comparatively unstable. With this method, the latent image can be localized with the least possible error and caution is needed becaused grain size is very small.

Staining

The gelatin may or may not necessarily be dissolved for staining after the photographic processing is completed. Staining can be performed even before applying the emulsion. A serious drawback of keeping the gelatin intact is the possibility of disruption of the grains through shattering of the gelatin layer by the electron beam. The removal of gelatin may be carried out by sublimation by gradually increasing the intensity of electron beams. In a modification of Grandboulan's technique, prior to staining, gelatin was first dissolved by acid hydrolysis.

In this schedule the grids are first floated for 30 min on the convex surface of distilled water, kept at 37°C and then transferred to 0.5 N acetic acid at 37°C and kept for 15 min, followed by rinsing in a stream of distilled water and then floated in a second change of distilled water at 22-25°C for 10 min. The staining is performed by first wetting the sections with drop of distilled water, staining in an inverted position in 7.5% aqueous uranyl acetate for 20 min at 45°C, subsequent drying on filter paper, wetting again with distilled water and then further staining with 0.2% lead citrate for 10-60 *s* to a few minutes. Uranyl acetate changes the properties of the macromolecules in such a way that they bind better with lead citrate. In order to avoid the possibility of removing or damaging the silver grains, following a method of staining prior to the application of the emulsion, a few drops of lead stain, or lead citrate staining for 5-30 min preceded by aqueous uranyl acetate or uranyl nitrate may be added to the sections on the slide. The excess stain is immediately flushed off with a stream of distilled water. Evaporation of carbon layer (5-10 nm Union Carbide SPK spectroscopic carbon) over the sections is necessary for screening the stained sections, and the emulsion, and also for providing a base for the emulsion layer.

Protocol for Spreading and Autoradiography

For short labelling, incubate the exponentially growing cells for a very short period with ^3H (80-100 μCi/ml, sp. act. 25-30/mmol CEA or Amersham and *for longer incubation* of 1 h and more, with tritiated precursor 50-30 μCi/ml. After labelling, transfer the culture to ice cold phosphate sucrose soln (0.2 mM Na_2PO_4, 0.1 M sucrose, pH 7.5). Centrifuge the cells at 304 g for mouse and 580 g for Drosophila cells and resuspend in 1 ml of the same soln. Add drops of 1 ml Nonidet P40 (NP40 Shell) in 0.2M EDTA, pH 7.4 with stirring. Dilute cell lysate with 50 ml of 0.2 M EDTA (pH 7.5) and keep in ice.

Layer 30 μl of lysate on the top of 4% formaldehyde, 0.1M sucrose soln., pH 8.5, in a translucide plastic chamber, the bottom of which contains freshly glow-discharged copper or gold EM grid coated with Formvar-coated membrane. Centrifuge the material at 2400 g at 4°C for 10 min. Remove the grid from the chamber and treat for 30 *s* in 0.4% Photoflo 600 (Kodak), pH 7.5-7.9. Air dry.

Stain with 1% PTA in 70% ethanol for 1 min. Dehydrate in 95% ethanol for 20 *s* and air dry. Rotary shadow the grids (angle 7°C-10°C) to obtain thin layer of platinum. Apply Ilford L4 emulsion using loop technique, prepared the previous day by dissolving gel shreds in water at 40°C and kept in cold. Develop in preparation following Elon ascorbic acid procedure following 2-8 months exposure.

Bring the material to room temperature and use safe light. Dip the holder into the intensification bath for 5 min. Transfer into dist. water bath for 10-20 *s*. Transfer into Elon ascorbic acid developer for 7.5 min. Transfer to fixing bath for 2 min. Wash in three successive baths of dist. water. Observe in EM using suitable objective aperture (30-40 μm).

CHAPTER 5

ELECTRON MICROSCOPY

The introduction of electron microscopy in the study of ultrastructure has helped to clarify chromosome structure at the submicroscopic level. The high magnification attained, coupled with good fixation and proper preservation of components *in vivo,* has resolved the inner details of chromosome and other cellular components. In electron microscopes, electrical or magnetic fields are adequately shaped to refract electrons, producing the desired image. The transformation of the electron image to a visible light image is effected on the fluorescent screen.

The basic constituents of the electron microscope are assembled in a vertical column. They are: an illuminating system, a specimen chamber, an objective lens, intermediate and projecting lenses, and a viewing chamber with facilities for photographing electron images. In the entire column, vacuum is maintained by oil and mercury diffusion pumps, at specific pressure. The source of the electron providing the illuminating system is generally constructed from a hot tungsten filament, serving as a cathode, and an anode with a surrounding Wehnelt cylinder, provided with a hole for the passage of the electron beam. The resultant image is detected either on a fluorescent screen or on a photographic plate. For details of instrumentation and the working of the microscopes, such as, RCA-EMU, Philips EM, Hitachi, Siemens ELMISCOP, Zeiss EM, Jeol JEM, AEI Corinth, Bendix-Akashi, etc., the reader is referred to Agar, Alderson and Chescoe (1974) and Meek (1976). To allow the passage of electrons ejected from a source or their penetration, and for good resolution of the object, ultra-thin sectioning in ultramicrotome is essential. For the preparation of materials to be observed under the electron microscope, the following steps should be adopted under strictly controlled conditions: (a) fixation, (b) dehydration, (c) embedding, (d) sectioning and (e) mounting.

FIXATION

With low temperature fixation, the life-like preservation of cells is maintained because of the absence of any chemical compounds. Frozen materials, after dehydration, either by sublimation of ice as in freezedrying, or by substitution with organic solvents as in freeze-substitution, have decided advantages in electron microscopy. These techniques are generally recommended for the study of enzyme localization and activity and are not applied directly to the analysis of chromosome structure. But the potential of their use in chromosome study cannot be ignored; especially in research involving the distribution of enzymes, such as alkaline phosphatase in chromosomes.

A chemical fixative should provide the natural milieu of a living system, especially with regard to pH, osmolarity and ionic concentration. This is necessary to prevent swelling, shrinkage or extraction of the material.

The osmolarity is normally adjusted by changing the buffer concentration or by addition of sodium chloride, polyvinyl pyrrilidone glucose, sucrose and such other non-ionic compounds. The chemical fixative consists of a fixing agent and a vehicle which is normally a buffer solution, with salts. *Of all the metallic fixatives* so far tried, osmium tetroxide (OsO_4) is the most widely used, and the buffered OsO_4 solution, with added calcium in certain cases (*see* Fawcett, 1964) is considered the most adequate one.

Of the non-metallic fixatives, formaldehyde is used to a certain extent.

An aldehyde with wide use in electron microscopy is glutaraldehyde, or more precisely, glutaric di-aldehyde, $(CH_2)_3$ CHO. The commercially available forms of glutaraldehyde are obtained as 25% or 50% solution in water. The monomeric glutraldehyde is the principal reactive compound but the commercial forms contain polymers as impurities. It is often necessary to purify the chemical by charcoal or distillation. Charcoal purification is performed by shaking a 25% solution of glutaraldehyde with 10% (w/v) activated charcoal at 4°C for 1 h before filtration.

In glutaraldehyde fixatives, pH should be kept below 7.5 to check polymerization. As with aldehyde fixatives in general, phosphate buffer is always preferable. For 2% solution, take 2.26% NaH_2PO_4. H_2O in water 64 ml; 25% glutaraldehyde in water 8 ml;2.25% NaOH in water for adjusting pH as required amount and add distilled water to make 100 ml.

For immersion fixation, the most suitable method is to take the tissue in a drop of fixative on a sheet of dental wax, cut into small pieces of required size with a blade, and transfer with the aid of pipette or tweezers to small glass vials containing the fixative. Several formulae for fixatives containing glutaraldehyde are now widely used in chromosome analysis.

Acrolein, otherwise known as 2-propenal or acrylic aldehyde, is another bifunctional aldehyde having the capacity of forming crosslinks between end-groups of proteins. The commercial form is liquid and by distillation the polymerized material can be removed. The fixative is usually prepared by 10% acrolein in 0.025 or 0.05 M phosphate buffer.

DEHYDRATION

Removal of water is usually performed through a sequence of solutions of ethanol or acetone. For embedding in polyester resins, the dehydration is carried out in acetone. For embedding in epoxy resins, dehydration is possible both through ethanol or acetone. As these resins react well with propylene oxide, the latter is often used at the last stage of dehydration. Propylene oxide can also be used as the dehydrating agent itself specially for Epon embedding. In that case, the materials are to be dehydrated through aqueous solutions of propylene oxide.

EMBEDDING

The principal criteria in the choice of embedding media are its stability in the electron beam, uniform polymerizing capacity, convenience in sectioning, low viscosity in the monomer form and solubility in the dehydrating agent. Of the three types of embedding media normally employed, i.e. epoxy resins, polyester resins and methacrylates, the former is the most widely used because it generally satisfies all the above conditions.

The embedding medium also contains an accelerator or activator for infiltration which is always freshly prepared. Though final embedding is carried out in polyethylene or gelatin capsules, the intermediate steps for the complete removal of the dehydraing agent are carried out through a sequence of solutions in the glass vials containing the fixed material. These intermediate solutions are different for different embedding media and are removed and replaced with the aid of a pipette and finally with pure embedding medium. The vials are kept overnight with lid open to accelerate evaporation of the dehydrating fluid.

For final embedding, gelatin or polyethylene (BEEM or TAAB) capsules are available of different sizes to suit the blocks. The ends of these blocks are shaped like a pyramid so that very little trimming is needed. The completely dried capsules are kept for a minimum period of overnight at 60°C oven before use, are half filled with final embedding medium and the specimens are gently transferred to the bottom of the capsules.

All epoxy resins chemically are polyaryl ethers of glycerol with terminal epoxy groups. Their viscosity may be different from one another and they are polymerised by bifunctional agents which criss-cross the epoxy groups forming ultimately a three- dimensional structure. The common epoxy resins are Araldites (CY212, 502, 6005, CIBA 506) and Epons (812, 815 Shell) which may be used separately or mixed together. All epoxy resin embedding media have, in addition to the resin, a hardener and an accelerator. The hardeners of the block are often softened by the addition of additives, plasticizers or flexibilisers. The infiltration time needed is directly proportional to the viscosity of the medium.

The common hardeners differing in viscosity from one another available in the market are: (a) dodecyl succinic anhydride (DDSA—290 cps at 25°C), (b) hexahydrophthalic anhydride (HHPA, m.p. 35°C, (c) methyl nadic anhydride or nadic methyl anhydride (MWA or NMA—175-275 cps at 25°C), and (d) nonenyl succinic anhydride (NSA—117 cps at 25°C). The common accelerators are: (a) 2,4,6-tridimethyl amino methyl phenol (DMP 30), (b) benzyldimethyl amine (BDMA) and (c) dimethylamino ethanol (DMAE). The additives generaly used are: (a) dibutyl phthalate—plasticiser (DBP), (b) carbowax 200 (polyethylene glycol 200), (c) triallylcyanurate (TAC), (d) polyglycol diepoxide—flexibiliser (DER 732, Dow), (e) diglycidyl ether of propylene glycol (DER 736, Dow), (f) long chain mono-epoxide, flexibiliser (Cardolite NC 513) and (g) polythiodithiol—liquid polymer of low viscosity, flexibiliser (Thiokot LP 8).

Some of the common epoxy embedding mixtures are

(i) Epon (812)—10 ml, DDSA—8 ml, BDMA 1%. (ii) Epon (812)—10 ml, HHPA—1 ml, MNA—8.5 ml, DMP 30—0.15/0.3 ml (iii) Thiokol at LP8 hardening takes about 4-7 days—1.5/3.0 ml (embedding at 26°C. Room temperature)

(iv) Epon (812)—10 ml, NSA— 13 ml, DMP 30—1.5/2%

(v) Epon (812) and (815) (ratio determining hardness)-10 ml, DDSA— 14-16 ml, DMP 30 or BDMA—2 ml

Schedule for embedding with Epoxy Resin Media

Drain off the solvent with a pipette. For flushing off, it is preferable to pour it down with a large volume of water in a fume chamber. If propylene oxide is used, keep the material slightly moist. Add a mixture of solvent and embedding medium (1:1) in the vial, shake for thorough mixing. Keep the vial for 30 min to 1 h at 24-28°C. Use pipette to remove the fluid and add the final embedding medium. Remove the cap from the vial and keep for 16-24 h at room temperature. Take several dry polyethylene (BEEM) capsules (dried overnight at 60°C), fit them in a cardboard box with punched holes of proper size, half fill it with embedding medium and transfer the material with the aid of tweezers into the capsules. Fill the capsules with the embedding medium and allow polymerization overnight or more at 60°C oven. Before sectioning, keep the capsules for a few days, if possible in the oven.

A number of water soluble embedding media have been devised, with the advantage that aqueous fixatives meant for cytochemical work can be used for the selective extraction of chromosome components, including Aquon-the water soluble constituent of epon.

Some of the common water soluble epoxy media are: (i) Durcupan, 5 ml; DDSA, 11.7 ml; DMP 30, 1 ml; (dibutylphthalate DP, 0.2 ml may be added). (ii) Durcupan, 100 ml, MNA, 120 ml; DMP 30, 1.5% (Thiokol LP8, 20-35 ml may be added in requisite proportion to secure adequate hardness). (iii) Aquon, 10 ml; DDSA, 25 ml; BDMA (benzyldimethyl amine), 0.35 ml; (iv) Epon 812, 20 ml; hexahydrophthalic anhydride (HHPA), 16 ml; BDMA 1.5%.

For aquon, the material is fixed in 10% formaldehyde-veronal acetate buffer (pH 7.3) at 3-4°C and washed. It is dehydrated through increasing concentrations of aquon in water to pure aquon at 4°C and finally kept immersed in the embedding mixture for 4 h. Curing is performed by transferring the material to a gelatin capsule with fresh embedding mixture and keeping it at 54-60°C for 4 days.

SECTIONING

For ultra thin sectioning (0.1-0.01 µm), *several models of ultramicrotomes are at present available.* In the ultramicrotome of LKB-Producter AB-Stockholm, fluctuation in section thickness is eliminated to a significant extent, the principle being based on a thermal advance system. A cantilever arm is the principal moving part. One end of the arm holds the specimen block, the other end is attached to a leaf spring joined to the base of the microtome. This spring causes the up and down motion of the bar. Thermal control of the cutting arm guides the advance of the block against the knife. The gravitational force controls the cutting stroke and a motor regulates the motion and the upward movement. An electromagnetic force, which acts during the return stroke, causes the flexing of the base below the knife holder necessary to ensure the bypass of the cutting surface and knife edge during the return stroke.

Of the different types of knives used, including hard steel, diamond and glass the latter is the most convenient for ultra thin sectioning and is, therefore, widely used.

A common method for preparing the glass knife is outlined below:

Make a 1.27 cm score mark with a sharp cutter on a clean 20 × 20 cm sheet of plate glass, at right angles to the base of the glass plate. Position the scored edge of the plate to overlap the edge of the working surface by about 0.63 cm. Keep the edge of the glass parallel to the edge of the table. Take a pair of glass breaking pliers with wide parallel jaws. (A narrow strip of adhesive tape is placed on the inner surface from the cutting edge to halfway to the middle of the bottom jaw. Two lateral strips are placed at the edges of the inner surface of the top jaw. Inner surfaces of both jaws are then covered with wide pieces of adhesive tape, smoothly.) Keep the jaws open and with the central piece of tape of the bottom jaw centred beneath the score mark on the glass, push the face of the bottom jaw flush against the table. Gently squeeze the pliers to produce a slow, even and straight break, with two smooth new surfaces, which will be free of artefacts except for the short line where the initial score was made. Turn the two pieces of glass through a right angle so that the smooth edges are away from the table edge. Score one piece of glass in the centre of the old long edge and repeat the procedure to have two 10 × 10 cm plates. Repeat the process till 2.54 × 2.54 cm squares are obtained, each with at least two smooth edges meeting at a 90 degree angle. Choose the best adjacent edges for the final break. Start a diagonal score 1 mm or so from the apex of the angle where the faces meet and extend to bisect the opposite corner. Carefully centre the pliers halfway along this line and gently increase pressure till the glass breaks to give a triangular knife. The good knife should have an even and straight cutting edge, an absolutely flat front surface and a back face with either a right or left-handed configuration when viewed from above. The part of the knife edge closest to the top of the arc formed by the back surface is best for thin sectioning. A good 45 degree angle knife is usually suitable for cutting tissue embedded in media of average hardness. An angle of 55 degrees has been recommended.

The trough is an integral part of the section-cutting equipment for diamond knives. With a glass knife, the usual procedure is to prepare a trough with adhesive-backed cloth or paper tape, which is disposable. The exposed adhesive surface of the trough is coated with paraffin to prevent contamination with the trough liquid. It is sealed to the glass with melted paraffin. The liquid in the trough should be able to detach the sections from the knife, eliminate all electrostatic charges, spread the sections through solvent action, and should have an adequate surface tension to penetrate the layer between section and knife facet.

For trimming and sectioning, the cutting edge of the knife, the embedding material, the cutting face of the block and operating speed are the principal controlling factors. In case of hard epoxy-embedded blocks, fine files and jeweller's saws are required for trimming, followed by a final finish with an acetone or chloroform-washed razor blade. To prepare the block for sectioning, the side walls of the portion delimited from the tissue should be trimmed—a surface area about 0.3 mm × 0.08 mm is usually desirable. The block should be oriented in the microtome with the long side parallel to the knife edge. After trimming, the final shape of the block should be that of a truncated pyramid or, in the case of larger materials, like a roof-top. Normally the cutting face should be square but some people prefer a trapezoidal block, having asymmetry, with the longer axis oriented towards the cutting face, the upper and lower edges being parallel. In general, one side of the pyramid, or the long face of the roof-top-shaped top is adjusted parallel to the knife edge. The blocks are dipped in a filtered mixture of Carnauba wax and paraffin (1:2) and kept at 80°C. The tissue specimen can be oriented by mounting the block on a holder made of a wooden dowel rod 5-16 mm in diameter, which fits well in a Porter-Blum microtome.

After the block is fitted in the chuck, the front edges of the jaws should clamp it strongly and the projecting portion alone should not be more than 3-4 mm. The knife should be tilted so as to have a 1-3 degree clearance angle and a rake (knife) angle of about 30 degrees. The entire process of sectioning should be performed very gently at uniform operational speed.

When the sections have been cut, the ribbon can be detached from the knife edge and transferred to a trough in which the level of the liquid is controlled with a hypodermic syringe fitted at the base with a plastic tube. The level of the fluid is generally maintained over the knife edge, forming a well-rounded meniscus, and the ribbon can be detached with a fine-hair brush. In order to estimate the thickness of the section correctly, it is always preferable to use reflection of interference colours while the sections are floating in the trough . The light should be adjusted to allow total reflection on the liquid surface. Peachey (1958) published a detailed account of the thickness of the sections and the corresponding interference colours, the former ranging form 60 to 320 nm and the latter from grey to yellow. A satisfactory method of observing sections by reflected light is by a fluorescent lamp.

MOUNTING

The removal of the ribbon from the fluid needs special care. It is mounted on a specimen grid with a backing film (Parlodion and Formvar are the common films used for this purpose because they provide good supporting media, being composed of light atoms). They dissolve quickly and become tough when the solvent evaporates. Parlodion is the trade name of nitrocellulose plastic (prepared by Mallinck Rodt Chemical Works, St. Louis). Polyvinyl formal plastic of Shawinigan Products Co., New York is called Formvar. Gay and Anderson's method (1954) is suitable for serial sectioning. The principal implement is a thin film of Formvar supported by a small wire, and these loops can be inserted in the liquid of the trough in a tilted position. By suitable adjustment, the ribbon is centred across the diameter, and when the loop is raised, the sections adhere to the Formvar, after which they are directly transferred to the supporting grids for examination. The grids are placed on a combination of transparent plastic discs, fitted on the top of an adjustable condenser in a standard microscope, and by lowering the condensers, the grid can be kept below the stage and the ribbon suitably arranged. Contact is achieved by lifting the condenser. Epon-embedded materials can be mounted on 300 mesh copper grids. For araldite-embedded materials, the sections do not require any support for mounting; even a carbon film may serve the purpose.

STAINING

Positive staining technique involves treatment with components which increase the weight density, whereas in negative staining, the material is surrounded with a structureless material of high weight density. Good negative staining can be obtained with sodium tungstate, uranium nitrate or disodium hydrogen phosphate.

The number of staining methods available is increasing gradually. Staining with uranyl nitrate may be performed, using a filtered aqueous saturated solution at pH 4.0. For combined staining, the section on the grid may first be moistened with a drop of distilled water, followed by staining (in an inverted position) in 7.5% uranyl acetate solution for 20 min at 45°C. The sections are dried on filter paper, again moistened and finally stained with 0.2% aqueous lead citrate solution for 10-60*s*.

IN SITU FIXATION AND EMBEDDING

Several methods have also been developed for fixation and embedding *in situ*, thus avoiding any distortion or displacement of the structure. A comparatively simple technique with least toxicity to the cell, applicable to monolayers, suspensions, ascites cells, cell smears, frozen materials, involves culturing of cell on cover slips sprayed with teflon. The fixation and dehydration of monolayers attached to coverslips are carried out in staining dishes before final embedding in epon or other media using a silicone rubber mould. Coverslips are then separated from the block by quick dipping in liquid nitrogen. The technique allows mass harvesting of cells as several coverslips with monolayers can be obtained from the single culture flask. Methods have been devised to use synthetic substrate for culture, which can be easily separated from embedding medium like falcon plastic petri dishes, silicone rubber membranes, vinyl plastic cups. For ultrastructural analysis of chromosomes at different stages of division, falcon dishes were used. The procedure involves glutaraldehyde fixation, dehydration in hydroxypropyl methacrylate and *in situ* embedding in epon. Polymerised blocks can be observed and photographed under light microscope and selected area embedded in BEEM capsules for sectioning. This technique allows rapid analysis of a large number of materials and is specially suited for chromosome analysis.

REPRESENTATIVE SCHEDULES

EM Chromosome Analysis of Slime Mold—*Echinostomum minutum De Bary:*

Fix plasmodia in a mixture of 3% glutaraldehyde + 1μM $CaCl_2$ buffer (pH 6.8) with 0.05 M Sörensen's phosphate buffer for 1 h. Post fix in 1% OsO_4 in 0.05M phosphate buffer (pH 6.8) for 1 h. Dehydrate in ethanol series (30, 50, 70, 95% and absolute). Transfer to propylene oxide and follow the usual procedure of embedding in Epon 812. Select metaphase cells under phase contrast microscope, cut the selected portion and remount in epon stubs. Cut 0.5 μm thick sections in ultramicrotome, and mount on Formvar-coated grids. Stain in 2% uranyl acetate in methanol and later in lead citrate for 45 and 20 min respectively.

N.B. It is desirable to apply a thin layer of carbon to the grids by a screened carbon source.

EM Chromosome Analysis of Nematode *(Ascaris lumbricoides* var. *suum)*

Fix the nematode in 2% glutaraldehyde solution in phosphate buffer (pH 7.2). Dissect ovary and testis and fix them again in the fresh fixative for overnight at 4°C. Post fix in Dalton's osmium chromic acid at 24-28°C for 2 h as used for yeast as well. Dehydrate as usual through ethanol and propylene oxide and embed in epon for sectioning. Stain with uranyl acetate and lead citrate.

EM Chromosome Analysis of Algae *(Acetabularia* sp)

Fix the algae in 5% glutaraldehyde solution in 0.1 M sodium cacodylate buffer (pH 7.2 at 5°C for 2 h). Rinse in cold buffer. Dehydrate through ethanol grades and follow the usual procedure for epon embedding. Cut 1.3 μm thick sections in ultramicrotome. De-eponise one set and observe following haematoxylin staining.

EM Analysis of Dinoflagellate Chromosomes

Centrifuge exponentially growing cells of dinoflagellates for 10min at 1200 rpm. They can then be sectioned or mounted whole. For sectioning, fix the cells for 1h at 4°C in 2% glutaraldehyde in sodium cacodylate-saccharose buffer and wash in same buffer for 30min at 4°C. Postfix in 1% osmium tetroxide in veronal acetate for 45min at 4°C. Embed in Epon, cut sections 80nm thick and stain with uranyl acetate and lead citrate.

For whole mounting, further centrifuge the cells at 1200rpm for 5min. Cover the non-disrupted pellet with 2ml of isolation buffer (1% citric acid, 1% Triton X-100, 6 mM $MgCl_2$) and keep at room temperature for 30min. To isolate chromosomes, suspend the cell pellet in the isolation buffer (2×10^4 cells/ml) in a Dounce homogenizer and give 10 strokes of a loose pestle. Centrifuge the homogenate at 1200rpm for 5min through a 2ml cushion of isolation buffer onto Formvar-coated grids. Wash in 50, 75 and 100% ethanol, keeping 10min in each. Immerse the grids in baths of amyl acetate twice for 5min each and airdry. Examine and photograph with a Zeiss 109 Turbo EM operated at 50KV. For negative staining and positive contrast of whole mounted chromosomes, place a drop of 2% phosphotungstic acid (PTA) at pH 7 on the grid with chromosomes after chromosome isolation and centrifugation through isolation buffer as described above. Keep for 40 *s*. Remove stain with filter paper. Air dry the grids at 37°C for 30 min. Positive contrast can be obtained by 2% PTA staining at pH 5 for 40 *s* (Costas and Goyanes 1987).

EM Analysis from Whole Mount Preparation of Protozoa

Permeabilize cells of live *Euplotes eurystomus* with the detergent—microtubule stabilizing buffer PHEM-Triton (60mM Pipes + 25mM Hepes + 10mM EGTA + 2mM $MgCl_2$, adjusted to pH 6.9 + 0.5% Triton X - 100). Mix 4ml *Euplotes* with an equal volume of 2 X PHEM-Triton buffer and incubate at room temperature for 5 or 30min. Fix samples for 15min with 8ml 1% glutaraldehyde in PHEM and centrifuge at 600rpm for 1-2 min. Resuspend the cell pellet in PHEM without Triton. Centrifuge and discard the supernatant. Suspend cells in 1% glutaraldehyde, 0.4% tannic acid in PHEM and fix for 25min in room temperature. Counterfix in 0.5% OsO_4 -PHEM for 30min, and then wash succesively four times with PHEM.

Centrifuge cells in a Shandon-Elliott cyto-centrifuge, (Sewickley, PA) onto subbed coverslips, wash in PHEM 4 times; fix in 0.5% OsO_4 -PHEM for 30min and again wash 4 times in PHEM.

Alternatively, fix cells as above, wash 4 times in PHEM and place a few drops of the cell pellet on a subbed coverslip, covered with a thin layer of 1.2% low melting agarose (Sea Prep tm 15/45; FMC agarose, Rockland, ME) in water. Allow the cells to settle and refrigerate the coverslips to gel the agarose. Dehydrate through an ascending series of ethanol, starting from 30% and then propylene oxide. Infiltrate with Epon according to the conventional schedule described earlier. Float gold grids, section downward, on the surface of a 5N HCl solution in a covered container for 25min at room temperature. Wash the grids thoroughly with water and place, section facing up, on a piece of filter paper. Prepare a 0.2% solution of the DNA-specific osmium ammine-B in water and bubble with SO_2 for 20 min. Float the sections in this solution, facing downward, in a well covered container for 30min at 37°C. Wash the grids thoroughly in water and air dry (Olins and Olins 1990).

EM analysis of Whole Mount Alkali Urea treated Polytene Chromosomes.

Pre-treat polytene chromosomes from salivary gland of *Drosophila melanogaster* in a mixture of 0.5 M NaOH and 10 M urea (1:1) for 5 min in the corner of a siliconized cover slip.

For centrifuging, take centrifuge discs (a modified type of Miller's disc), provided with freshly carbon coated grids, and fill with 10% formalin in a borate buffer (pH 9.2—spreading solution). Centrifuge for 10 min at 3000 r/min and handle the grids as usual. Dehydrate in 30, 50, 70, 95% and absolute ethanol for 30 *s* each, by submerging the grids in solution. Stain for 30 *s* in cold 2% uranyl acetate in methanol and rinse for 30 *s* each in ethanol I and II. Dry the grids in air and observe under the electron microscope. The presence of axial fibrils in spread polytene chromosomes can be demonstrated by this method. The whole mount method permits the retention and identification of basic structural features for kinetochrores, interchromosomal fibres, centromeric heterochromatin as well as basic chromatin fibre (Jeffrey and Geneix, 1974).

Study of Synaptonemal Complex

The synaptonemal complex is recognized as the ultrastructural manifestation of meiotic chromosome pairing. The sectioning permits a study from one end of the chromosome to other. Development of the technique of spreading the complex has facilitated its analysis to a great extent (Dresser and Moses, 1980; Gillies, 1983). The surface spreading method enables a detailed study of the configuration through wide spreading of the skeleton revealing details of architecture. It involves spreading the material in dilute solution of sucrose followed by transfer to plastic coated slide and fixation and hardening in formaldehyde added fixative. The fixation within a short period checks overspreading of the material. After air drying, the slides can be retained and plastic discs can be mounted on copper grids for observation under electron microscope.

A. Animal Synaptonemal Complexes

Mix testes samples of mature bulls in phosphate-buffered saline pH 7.3. Centrifuge at 500rpm for 5min. Discard supernatant with excess spermatids. Resuspend cell pellet in a small quantity of PBS. Spread two drops of cell suspension on a watch glass full of 0.2M sucrose solution. Pick up immediately on a plastic coated slide. Fix for 5min in 4% paraformaldehyde, containing 0.03% sodium dodecyl sulphate, 5min in 4% paraformaldehyde solution. Dip the slide in 0.4% Photoflo solution and air dry. Stain with silver nitrate or with 1% ethanol-phosphotungstic acid (PTA). Transfer to 50 x 75 mesh electron microscope grids, examine with a JEOL 100S microscope and photograph using Kodak Ortholith high contrast 35 mm film (after Dollin *et al* 1989).

B. Plant Synaptonemal Complexes

Dissect out fresh anthers. Place in a drop of spreading medium (0.1% bovine serum albumin and 2mM disodium EDTA salt in Eagle's minimum medium, pH 7.7). Cut each anther in half, squeeze out meiotic cells and remove anther debris. Take up the suspension of cells in micropipette and place one drop on the convex surface of 0.5% NaCl solution in a black dissecting dish, of about 35 mm diameter. Pick up the meiotic cells spread out on the surface of the NaCl solution by touching the surface with a plastic coated slide (prepared previously by dipping in a solution of 0.6% Falcon optilux petri dish dissolved in chloroform). Fix slide with attached spread cells for 5 min in 4% paraformaldehyde (pH 8.2) containing 0.03% sodium dodecyl sulphate (SDS); then in 4% paraformaldehyde without SDS for 5 min and then rinse for 20 *s* in 0.4% Photoflo (Kodak pH 8.0) and air dry in a vertical position. Cover dried slide with coverslip and observe under phase contrast to detect SC and nucleolus at 40X. Stain suitable slides for electron microscopy with either phosphotungstic acid (PTA) in 75% ethanol for 30min or ammoniacal silver nitrate solution (after Gillies 1983).

From Allium cepa and A. fistulosum

Tap out pollen mother cells from fresh anthers in prophase I in a digestion medium (0.1g snail gut enzyme LKB cytohelicase; 0.375g polyvinyl pyrrolidone; 0.25g sucrose, in 25ml sterile distilled water) and keep for 5min. Transfer a single drop onto a single drop of detergent solution (0.5% Lipsol) on plastic coated slide. Add after 5 min, 5 to 6 drops of paraformaldehyde and dry at 20- 25°C for 6h. Rinse and air dry. For PTA staining, immerse slide in 1% ethanolic PTA for 10min, rinse in 95% ethanol and air-dry. For silver staining, place a few drops of $AgNO_3$ solution on slides covered with patches of nylon cloth (instead of coverslip) and incubate in a moist chamber at 60°C for 40- 45min. Transfer suitable surface spreads to EM grids by scoring and floating the plastic film on a water surface. Mark the positions of nuclei on underside of slide, and place a grid over each before floating the film off .Examine and photograph spread and stained nuclei using a Philips EM301 (after Albini and Jones 1987).

EM Study of Nucleosome

In the eukaryotic chromosomes, the structure is beaded in appearance, in which DNA is packed into repeating chains of nucleosome subunits made up of octamers of histones complexed with almost 200 base pairs of DNA. Of these base pairs, nearly 140-160 bp remain associated with H2A, H2B, He, H4 around a central histone octamer and further 20 base pairs remain associated with H1 in the linker region adjoining the core. To some extent, the heterogeneity of the nucleosome is achieved by the histone variants but the relative amount of nonhistone protein is more variable then the histons. In addition to the function of nonhistones in gene action, their role in chromosome structure is established (Kornberg, 1981). Two proteins in the nucleus are deeply involved in the nucleosome assembly, namely DNA Topoisomerase I, which is a nicking closing enzyme interacting with DNA followed by addition of histone, and the other nucleoplasmin which promotes also histone-histone-interaction (Laskey and Earnshaw, 1980). The nucleosomes can be visualized by utilizing spreading technique as well as electron microscopy. The spreading technique of Miller and associates permits the study of chromatin from a variety of chromosomal types. In principle, it involves swelling and dispersion of nucleoplasmic chromosomal material with low ionic solution followed by centrifugation. Ultimately, the three dimensional structure becomes two dimensional on fitting over the surface of electron microscope grid. In order to unwind and disperse the chromatin, low ionic solution, such as distilled water or water at alkaline pH is employed which removes some chromosomal protein including H1. This permits the higher order fibres to unwind into basic strand of beads representing nucleosomes.

Study of Chromatin in Animals, ("Miller" Spread)

Initially devised by Miller and his associates, modifications of this technique have been applied to the study of chromatin from a wide variety of chromosome and cell types. The basic principle is the swelling and dispersal of nucleoplasmic chromosomal material in a low ionic environment followed by centrifugation of these three-dimensional structures onto a two-dimensional flat electron microscope grid surface. By this method, long stretches of continuous chromatin fibres can be traced.

Schedule for oocytes

A sample technique is given for *Xenopus* or *Triturus* oocytes.

Place a small bunch of freshly dissected oocytes in 5:1 K/NaCl. Isolate two or three germinal vesicles (GV) using fine forceps and a dissecting microscope with reflected light on a black surface. Transfer to fresh K/NaCl with a straight micropipette. Clean adherent yolk by sucking in and out of the micropipette. Remove the nuclear envelope carefully with the help of forceps and tungsten needle. Immediately pick up the translucent nuclear content in the tip of a small bore micropipette with a minimum of saline and transfer to a drop of pH 9 glass-distilled water in a clean plastic Petri dispersal dish. Cover the dish and allow the nuclear contents to disperse (5-20min).

Prepare two wells in the microcentrifugation chamber (MCC) as described later under metaphase chromosomes in this chapter and fill each with 0.1M sucrose in 4% formaldehyde, pH 8.5. Take a carbon-coated hydrophilic copper electron microscope grid with a pair of forceps, rinse with 95% ethanol, and then sucrose-formaldehyde solution. Place it inside the well to settle at the bottom with carbon side uppermost. Remove three-fourth of the sucrose-formaldehyde solution. Gently pick up the drop of dispersed GVs with a braking micropipette, mixing the contents. Gently layer the nuclear contents on the top of the sucrose-formaldehyde in the two wells of the MCC. Place a round coverslip on the top of the MCC. Blot excess fluid. Place the chamber in centrifuge tube or bucket adapted for the purpose, as described later in this chapter. Centrifuge at 2500-3000g for 20min.

Then remove the MCC from its holder. Remove the coverslip with a pair of forceps and round up the meniscus with a drop of formaldehyde-sucrose. Invert the MCC. The grid will fall into the meniscus. Lift it up with curved forceps. Rinse it immediately in glass distilled water pH 9 and then pH 9 water + 0.5% Photoflo 200 solution, keeping 7 *s* in each. Remove excess fluid by pulling the edge of the grid across a piece of vellum tissue. Air dry.

To stain the grids, float them face down on a drop or 1% phosphotungstic acid in a clean petri dish. Alternatively, immerse each grid in the same stain in the well of a microtitre plate with a pair of forceps. Stain for 30 *s*. Rinse in 95% ethanol twice and once in pH 9 water + Photoflo 200. Dry the grid with vellum paper. To enhance the contrast of DNP and RNP, the grids can be stained also with 1% uranyl acetate and can be rotary shadowed with platinum or palladium in a vacuum coating unit.

In a modified method used entirely for lampbrush chromosomes, which are easily broken during the long process given above, a shorter schedule is used as follows:

In a MCC, remove half the formaldehyde - sucrose solution and add the "Joy" solution to make two distinct layers (0.025%"Joy" lemon-fresh washing up liquid, Proctor & Gamble Co, Cincinnati, Ohio, USA and glass distilled water). Isolate two GVs and wash off the yolk. With a braking micropipette, rinse the GV several times in pH 9 water and place it in the upper layer of a well of the MCC. The nucleus ruptures and the nuclear contents disappear from view on coming in contact with the detergent "Joy" in the chamber. Keep the MCC in a moisture chamber for 15 min to allow the nuclear contents to disperse. Cover with a coverslip and centrifuge at 2500-3000 g for 20 min. Remove the grid from the well, rinse, dry and stain. The preparation should be rotary shadowed with 80% platinum or 20% palladium.

Schedule for embryonic cells:

Prepare an MCC as described earlier. Load only with formaldehyde- sucrose solution and a coated grid. Do not put water on top. Isolate and wash the embryos in the saline suitable for the species, like modified Locke's insect Ringer for *Drosophila*. Mince embryo in a 100 µl droplet of 0.05% "Joy" at pH 9 with a NaOH-borate buffer on the surface of a siliconized coverslip, using fine forceps. Stir droplet gently for one min. Add 100 µl of a solution containing formaldehyde - sucrose solution pH 8.5. Keep for 5 min. Carefully layer the 200µl droplet on the top of the formaldehyde - sucrose layer in the MCC. Centrifuge. The next steps are the same as the first schedule.

Schedule for tissue culture cells:

Culture the cells till they are in the log phase of growth (10^6 - 5 X 10^6 cells/ml). In spinner cultures, gently pellet the cells at 2000 rpm for 3 min. With cells grown on plastic surfaces, collect by trypsinization, wash once in the media and twice with a balanced salt solution. To the cells, obtained by either process, add an equal volume of 0.1 to 0.5% "Joy", pH 9, to lyse the cells and nuclei and moniter under a phase contrast microscope. When most nuclei are ruptured, add 10 vol of pH 9 glass distilled water and allow to disperse. Moniter under phase contrast. When nuclei are dispersed, place the dispersal vessel on ice. Add 0.1 vol of formaldehyde-sucrose solution to stabilize the chromatin. Layer the mixture into the well of the MCC, already filled one quarter with formaldehyde-sucrose solution. Centrifuge and process for electron microscopy as described earlier. Check in the EM for concentration of material. Stain with PTA and uranyl acetate if needed. Rotary shadow if needed.

EM Study of Spheroid Chromatin Units (v Bodies)

Isolate interphase nuclei from tissues like fresh rat thymus, rat liver and chicken erythrocytes. Wash and centrifuge twice in CKM buffer and once in 0.2M KCl, suspend in 0.2M KCl at a concentration of approximately 10^8 nuclei per ml and dilute 200- fold into distilled water. Allow nuclei to swell for 10-15 min. Make 1% in formalin (pH 6.8 to 7). Fix for 30 min. These steps are carried out in cold (0-4°C). Centrifuge aliquots of the swollen and fixed nuclei through 10% formalin (pH 6.8 to 7.0) onto carbon-covered grids, rinse in dilute Kodak Photoflo and air-dry. When examined after positive staining, chromatin fibres are observed to stream out of ruptured nuclei (Olins and Olins 1974).

CKM buffer = 0.05 M sodium cacodylate, pH 7.5; 0.025 M KCl; 0.005 M $MgCl_2$ and 0.25 M sucrose.

EM Study of Scaffold

The depleted chromosome in metaphase represents a scaffold or core which has the shape characteristic of chromosome and a halo of DNA loops anchored in the scaffold (Paulson and Laemmli, 1977; Lewis and Laemmli. 1982). It has also been defined as those intergenic structures of chromosomal filament which serve as structural and functional entities by spatial ordering of axis and a linear structure (Haapala, 1985). The loop and core architecture of metaphase chromosome was first suggested by Stubblefied and Wray (1971) following treatment of isolated chromosomes with NaCl and urea, removing protein and DNA. The core structure resembled chromosomes in shape. There are two ways through which the core structure can be visualized. The study of histone depleted chromosomes after isolation and spreading principally involves the observation of non-histone protein cores. The other method is the study of the core in intact chromosome by silver staining. For greater resolution and study of details, a technique has been developed based on squash silver staining as well as correlated light and electron microscope analysis of the fine structure of mitotic chromosomes. It has been shown that the core is basically a compact network of fibres rather than a simple coiled structure. The axial element has even been localized as specific segments interspersed along the chromosomal filament when assembled within scaffolding proteins (Haapala, 1985).

Study of Chromosome Core of Trilophidia

Dissect out testes and place in 2% sodium citrate. Further tease out seminiferous tubules in 0.7% sodium citrate and keep for 90 to 120min. Transfer a single tubule to a drop of 45% acetic acid on a slide. Keep for 5 to 10min and squash. Remove coverslip by freezing with liquid nitrogen and air dry the slide. Fix the preparation in a mixture of acetic acid-methanol (1:3) for 30min and rinse in distilled water. Incubate the slides in 2 x SSC solution at 70°C for 60 to 80min. Rinse slides in distilled water, air dry and stain with silver nitrate. Photograph under light microscope. Select preparations with well dispersed chromatin. Dip in 0.6% Folcan solution dissolved in chloroform to coat the preparations with a thin layer of Folcan. Re-examine by light microscope and mark the positions of desirable cells. Float off the Folcan film with specimens onto the surface of aqueous 1% hydrofluoric acid solution. Transfer pieces of the film to the surface of distilled water and keep for 1 min. Pick up the pieces of the film with 50mesh grids. Dry at 45°C. Observe with an Hitachi-600 electron microscope. The core is seen to be a basically compact network of fibres (Zhao *et al* 1991).

Scanning Electron Microscopy

For the study of chromosomes, transmission electron microscopy is normally applied. But an analysis of the details of chromosome structure through TEM is fraught with the limitation of reconstruction from serial sections. The whole mounts of chromosomes suffer from problems of high electron density. In order to obviate some of these difficulties scanning electron microscopy has been adopted from conventional acetic-fixed chromosomes utilizing osmium tetroxide impregnation.

A. For Human Chromosomes

Prepare chromosomes from human blood cultures, stimulated by phytohaemagglutinin, treated with 0.075M KCl and fixed with acetic acid-methanol (1:3) following the usual schedule. Digest the slides or 22mm square coverslips with chromosome spreads for 5 *s* in trypsin (Difco Bacto-trypsin, reconstituted according to the manufacturer's instructions and then diluted 100fold with distilled water). Then wash thoroughly with distilled water and place in glutaraldehyde (2.5% in 0.1M cacodylate buffer, pH 7.4, containing 0.1M sucrose) for 30min or overnight. Next wash the preparations thoroughly with tap water and immerse for 5 min in freshly prepared 1% osmium tetroxide in distilled water. Again wash thoroughly in tap water. Immerse for 5min in a freshly made 0.5% solution of thiocarbohydrazide in distilled water. Alternate the treatments with osmium tetroxide and thiocarbohydrazide to give a total of upto 11 stages, always finishing with osmium tetroxide. Wash thoroughly between treatments. Dehydrate and dry at critical point. Break the slide or coverslip bearing the chromosomes into appropriately sized pieces and mount on a stub. Finally lightly coat the specimens with palladium or platinum. Observe with a Cambridge Stereoscan S-180 microscope or latterly with a Hitachi S-800 field emission SEM (after Harrison *et al* 1987, Sumner 1991).

B. For Plant Chromosomes:

Pretreat roottips in a suitable pretreating chemical. Fix in acetic acid-ethanol (1:3). Squash in 45% acetic acid. Remove coverslips by freezing in liquid nitrogen or solid CO_2. Immerse the slides in a 2.5% glutaraldehyde fixative buffer (either 50mM cacodylate, 2mM $MgCl_2$, pH 7.2 or 100mM sodium phosphate, pH 7). Wash three times in buffer and then once in distilled water. Treat with 1% osmium tetroxide and a saturated solution of thiocarbohydride (or 1mM dithioerythritol), followed by osmium tetroxide (twice). Wash with distilled water between steps. Then dehydrate the specimens through a graded acetone series (20-100%). Critical point dry with liquid CO_2 and sputter with 3-5 nm of gold or palladium. View with a Hitachi S-800 FESEM (after Wanner *et al* 1991).

EM Study following *In Situ* Hybridization to Metaphase Chromosomes in Animals.

Preparation of whole mount metaphase chromosomes:

Arrest subconfluent cultures of mouse cells in metaphase by treatement with colcemid (50-80ng/ml, Gibco) for 6-12h. Collect and pellet the cells by centrifuging at 2000rpm for 10min. Resuspend at 25°C in 0.1 to 0.2 times the original volume of culture medium. For cell lysis, place one drop of the cell suspension in one drop of 1% Nonidet-P 40, adjusted to pH 8-9 and filtered through Millipore filters of 0.22μm pore size, on a clean microscope slide. Mix carefully. Most cells release the chromosomes within 5min. Adjust cell suspensions to a final concentration of about 1×10^5 cells/ml prior to lysis. All chemicals are filtered through millipore filters. Prepare microcentrifugation chambers (MCC) from clear plastic material like Perspex, of 1cm thickness. Cut a central hole and two peripheral grooves into the disc. The outer diameter of the disc depends on the diameter of the buckets in the swing-out-rotor of the centrifuge to be used. The diameter of the central hole is slightly wider than the diameter of the electron microscope grids to be used. The peripheral grooves help in picking up the chamber with a pair of forceps. Centrifuge the chamber using the swing- out-rotor of a refrigerated centrifuge. Make solid adaptors to present a level platform near the top of the bucket of the rotor. The chambers are seated in the buckets. The adaptors can be made as Araldite plugs moulded in plastic centrifuge tubes that fit the buckets (see Sommerville and Scheer 1987).

Deposit the chromosomes on 400mesh gold EM grids, covered with a Parlodion-carbon support film. Place grids in 95% ethanol for 1min to render the support film hydrophilic. Dip the grids briefly in 1M sucrose pH 8.5 (Millipore filtered) and drop them with the carbon film side up into microcentrifugation chambers filled with sucrose pH 8.5. Remove half of the sucrose, thus forming a cushion in the chambers.

Carry out cell lysis in a siliconized test tube. Mix cells with 1% NP-40 (1:1). Keep at room temperature for 1 min. Mix the lysate carefully to suspend released chromosomes and layer 35-50µl on top of the sucrose cushion in each chamber. Centrifuge the preparations for 5min at 3200 rpm in 50ml swinging buckets in a Sorvall GLC-2B bench centrifuge. Add sufficient 1M sucrose to each chamber to form a rounded meniscus. Invert the chamber and remove the grids when these fall into the hanging droplet of sucrose. Rinse the grids in Kodak Photoflo 200 solution, pH 7.0. Air dry. Moniter for chromosome density and morphology by phase contrast microscopy.

Prepare fixative by diluting 1.25ml of 8% glutaraldehyde (EM grade) with 100ml of freshly prepared 2X SSC,pH 7.0 in a beaker on a magnetic stirrer. Immerse grids in single grid holders in the fixative. Remove any bubbles formed. Fix for 20min at 25°C with gentle stirring. Rinse the grids in Photoflo solution and air dry. Glutaraldehyde is more suitable than formaldehyde as a fixative and specimens can be stored at 25°C for at least 3 months.

Denaturation:

Add 24 drops of 10M NaOH to 100ml of freshly prepared 2 X SSC in a beaker on a magnetic stirrer. Mix to get the denaturation solution (pH 12.0). Immerse specimens in single grid holders in the solution for 2-10min, avoiding formation of air bubbles. Rinse for 5 *s* in Photoflo solution and air dry.

Hybridization should immediately take place after denaturation. for nick translation reaction, the stock solution are:

10 X Reaction mix = 0.2mM each of dATP, dCTP and dGTP + 0.5mM Tris-HCl, pH 7.8 + 50mM $MgCl_2$ + 100mM 2-mercaptoethanol + 100µg/ml BSA (BRL, nuclease-free).

10 X Enzymes = 0.4 units/µl DNA polymerase I + 40pg/µl DNase I + 50mM Tris-HCl, pH 7.5 + 5mM magnesium acetate + 1mM 2- mercaptoethanol + 0.1mM phenyl methyl sulphonyl fluoride + 50% (v/v) glycerol + 100µg/ml BSA (BRL, nuclease free).

BRL nick translation kits are available, containing these mixtures.

Hybridization

Label probes by nick translation in the presence of biotinylated nucleotides as follows:

Dissolve 2µg of DNA for each reaction in 10mM Tris-HCl (pH 7.6) and 1mM EDTA (TE). Monitor incorporation of biotinylated nucleotides by dosing the reaction with (^{3}H) dATP. Keep a parallel reaction without biotinylated nucleotide as control for the efficiency of nick-translation with the modified nucleotide. Lyophilize 20μCi of (^{3}H) dATP (Amersham International TRK 347, 17Ci/mmol) per reaction in Eppendorf tubes. Place on ice. Set up both reactions on ice (both containing 10μl of 10 X reaction mix + 2μg DNA + distilled water to 90μl; solution without bionucleotide contains 2μl of 1mM dTTP instead of 3μl 0.7mM bio-dUTP added to the solution with bionucleotide). Mix well.

Add 10μl of DNA polymerase - DNase 10X stock solution to each tube. Mix well and incubate at 15°C for 90 min. Terminate the reaction by adding 10μl of 300mM EDTA, pH 8.0 to each tube. To determine the specific activity, place a 5μl sample from each reaction on a GF/C Whatman filter. Dry the filter and keep in cold 5% TCA for at least one h. Wash with 5 changes of cold 5% TCA followed by two rinses in 95% ethanol. Air dry. Estimate the specific activity (cpm/μg DNA) of the reaction by liquid scintillation counting. Remove unincorporated nucleotides from the filters while drying by Sephadex spin column chromatography using the spin column procedure after equilibrating the Sephadex G-50 used for the columns with 100mM NaCl, 1mM EDTA and 10mM Tris-HCl pH 7.6. Measure the volume of the column eluate and determine the amount of incorporated radioactivity by liquid scintillation counting. Determine the amount (μg) of DNA recovered in the column eluate by dividing the total radioactivity by the specific activity. Dilute the probe to a concentration of 4μg/ml in either 11mM NaCl+1mM EDTA + 10mM Tris (pH 7.6 or distilled water). Store probes at -20°C for a year. Biotinylated deoxyribonucleotides and nick translation kits are available commercially from several companies including BRL, Enzo Biochemicals, Amersham International. Such nucleotides have carbon linkers of different lengths, which bridge the biotin moiety and the base. Slight variations in labelling have been recorded (Narayanswami and Hamkalo 1987).

Make up the bybridization buffer (HB) as follows:

Dextran sulphate (Pharmacia) 200mg; formamide (BRL, 2 X, recrystallized) 1ml, 0.5M EDTA (neutralized) 4μl; 0.5M Tris- HCl, pH 7.6, 40μl; 2% Ficoll-400 (Pharmacia) 20μl; 50mg/ml BSA (BRL, nuclease free) 40μl; 2mg/ml DNA (E. coli or salmon sperm) 40μl; 5M NaCl 240μl. The DNA used should be pretreated with RNAse A and with proteinase K, then extracted with phenol and precipitated with ethanol. Add to a siliconized Eppendorf tube 79μl of hybridization buffer, 1μl of 7mg/ml yeast tRNA (extracted with phenol, precipitated with 2 vol ethanol and dissolved in distilled water) and 20μl of probe DNA (4μg/ml). The final concentration of the probe is 800ng/ml. Hybridize the specimens, two grids per vial, in siliconized reactivials (Kontes Cat #749001-0000) with 50μl of HB per vial. Heat the HB for 5min at 100°C to denature the probe and rapidly cool on ice for 2min. Spin briefly in a microfuge at 15000g for 5 *s* and place on ice. Transfer

50µl aliquots of HB into vials. Place two grids, back-to- back, in each vial. Cap the vials and incubate at 30°C for 12-15h in a water bath. After hybridization, rinse grids three times successively in 2XSSC at 25°C in petri dishes with agitation, to remove unreacted probe. The period of hybridization and the temperature required depend upon the reiteration frequency of the sequence and probe concentration.

In order to detect hybrid, first incubate with rabbit anti-biotin antibody. Place grids, film side up, in 50µl droplets of affinity-purified rabbit anti-biotin antibody (0.8µg/ml in PBS supplemented with 250-500mM NaCl in Petri dishes). Incubate for 4h at 37°C in a moist chamber (tightly fitting perspex boxes lined with moist filter paper). Rinse the preparations three times consecutively with PBS/NaCl for 10min each at 25°C to remove unbound antibody.

For incubation with secondary antibody-colloidal gold, soak and wash all glassware in a hot 1% solution of Micro (Fullam, Cat #1547), rinse 10 times with distilled water and autoclave at 200°C for 12h. Centrifuge gold sols and store in polycarbonate centrifuge tubes (Sorvall). Pellet the preparations of gold particles labelled with appropriate secondary antibodies, like goat anti-rabbit IgGs, immediately before use to remove unadsorbed protein. Centrifuge 0.5ml of antibody-labelled gold stock (10-20nm particles) at 10000rpm for 15min at 4°C in a Beckman JA20 rotor. Alternatively 5nm gold sols are centrifuged at 14000rpm for 1h at 4°C. Discard the supernatant. Resuspend the pellet in the same volume of 1% BSA buffer (1% bovine serum albumin Sigma fraction V, 0.9% NaCl, 0.02M sodium azide, 20mM Tris-HCl, pH 8.2), filtered through Millipore filter (8.0µm pore size followed by 0.22µm pore size). Dilute the gold probe in 1% BSA buffer including 500mM NaCl. Centrifuge at 400- 700g for 10-20min at 25°C to remove aggregates. Dilutions may vary from 1/10 to 1/100 of a gold stock which has an absorbance of 520nm of 10.0.

Place grids individually in 50µl droplets of gold probe in clean Petri dishes. Incubate at 25°C in a moist atmosphere upto overnight. Place preparations in single grid holders and rinse thrice, for 20min each time in 1% BSA buffer at 25°C with continuous stirring. Rinse grids in Photoflo solution and air dry. View the electron dense chromosomes directly or after positive staining or heavy metal shadowing.

Reagents for EM:

Sodium hydroxide-borate buffer: 3.1g of boric acid in 250ml of glass-distilled water to give approximately 0.2M solution. Adjust pH to about 9.5 by adding 30ml of 1M NaOH.

Formaldehyde-sucrose: Dissolve 4g of paraformaldehyde powder in pH 8.5-9 water to a final volume of 100ml. Stir and warm to dissolve. Do not boil. Cool and add 3.4g of sucrose (Grade I, Sigma Cat.no. S9378) to make an approximately 0.1M solution. Filter through nitrocellulose bacterial filter.

pH 9 water: Bring glass distilled water to pH by adding NaOH- borate buffer. Prepare just before use.

Phosphotungstic acid (PTA):, 0.2g (Polaron Equipment Ltd, Cat. no. NC 3009) dissolved in 5ml of glass distilled water. Filter through nitrocellulose filter. Dilute one part stock with 3 parts 95% ethanol immediately before use.

CHAPTER 6

STUDY OF CHROMOSOMES FROM TISSUE AND PROTOPLAST CULTURE, CELL FUSION AND GENE TRANSFER IN PLANTS

A. TISSUE CULTURE

Tissue culture technique allows the cultivation of cells, tissues and organs *in vitro* in natural medium or in artificial medium, but reproducing as far as is practicable the conditions *in vivo*. The technique has immense application in the study of biology in general, and cytology and cytochemistry in particular.

In the study of cytogenetics in relation to plant breeding cell, tissue, and embryo culture has acquired the status of a routine technique in the laboratory. The importance of tissue culture lies principally in its capacity to secure rapid propagation of disease-free clones and of embryo culture in the production of interspecific hybrids and in overcoming the incompatibility barrier. The potentiality of protoplast culture in foreign gene incorporation and of pollen culture in the production of haploids is immense. The significance of the totipotency of plant cells in regeneration of whole plants can hardly be overestimated.

Principal Steps involved in Tissue Culture

Due to the restricted nature of growth in certain specialized and local regions, not all the tissues and organs form convenient material for cultural studies. The highly active growing points, such as apices of stem and root, buds, lateral cambium, intercalary meristems, leaf meristems, phloem tissues, etc., provide excellent materials for tissue culture. Pollen, embryos and endosperm of higher plants, as well as spores and prothalli of lower groups, do not present much difficulty in culturing in a suitable medium. Even flowers, fruits, excised leaves and seed primordia have been cultivated. Phloem tissue of carrot and tomato root, *Haplopappus* stem and pith callus of tobacco, provide ideal materials. Dicotyledons appear to be more responsive to callus initiation from various organs than monocotyledons, gymnosperms and bryophytes.

One of the important factors in tissue culture is the apparatus used for cultivating the tissue. The techniques adopted for the explants are based on the same principle, the chief differences being the type of container employed for the purpose.

They can be classified under three categories, (a) hanging drop, (b) flask and (c) watch-glass and petri dish techniques. All three methods are practised for different animal and plant materials and each has separate advantages and disadvantages. Modifications of the different methods have been made on the basis of the requirements of long-or short-term growth or the convenience of observation. Sub-culture from the primary explant is an essential step in tissue culture.

All nutrient media for tissue culture contain essential macro-and microelements, sucrose as carbon source, supplemented by vitamins, auxins, amino acids, cytokinins, etc. along with in some cases, antibiotics. The nature of the medium is also an important factor, as a mixture which is effective for callus formation in a liquid medium may not be so in a solid medium. Some commonly used media are listed at the end of this section. In general, there is a high concentration of mineral nutrients in Murashige and Skoog and Nitsch and a lower concentration in Gautheret's medium.

Callus culture can be done either in solid or liquid media. For solid media, the most convenient and simple ingredient used for solidification, is agar. The liquid media may be either stationary or agitated, the latter being of wide use as adequate aeration is ensured.

For ultrastructure analysis of chromosomes of callus, the most adequate method of fixation is overnight fixation with 6% glutaraldehyde in 0.1 M phosphate buffer (pH 6.9) at 5°C (initial 2 h being at 25-28°C) followed by washing in phosphate buffer for 3 h and then post-fixation for 1 h in 1-2% buffered osmium tetroxide. Dehydration in ethanol, embedding through propylene-araldite to araldite and staining in uranyl acetate are recommended.

In suspension cultures, cells or colonies are grown in a liquid medium through dispersion and movement. There is an optimal size of the colony formed following incubation, after which it is desirable to subculture the cells in fresh nutrient solution. Materials for suspension culture may be obtained from pieces of friable callus or by grinding the tissue or embryo in a homogeniser. In the latter case, after first suspension passage, it is desirable to pipette out the fine suspensions for sub- culturing bearing the macromolecular particles. The cell suspension after a period of growth may be plated on agar when colonies from single cell can be obtained.The suspension culture filtrate may be subjected to cellular count after pectinase treatment for separation by low speed centrifugation and removal of the supernatant. The 2 ml suspension of the required density is mixed with a sterile medium in 0.6% agar, pre-cooled to 35°C and then finally plated in 9 cm sterile petri dishes.

In cell culture, the minimum effective density or rather the minimum size of the clone which would allow the growth of the culture, is of prime importance and is controlled by the nature of the culture, the nature of the medium and the period and conditions of incubation. The growth of cell colonies is more rapid from aggregate of cells in suspension plates rather that from single cells.

The necessity of maintaining strictly aseptic conditions at all phases of tissue culture, including sterilization of the object, tools, medium, etc., is obvious. Contamination at any stage may not only spoil the culture but also give erroneous data.

Several chemical agents are used for tissue sterilization, the most common ones being calcium hypochlorite (9-10%), sodium hypochlorite (2%), mercuric chloride (0.1-1%), hydrogen peroxide (10-12%), bromine water (1-2%), silver nitrate (1%) and antibiotics (4-50 mg/l). To accelerate sterilization, occasionally light detergent solution (0.05% Teepol or Lissapol F) may be employed.

Chromosome Analysis

For chromosome analysis from liquid medium if necessary, the medium should be centrifuged and pellet of cells taken. For callus, it is desirable to take (50 mg or so) samples from different parts, placed in fresh medium for growth for a week or more. The period of maximum frequency of division should be chosen for fixation. Usual schedules may be followed for fixing, hydrolysing and staining of the cells (Sharma and Sharma, 1980), the most common fixative used being acetic ethanol (1:3). For Feulgen staining, hydrolysis is preferably carried out in 5 N HCl for 1 h at 24-28°C. Dry ice technique may be adopted for making permanent preparations of squashes followed by mounting in euparal.

For softening of the tissue, treatment with mixture of pectinase and cellulase (0.5% of each in 0.1 M sodium acetate buffer, pH 4.5) for 1-2 h has been recommended. For securing a high frequency of division and large number of metaphases, placing the culture for 12-14 h in darkness at 15°C followed by 11 h at 27-29°C and application of colchicine 5 h before harvesting have also been suggested.

Determination of the Cell Cycle

The term 'cell cycle' implies the sequential occurrence of different phases of the cell, initiated from a division till the completion of the next cell division. In the actual method of analysis, it involves the time span between one point in a cycle to the same point in the next cycle.

At the time of counting in culture, cells are in the logarithmic phase of growth, in which every cell is active. In this type of culture, the cell count at each stage is nearly proportional to the time taken by the cells to complete this phase. For example, if the generation time is 15 h and the frequency of dividing cells is 5%, the time required for mitosis is $15 \times 0.05 = 0.75$ h. This method of calculating the generation time also depends on the nature of the medium and the type of the cell.

Mitotic indices can also be utilized in working out the duration of the cell cycle. The technique is based on the fact that the generation time (T) is inversely proportional to the division frequency of cells per unit time. If M is the mitotic index and d the period or duration of mitosis, then the number of cells entering into mitosis per hour is M/d. The practice of considering the generation time as $T = \frac{d}{M}$ ignores the fact that there is a continuous increase in the total number of cells during the cell cycle and so there is an error of nearly 30%. The following formula for working out the generation time fits well with the observational data

$$T = \log 2\frac{d}{M} = 0.693\,\frac{d}{M}$$

Mitotic index at any given time is,

$$M(t) = \frac{n(t+d) - n(t)}{n(t)}$$

when n is the number of cells at the time t, and d is the mean duration of mitosis. In the logarithmic phase of cell multiplication, as in cultures, the number of cells at any time t is also $n(t) = n_0 \,.\, 2^{\frac{t}{T}}$ where n_0 represents the number of cells counted at the beginning of this phase.

To determine the duration of mitotic cycle, germinating seeds or bulbs with 5-10 mm. long roots are heated with 0.1% caffeine solution for one hour. After washing, the seedlings or bulbs are transferred to Knop's solution. At successive hourly intervals, root-tips are fixed to observe the development of binucleate cells induced by caffeine. The criterion by which the duration of mitotic cycle is calculated, is based on the first observation of telophase in a binucleate cell, since binucleate cells are obtained through inhibition of cytokinesis in telophase of the previous mitotic division. Confirmation of the mitotic cycle time can be made with a second treatment of caffeine when quadrinucleate cells form.

The entire mitotic cycle is divided into four phases, namely, G_1—growth phase, post telophase; S = DNA synthetic phase; G_2—post-synthetic growth phase, and M—rest of the mitotic phase. Species differ with respect to the durations of these different phases, G_1 phase in general being variable and taking the maximum and G_2 minimum (except in HeLa cells) periods of interphase. The method of calculating the duration of the different phases in culture is based on autoradiographic procedure. After pulse labelling or short treatment with tritiated thymidine and by varying the time between labelling and fixation, the duration of the three different phases can be worked out. If the cells are fixed after a long interval, only those which were at the S phase at the time of treatment show labelling.

Representative Schedules for Plant Tissue Culture

Preparation of Media

For liquid nutrient medium (a) Add 100 ml of 0.005% aqueous ferric sulphate solution to 500 ml of 8% aqueous sucrose solution, 200 ml of standard salt solution, 2 ml of organic accessory solution and add distilled water to prepare 1000 ml of *stock* solution. (b) Further add 100 ml of distilled water to the stock solution. Distribute in 50 ml portions in flasks and in 10 ml portions in test-tubes. Plug with sterile cotton wool and autoclave at 18 lb pressure for 20 min.

For agar nutrient medium (a) Prepare stock liquid nutrient medium as described in (a) above. Add an equal quantity of 1% hot agar solution in water to the liquid nutrient, mix, keep half of the mixture thus prepared in portions of 15 ml in test-tubes. (b) To the remaining half, for every 100 ml of agar nutrient, add 1 ml each of calcium pantothenate, biotin and naphthalene acetic acid solution, mix, divide into test-tubes, plug and autoclave.

Culture of Tomato Roots

Technique 1:

Wash and dry a ripe healthy tomato, then cut it into four quarters with a shallow incision with a sterile scalpel and open it to expose the seeds, without touching them. Transfer selected well formed seeds by forceps to a petri dish on a sterile filter paper moistened with sterile water and germinate in the dark. After germination remove healthy roots, 2-3 cm long, with a sterile scalpel and transfer each to a flask of nutrient medium. After a week, cut out 1 cm long healthy root tips from the developing root system with pair of scissors and transfer individually to fresh nutrient. The roots can be grown indefinitely.

Technique 2:

Prepare cuttings from a healthy tomato plant, remove the leaves, wash in a mild antiseptic solution, followed by sterile water. Bore holes in a stiff paraffinized cardboard and cut it to size slightly larger than the mouth of a battery jar. Push each cutting through the holes in two sheets of this cardboard with stem side up. Place the cardboard covers at the mouth of a battery jar previously lined with sterile blotting paper and containing a thin layer of sterile water at the bottom. The bottom end of the stem protrudes about 25 cm through the cardboard into the moist chamber formed by the battery jar. Roots develop on this end. Keep in dark. When roots develop, cut out healthy root tips, 1 cm long, with sterile scissors and transfer to a flask containing nutrient medium. Some roots will develop into healthy clones.

Culture of Carrot Callus

Insert long narrow strips of sterilized filter paper twice along the breadth for thickness into test-tubes containing liquid nutrient medium. Wash and dry a healthy carrot, about 15 cm long and break in the middle. Remove a series of cores with a sterile cork borer from the middle with the cambium traversing them lengthwise, and put them on a sterile petri dish. Cut the cores into discs 1 mm thick. Transfer two discs to each test-tube, placing them side by side on the strip of paper. Keep the tube tilted at an angle of 30 degrees, so that the paper is kept moist by dipping in the nutrient medium but the discs are not immersed in it. The discs develop into callus tissue within fortnight. They can be transferred to agar nutrient medium with naphthalene acetic acid and will develop into clones, which can grow indefinitely.

Micro Culture for Observation of Chromosomes under Phase Contrast

Prepare hybrid tobacco (*Nicotiana tabacum X N. glutinosa*) single cell clones, isolated from stem callus and grown in liquid 'tobacco' supplemented by coconut milk (150 ml/l), calcium pantothenate (2.5 mg/l), naphthalene acetic acid (0.1 mg/l) and 2,4-dichlorophenoxyacetic acid (6.0 mg/l), in tubes within a shaker. Place a drop of paraffin oil near each end of a standard microscope slide. Lower a No. 1, 22 mm square cover slip on to each droplet to form risers for a shallow central chamber on the slide. Put a drop of mineral oil in a rectangle on the slide, connecting the two cover slip risers and covering the inner end of each. Place a droplet of liquid medium at the centre of a third square cover slip. Isolate a single cell or a small cluster of cells from a culture-tube with a pair of flattened teasing needles under a dissecting microscope and transplant it in the droplet of liquid medium on the third cover slip. Invert the cover slip over the rectangle of mineral oil on the slide in such a manner that the mineral oil surrounds the liquid medium with its enclosed cells and the ends of the top cover slip lie upon the inner ends of the cover slip risers. The culture thus lies in a liquid medium in a tiny micro-culture chamber filled with liquid paraffin. Observe the micro-cultures directly under the microscope. Keep them in sterile petri dishes in the dark at 26°C at controlled humidity. By this method, the different cytological changes during the growth of the culture in living cells can be observed under both ordinary and phase contrast microscopes (after Jones and colleagues 1960).

B. ISOLATION AND CULTURE OF PROTOPLASTS

Lately, isolation of protoplasts from the cell and culturing them *in vitro* has become a convenient tool, in the study of the property of the membranes, in securing cell fusion, in the study of photosynthesis, as well as in the incorporation of foreign genes of DNA in the protoplast system.

The objective is to isolate protoplast without causing any irreversible damage to its structure. The maintenance of a correct osmotic level without causing excessive plasmolysis is an important factor in protoplast isolation. The actual concentration of osmotic stabilizer varies from tissue to tissue. At present for isolation, enzyme preparations, such as pectinase (Macerozyme) derived from the fungus *Rhizopus,* cellulase (Onazukacellulase) derived from *Trichoderma viride*, driselase—a cellulose enzyme complex derived from a basidiomycete as well as hemicellulase are widely used. Helicase from snail was also used. In view of the fact that enzyme preparations often contain some low molecular weight chemicals, which may cause toxicity by affecting osmotic concentration, it is often desirable to desalt the enzymes through centrifuging in a salt solution such as 15 g of cellulase in 30 ml of 0.5 M NaCl. Moreover, to free the enzymes from contaminants antibiotics are also added.

There are essentially two methods for protoplast isolation based on the principle whether the enzymes are to be put simultaneously or sequentially. For *protoplast* fusion *see* chapter on Somatic Cell Fusion.

For protoplast isolation, it is always desirable to grow plants under controlled growth conditions such as 1000-10 000 lx light intensity for 16 h at 22-25°C. For best yield of protoplasts, Cassells and Burlass (1978) suggested growth under low light intensity (2.52-10.8 MJ/m^2/day for 15 h a day) with balanced fertilizer containing calcium nitrate for continued supply of mesophyll cells. The entire isolation technique should be carried out before a laminar air flow cabinet by treating with 5-7% sodium hypochlorite (Domestos).

Techniques for Protoplast Isolation

Technique 1:

Sterilise the leaf surface, immersing in 70% ethanol for 30 *s* followed by 2.5% sodium hypochlorite for 30 min and thorough washing with sterile water. Peel off the lower epidermis of the leaf in strips (4-6 g) and place in 20 ml of maceration medium (0.5% macerozyme, 13% mannitol and 1% potassium dextran sulphate).

Adjust the pH to 5.8 with 2 N HCl, before filter sterilization. Shake the materials in the maceration medium at 25°C in a reciprocal shaker (100-120 cycles/min, 4.5 cm stroke). Isolate cell fractions after 30 min, 1 h, 1-15 min, 2 h and 3 h, replacing the macerating medium each time. Centrifuge the last two fractions which are almost pure (100-200 × g, 3 min) and resuspend twice in fresh 13% mannitol and centrifuge again. Put the isolated cells in 40 ml, 4% 'Onazuka' cellulase in 13% mannitol,—pH 5.2 (adjusted with 2 N HCl before filter sterilization). Incubate the suspension at 36°C for 3 – $3\frac{1}{2}$ h with gentle swirling. Harvest the protoplasts from the medium by slow centrifugation (100 × g) for 1 min. Resuspend twice for washing in 13% mannitol with 0.1 mM calcium chloride and centrifuge again.

Suspend the protoplasts in 5 ml, 13% mannitol, count a sample in counting chamber and sediment again by centrifugation. Resuspend in fresh 13% mannitol to get a concentration of 1- 4 x 10^6 protoplasts/ml.

Technique 2

Surface sterilise the leaves of *Nicotiana* sp and wash in sterile water. Take slightly flaccid leaves, remove the lower epidermis. Cut pieces of peeled areas and float exposed surface downwards in mixture of 13% mannitol and CPW salts (KH_2PO_4, 27.2 mg/l; KNO_3, 101.0 mg/l; $CaCl_2.2H_2O$, 1480 mg/l; $MgSO_4$. $7H_2O$, 246.0 mg/l; KI, 0.16 mg/l; $CuSO_4.5H_2O$, , 0.025 mg/l, pH 5.8) in a petri dish (14 cm in diameter) for 1-2 h. Remove the mixture and replace with 20 ml (approx) filter sterilised enzyme mixture (4% w/v meicelase, 0.4 w/v macerozyme, 13% mannito, CPW salts, pH 5.8, adjusted with 5 N HCl) and incubate for 18 h in the dark. With forceps, agitate slowly to release protoplasts, tilt the dish and allow the protoplasts to settle for 30 min. Pipette out the enzyme mixture. Transfer the protoplasts to screw capped tubes, suspend the mannitol/CPW salts and centrifuge at 35 x g for 10 min. Remove the supernatant, suspend the protoplasts again in 20% sucrose solution containing CPW salts and centrifuge at 50 x g for 10 min. Remove the protoplast from the supernatant with a pipette and resuspend in 10 ml mannitol/salts. Count the sample. Sediment the protoplasts for 5 min at 35 x g and resuspend in nutrient medium to get a final concentration of 1 x 10^5/ml. The technique may be modified specially in relation to enzyme level and plasmolyticum to suit different species. Protoplasts can also be isolated from cultured cells. The friable nature or the cells allows rapid penetration of enzymes. Regeneration of protoplasts from cultured cells is most convenient in view of the fact that the parent material is the product of *in vitro* growth.

The protoplasts can conveniently be cultured to yield cell colonies and ultimately lead to plant regeneration. In liquid culture, it is preferable to keep protoplasts in a stationary phase or with slight shaking to prevent breakage. Shallow 9 cm plastic petri dishes may be used with 6 ml of protoplast suspension. To prevent bursting, 0.04-0.02% Tween 80 may be added. Hanging drop culture is also possible. The other method of culture is to have a thin layer of protoplast over a layer of nutrient medium solidified with agar to secure a high plating efficiency. Colony formation from individual protoplasts can be recorded.

Chromosome Analysis

For chromosome analysis, it is desirable to centrifuge the protoplasts and take the sediment with nuclei. Pre-treatment before fixation can be performed with saturated solution of aesculine in water for 30 min at 12°C before fixing in ethanol acetic acid (3:1) for 2-3 h. The next step is to warm in a mixture of acetic- orcein/N HCl (9:1) for a few seconds and after 1 h, squashing in 1% acetic-orcein or 45% acetic acid. Carbol fuchsin stain can also be employed.

C. ANTHER AND POLLEN CULTURE

The anther and pollen culture technique is adopted to induce embryoids as well as haploids. It is always desirable to choose anthers, where the pollen is undergoing first mitotic division, which may be checked before selecting the anther.

The schedule for anther culture involves surface sterilization of unopened flower buds at proper stage, removal of sepals and petals, and plating the anthers immediately in basal agar medium of Murashige and Skoog (1962) or floating on liquid medium. Several species may require hormones and cytokinins in conjunction with sucrose for their growth.

The schedule for pollen culture involves initially the gentle grinding of anthers of suitable size in a glass homogenizer containing liquid culture medium as in anther culture. The suspension is filtered through nylon sieve to hold the large particles allowing the pollen suspension to be separated as filtrate. The filtrate is centrifuged at 100 × g for 4 min and the pellet is washed repeatedly in culture medium. After the addition of adequate medium, the suspension is transferred to culture vessel and may be gently agitated for aeration, if necessary. The suspension can also be cultured in agar medium.

D. FUSION BETWEEN PLANT CELLS

In plants, the principal limitation in securing somatic hybridization is the presence of cell wall and middle lamella which are to be digested by specific enzymes acting on cellulose and pectin. Following such digestion, the protoplast can be isolated and cultured, according to the methods described earlier.

For *protoplast fusion in plant system,* in contrast to viruses or animals, several inducing agents such as polyethylene glycol or sodium nitrate are needed to bring the protoplasts of different species or variants in contact with each other.

As compared to animal species, selection systems in plants are quite different for the detection of hybridity or heterocaryon nature. Refined chromosome techniques are essential for checking fusion. Carlson, Smith and Dearing (1972) utilized nutritional requirements as selection criteria of *Nicotiana glauca* and *N. langsdorfii*. Chlorophyll deficient mutants have been used as markers in varieties and mutants of *Nicotiana*. In *Sphaerocarpos*, Schieder's method (1975) of utilising auxotrophic mutants is analogous to the method adopted in animal system. In maize, mutant strains have been used (Giles, 1974). Sensitivity to drugs, such as actinomycin D, has also proved to be an effective marker. The advancement in the method of banding pattern analysis of chromosomes is an effective method of detection of somatic hybridity. Application of refined techniques for chromosome analysis available has immense potential in the identification of hybrids.

Techniques for Cell Fusion in Plants

(1) *Sodium nitrate method* (Power, Cummins and Cocking, 1970): Suspend isolated protoplasts in a mixture of 5.5% $NaNO_2$ in 10% sucrose solution for 5 min at 35°C. Centrifuge at 200 × g for 5 min. Transfer the pellet to a water-bath and keep for 30 min. Slowly replace the supernatant (mixture) with Murashige and Skoog's medium (M/S) with an addition of 0.1% $NaNO_2$. After some interval, wash twice with the medium and plate.

(2) *Calcium ion method* (Keller and Melchers, 1973): Isolate protoplasts and centrifuge in a mixture containing 0.05 M $CaCl_2$, $2H_2O$, in 0.4 M mannitol, pH 10.5 at 50 × g for 3 min. Keep the tubes in a water-bath for 40-50 min at 37°C, when fusion can be noticed in a maximum of 50% protoplasts.

(3) *Polyethylene glycol method* (Kao and Michayluk, 1974): Mix two types of protoplasts in equal quantities. Take a few microdrops (100 μl each) on petri dishes, preferably on cover slips placed at the centre of petri dishes (Bajaj, 1977). Keep for 5-10 min at 24-26°C. Add 2-3 drops of PEG (50 μl) to the periphery and incubate for 30 min. Addition of calcium in PEG accelerates fusion. Replace the PEG in the suspension with culture medium by gradually washing the protoplasts. Plate and stain to note fusion.

(4) *Method followed by Cocking and his group*
Isolate leaf protoplast of *Petunia parodii* and *P. hybrida* as outlined under tissue culture. Suspend the protoplasts in 9% (w/v) mannitol solution containing inorganic salts. The use of calcium ions at pH 8.5 is also recommended (Reinert and Bajaj, 1977). Keep the samples in screw- capped 8 ml tubes and use equal volume of each species for fusion. Centrifuge the tubes at 80 × g for 10 min and remove the supernatant. To induce fusion, add 2 ml 15% (w/v) polythylene glycol (molecular weight 6000), 4% (w/v) sucrose and 0.01 M $CaCl_2$ in the tubes. Re-suspend protoplasts and keep at 25°C for 10 min. Add after every 5 min, M/S medium 0.5, 1.0, 2.0, 3.0 and 4.0 ml, continually re-suspending the protoplasts after each addition. Centrifuge at 60 × g for 15 min. Remove the supernatant and add 8.0 ml M/S medium in each tube. Keep the tubes for 1 h before plating (A sample count may be taken at this stage which shows nearly 4% of the nuclei in a fused state).

Culture the protoplasts on liquid agar medium in the following way: Take 9 cm plastic petri dishes and add 8 ml M/S medium with actinomycin D (1.0 μg/ml) solidified with agar (0.5% w/v). Add 4 ml M/S medium with actinomycin D (2.0 μg/ml) and 4 ml protoplast suspension (2.0×10^5/ml) in M/S medium without actinomycin D on the surface of the agar. After this dilution the concentration of actinomycin D on the liquid becomes 1.0 μg/ml at a protoplast density 1×10^5/ml. Keep the cultures at 27°C using daylight fluorescent tubes for 28 days. Transfer the cultures to M/S medium with 3% mannitol solidified with 1% agar, without actinomycin.

After 60 days, transfer the cultures to M/S medium without mannitol for the formation of callus. After 10 weeks, transfer the fused hybrid callus to M/S medium with IAA (2.0 μg/ml) and 6-benzylaminopurine (1.0 μg/ml) for shoot regeneration. After shoot formation, transfer to M/S medium with NAA (0.1 μg/ml) and 0.3% agar. After the formation of plantlets of suitable size, transfer to pots to grow till maturity. Following this method, hybrids of *P. parodii* and *P. hybrida* were obtained with purple flowers and chromosome numbers ranging between 2n = 28 and 2n = 24 showing tetra and hypotetraploid constitution.

E. SOMATIC CELL FUSION AND GENE TRANSFER

In somatic cell fusion, the present goal is to achieve limited gene transfer through the production of asymmetric hybrids and cybrids. Through the accelerated elimination of non-viable chromosomes and judicious application of selective pressure, it may be possible to develop asymmetric hybrids of importance (Climelius 1988).

The scope of controlling chromosome elimination in somatic hybrids offers great possibility in gene transfer. There are two methods which have been found to be useful in securing asymmetric hybrids.

The fusion between dividing and mitotically inactive protoplasts has been successful in achieving such hybrid. The protoplast of fast growing suspension of albino mutant of carrot could be fused with leaf protoplasts of *Aegopodium prograria*. The plant had 2n=18 chromosomes like the carrot in which sub-chromosomal fragment from *Aegopodium* was integrated. It was responsible for showing several specific *Aegopodium* characters. The other approach, namely, irradiation of one of the protoplasts, has also been successful in securing asymmetric hybridization and gene transfer, as in case of several Solanaceous species (Dudits 1987, Bales *et al.* 1988).

Methods for Gene Transfer

There are at present a number of physical methods of transformation of plant cell through transfer of foreign DNA, chromosome fragments or chromosomes, involving cell cultures.

The insertion of cloned DNA though a suitable vector, specially Agrobacterium plasmid, is a convenient method but the transformation efficiency is not very high in all groups, specially in monocotyledons. In order to overcome its limitations, several chemical and physical methods such as PEG and electroporation have been devised.

Microinjection has been found to be effective for direct delivery of DNA into protoplasts or intact cells. Microinjection was first used in rapeseed, where injection into the cytoplasm of pollen- derived embryoids was carried out with an equal mixture of linear and supercoiled DNA. Several chimeras were obtained. The other biolistic approach is to have the particle or shot gun method, which is gradually becoming very popular. In this method micron size tungsten or gold particles are accelerated in the barrel of a gun to a high velocity but non lethal to membrane or cell wall. Tungsten is chosen as carrier of nucleic acid because of non-toxicity and adequate density for rapid penetration into target material. For absorption of DNA in the tungsten particles, spermidine and $CaCl_2$ have been found to be very effective at concentrations of 8 mM and 0.2-1.9 M respectively. The best result was secured with 1.2 to 2.5 µl/2 µg/mg DNA - tungsten preparation (see review by Oard 1991).

High voltage discharge delivered to a small droplet of water has also been utilized. Following vaporization, the released energy is directed to propel DNA-coated gold spheres to target cells.

Electrofusion/Electroporation

The principle involves the generation of high voltage electric field pulses to increase the permeability of cells. Electrically induced protoplast fusion has been used to produce somatic hybrids in *Nicotiana* and other crops (Bates and Hazenkampf, 1985).

The fusion is achieved in a multielectrode fusion chamber (500- 700 gel capacity) containing a mixture (1:1) of mesophyll protoplasts of both species. Following an alignment of protoplasts induced by an AC field of 125 v/cm and 1 Mdz, fusion was initiated by an exposure of protoplast samples to a train of 3- 4 DC pulses of 1.2 Kv/cm each 20 ns. The fusion rate was estimated as 20-40%, 30% being binary fusions. A large number of somatic hybrid plants were recorded following fusion experiments.

Irradiation

X-radiation or gamma radiation of protoplasts is often used in the transfer of foreign traits through cell fusion. Similarly, UV laser light of nanosecond pulse length has been used as apical needle to prick cells. It causes a self-healing perforation in the membrane, through which foreign DNA kept in the medium can pass. Irradiation prevents colony formation in protoplasts and irradiated nucleus often undergoes elimination in the fused cell. As such, the donor cell can be irradiated and the recipient can be treated with iodoacetate for inactivation of protoplast. This method of fusion leads to production of *cybrid*. The cybrids are particularly useful while dealing with cytoplasmically inherited cells such as cytoplasmic male sterility (Ichikawa *et al.* 1988). The gameto-somatic hybrids can be obtained by fusing protoplasts of pollen culture with those of leaf protoplasts though PEG technique (Lee and Power 1988).

Biolistic Approach or "Shot Gun" Method

The biolistic approach involves direct delivery of DNA into tissue facilitating gene transfer. The instrument of shooting DNA into cells includes microprojectiles which go deep into the tissue. The foreign genes remain active when put into microprojectiles and their entry into the tissue can be confirmed through reporter markers such as glucouronidase - GUS gene. It involves the assembly of a gun holding cartridges and a barrel to hold the microprojectiles. There are different devices of particle delivery system as applied to different species. Even helium driven biolistic devices are available.

Microinjection

Microinjection is an effective method for injecting chromosome fragment or DNA in the recipient (Detaat and Blaas 1989). Successful examples of the use of this method are introduction of subchromosomal fragments containing nuclear genes into the recipient genome in mammalian systems (Klobutcher and Ruddle 1981) and of chromosomes of *Petunia alpicola* into protoplasts of *P. hybrida* (Griesbach 1987). One method is to separate nuclei and cytoplasm before microinjection and then to grow nuclei individually or after rejoining with cytoplasm by electrofusion. In injecting the chromosome with microneedle of micromanipulator, the first step is to obtain evacuolated plant protoplasts.

Protocol for securing evacuolated protoplasts

Isolate mesophyll protoplasts from leaf tissue (1 × 10 mm) by overnight treatment with 0.8% cellulysin, 0.4% macerase, 10% mannitol, 5 mM-MES (2N morpholino-ethanesulfonic acid) and 10 mM $CaCl_2$, pH 5.7. Clean the protoplast suspension by layering on top of 30% sucrose soln and centrifuging at 200 × g for 15 min. Wash the cleaned protoplast band at interface twice in 5 mM-MES, 10 mol $CaCl_2$ and 10% mannitol, pH 6.0 by centrifugation at 100×g for 5 min. Mix 0.5 ml of 2×10^6 protoplasts/ml with 45 ml Percoll containing 100 mM $CaCl_2$, 5 mM N-2 hydroxyethyl piperazine - N2 ethanesulphonic acid (HEPES) pH 7.0 and 8% mannitol. Introduce Percoll/protoplast suspension through pipette into a 5 ml polycarbonate centrifuge tube and centrifuge at 40000 rpm for 60 min at 23°C (Beckman SW 56 rotor). Aseptically remove the band of evacuoalated protoplasts with a Pasteur pipette. Wash three times with MS regeneration medium with 2g/1 bactopeptone, 20g/1 sucrose, 100g/1 mannitol, 1 mg/1 NAA and 2 mg/1 BAP supplement at pH 6.0.

Protocol for uptake of chromosomes in protoplast

Stain isolated chromosomes in 0.1% 4.6 - diamidino-2-phenylindole (DAPI) for 1 h in dark. Wash off the unbound stain through centrifugation in protoplast buffer

(1 mM $CaCl_2$, 5 mM/ 2 (n- morpholino) ethanesulfonic acid (MES) and 10% mannitol, pH 6.5. Prepare mesophyll protoplasts of recipient species by overnight incubation in enzyme mixture (2% cellulysin Calbiochem, 1% macerase and 10% mannitol, pH 5.5). Wash off the enzymes through centrifugation and suspension. Incubate 1×10^6/ml of protoplast for 20 min and 1×10^7/ml of isolated chromosomes in 35% PEG 4000, 2% mannitol, and 12 mM $CaCl_2$, pH 6.0. Add 4 vol of 50 mM $CaCl_2$ and 10% mannitol, pH 8.5 to stop the reaction. Collect protoplasts by centrifugation and wash in protoplast buffer. Use fluorescence microscope for visualization of stained chromosome in protoplast. DAPI stained chromosomes fluoresce yellow green against red background.

Table 1: Plant tissue culture media (Nitsch and Nitsch, 1956)

Constituent	mg/l	Constituent	mg/l
KCl	1500	$CaCl_2$	25
$MgSO_4$, $7H_2O$	250	$MnSO_4$, $4H_2O$	3
NaH_2PO_4, H_2O	250	$CuSO_4$, $5H_2O$	0.025
KNO_3	2000	H_3BO_3	0.5
IAA	0.18-1.8	Na_2MoO_4, $2H_2O$	0.025
Sucrose	34 000	$ZnSO_4$, $7H_2O$	0.5

Table 2: Plant tissue culture media (White, 1954)

Constituent	mg/l	Constituent	mg/l
KCl	65	H_3BO_3	1.5
$MgSO_4$, $7H_2O$	720	$Fe(SO_4)_3$	2.5
NaH_2PO_4, H_2O	16.5	Sucrose	20000
KNO_3	80	Glycine	3
Na_2SO_4	200	Cysteine	1.0
$Ca(NO_3)_2$, $4H_2O$	300	Vit B_1	0.1
$MnSO_4$, $4H_2O$	7	Vit B_6	0.1
KI	0.75	Nicotinic acid	0.5
$ZnSO_4$, $7H_2O$	3	Ca D-pantothenic acid	1.0

Table 3: Plant tissue culture media (Murashige and Skoog, 1962)

Constituent	mg/l	Constituent	mg/l
$MgSO_4$, $7H_2O$	370	$CuSO_4$, $5H_2O$	0.025
$CaCl_2$,$2H_2O$	440	H_3BO_3	6.2
NH_4NO_3	1650	Na_2MoO_4, $2H_2O$	0.25
KH_2PO_4	170	Sucrose	30000
KNO_3	1900	Glycine	2
$FeSO_4$, $7H_2O$	27.8	Myo-Inositol	100
$MnSO_4$, $4H_2O$	22.3	IAA	1-30
KI	0.83	Vit B_1	0.1
$CoCl_2$, $6H_2O$	0.025	Vit B_6	0.5
$ZnSO_4$, $7H_2O$	8.6	Nicotinic acid	0.5
EDTA (disodium salt)	37.3	Kinetine	0.04-10

CHAPTER 7

CHROMOSOME ANALYSIS FOLLOWING SHORT-AND LONG-TERM CULTURES IN MAMMALIAN AND HUMAN SYSTEMS, FROM NORMAL AND MALIGNANT TISSUES

A. CHROMOSOME STUDY FROM NORMAL TISSUES

The remarkable progress in the study of cytogenetics in lower species encouraged a comparable study of mammalian, and later of human cytogenetics. There are now many techniques for studying mammalian chromosomes, particularly human ones, each laboratory having evolved its own variant of different published methods. Most of them have been devised in quick succession , as a result of improvements like: development of the tissue culture schedule to obtain cells *in vitro*, either suspended or forming a monolayer; pre-treatment by colchicine or its derivatives to accumulate a large number of mitotic figures; hypotonic solution treatment causing swelling of the cells to aid chromosome scattering and air-drying to force the chromosomes to lie in one plane. The techniques, in general, aid in the study of the karyotype, of the meiosis, and of the sex chromatin. A suitable method for karyotype study must fulfil three demands: (a) adequate scattering of chromosomes, (b) minimum distortion and (c) flattening of the cells. The three principal schedules, with variants, utilized in karyotype analysis, are the short-term peripheral blood culture, the direct bone marrow technique, and the long-term fibroblast culture. For meiotic studies, methods have been developed for use on gonadal tissues. Sex chromatin studies are done from buccal smears, tissue sections and peripheral blood neutrophiles.

As the methods were first developed either for other mammalian materials and then applied to human chromosomes, or *vice versa*, and because the method employed is decided by the nature of the tissue used, the different schedules will be discussed in relation to the tissue utilized, irrespective of their origin.

The dispersal of chromosomes by hypotonic shock, as discovered by Hsu (1952), led to the evolution of most of the methods on mammalian chromosomes. The simple air-drying technique, as devised by Rothfels and Siminovitch (1958), gives very satisfactory flattening of the chromosomes and is a common feature of most techniques. Gonadal tissues are usually obtained through biopsy and cut into small cubes. For mitotic prophase and telophase figures of germ cells, repeated suspension in a cold isotonic solution like Medium 858, or Hanks basal salt solution, at 3°C is adequate. It is also suitable for first meiotic prophase figures of male germ cells, obtained by testicular biopsy. In female cells, the divisional figures are numerous in fetal ovaries between the fourth and eight months. The ovary is immersed in isotonic solution, dissected, incubated at 37°C trypsin solution for 30 min, followed by three 10 min rinsings in isotonic solution. Swelling needed for mitotic metaphase and anaphase stages can be acquired by immersion for 10 min in neutral (pH 7) double distilled water, at room temperature, two or three times.

In the male, meiotic metaphase figures are greatly improved by hypotonic pretreatment. Individual seminiferous tubules are dissected out from the tissue immersed in isotonic solution, transferred to double distilled water at room temperature and kept for 30 min prior to fixation. In the human female, first and second meiotic metaphase figures are difficult to obtain, since they occur only in the mature follicle in the ovary of woman in her reproductive span, on the day of ovulation. The mature follicle is punctured in other mammals and the ovum dissected out and incubated at 37°C. It is then transferred to the fixative (50% acetic acid), kept for 15 min and covered gently to separate the bivalents (see Ohno, 1965 and also Edwards 1962).

The period of fixation ranges from 15-45 min, a cube of material (2 × 2 × 2 mm) being placed on a slide in 1 ml of fixative and tapped to release the free cells. It is covered with a cover slip and squashed with straight, uniform pressure. Later, the slide is immersed for 1 min in a mixture of dry ice and methanol, dried in air, treated in methanol for 15 min to remove fatty substances, dried, washed in water, hydrolysed in N HCl at 60°C for 15 min, and then stained, dried and mounted in synthetic balsam. Staining for 3 h with Feulgen reagent or 5 min with Giemsa or 1 min with 0.25 basic fuchsin solution gives good results.

Study of Chromosomes from Male Germ Cells

Schedule for Human Chromosomes

Germinal cells are usually collected by testicular biopsy, preferably undertaken under general anaesthesia to avoid any interference by injecting an anaesthetic agent locally. Tissue of the size of a safety match head provides enough material in most cases.

Two types of schedules are available - for squash and for air or flame-dried preparations (Hamerton, 1971; Sharma, A. and Talukder, 1974).

Squash preparations

The type of pre-treatment required depends on the developmental stage of the germ cells to be studied. The later steps of fixing, squashing and staining are similar.

First meiotic prophase figures

Gonadal tissues, obtained by testicular biopsy, are cut into cubes of 2 mm^3 and suspended in an isotonic solution, like Medium 858 or Hank's BSS (Difco) at 3°C for 10 min. Two changes are given in the isotonic solution in cold at intervals of 10 min each. The tissue is then transferred to fresh fixative (glacial acetic acid and distilled water in equal proportions), 50 vol of fixative to 1 vol of the tissue. It is kept for 15 to 45 min. A cube of the tissue is transferred to a clean grease-free glass slide with about 0.1 ml of fixative. It is tapped gently to release the free cells. The stringy connective tissue in the free cell suspension is removed. The material is covered carefully with a long cover slip. The slide is placed between four layers of filter paper and the material squashed by applying firm uniform pressure vertically on the cover slip. The slide is kept for 1 min in a mixture of dry ice and methanol in a beaker to freeze the fixative. The cover slip is then removed by inserting a razor blade. The slide is dried in air and then kept in methanol for 15 min to dissolve the fatty contents of the gonadal tissue. It is again dried in air, rinsed in tap water and hydrolysed in normal hydrochloric acid at 60°C for 15 min to remove most of the RNA from chromosomes and cytoplasm. The slide is stained using either Giemsa solution (5 min) or Feulgen reagent (3 h) or 0.25% basic fuchsin solution (1 min). It is dried in air, a drop of synthetic balsam is placed on the material and covered with a coverslip. Pachytene configurations with a heavily condensed XY- bivalent and 22 autosomal bivalents showing chromomeres can be observed. This schedule can be utilized to study mitotic prophase and telophase figures of germ cells as well (Ohno,1965).

First and second meiotic metaphases

A piece of fresh tissue, obtained by testicular biopsy, is immersed in isotonic solution. With the help of small forceps under a dissecting microscope, seminiferous tubules, each at least 2 cm long, are isolated. A long piece of tubule is transferred to 2 ml double distilled water on a clean slide and allowed to stay in it for 30 min at room temperature. This hypotonic treatment causes the tubule to swell to nearly three times its original size. The excess moisture is blotted off with filter paper. 0.5 ml of 50% acetic acid is added to the tubule and fixed for 15 min. The tubule is then teased into small pieces, covered with a coverslip and squashed and stained according to the schedule described for meiotic prophase. The first meiotic metaphase shows 22 autosomal bivalents with chiasmata and the XY-bivalent, associated end to end. In the second metaphase, 22 autosomes with the chromatids held together at the centromere and a heavily condensed sex chromosome, either X or Y, are seen.This method can be applied in observing mitotic metaphase and anaphase stages in germ cells as well (Ohno,1965).

Alternative squash method for study of pachytene stages

A piece of testicular tissue is placed in 0.3% aqueous sodium citrate solution for 1 to 6 h depending on the size of the tissue. A few tubules are dissected out, stained and squashed to confirm that the divisional stages required are present. The remaining tissue is kept in a 3 M solution of glucono-deltalactone for 2 h for softening. It is then stained in acetic or propionic-carmine (1%) solution for a period of 10-12 h. The tissue is washed in four successive changes of 70% ethanol to remove excess stain. It is transferred to a large watchglass or petri dish containing acetic acid- ethanol (1 : 1) mixture and minced into a thick suspension. The cell suspension is filtered through several layers of cheese cloth. The filtrate is centrifuged at 250 r/min for 15 min. The upper half of the supernatant is discarded. The fluid remaining just above the coarse precipitate is pipetted off. It contains most of the spermatocytes. One drop of the cell suspension is dropped from the pipette on to a siliconized slide. A drop of water-soluble mounting medium is added to the suspension and mixed with it. A coverslip is placed on the material and the excess fluid removed by blotting with filter paper. The slide is warmed on a hot plate to flatten cells and attach them to it. It is now inverted on several sheets of absorbent paper and squashed, using uniform pressure, with the thumb (Gardner and Punnett,1964).

Squash method for the study of stages other than first meiotic prophase

The fresh testicular tissue is placed in hypotonic solution (pH 7.0; 50 % potassium glycerophosphosphate, 93.0 ml; 20% glycerophosphoric acid, 24.92 ml and distilled water made up to 200 ml). The tubules are teased out and kept at room temperature for 3 min. The material is fixed in acetic acid-ethanol (1 : 3) for 1 h and then treated in 30% ethanol and distilled water successively for 3-5 min for hydration. It is hydrolyzed in normal hydrochloric acid at 60°C for 8 min, stained in Feulgen reagent for 1 h, rinsed successively in SO_2 water and cold 45% acetic acid, keeping 3-5 min in each. The tubules are transferred to 45% acetic acid and squashed (Ford, 1961; Turpin and Lejeune,1969).

Modified squash method

Hypotonic solution used is tri-sodium citrate in concentrations varying from 0.55 to 1.12% for 15 min or more. The specimen is then transferred to a fresh 2% acetic-orcein solution and cut into smaller bits. It is kept in this solution from 15 min to 2 h. The tissue is then transferred on to a clean grease-free slide, teased out; the debris is rejected and a cover slip put on the material. The excess fluid is removed by blotting under filter paper. A gentle pressure is applied to squash the chromosomes and the preparation sealed by Kröning's cement.

Air dry method for the study of pachytene chromosomes

The tissues were mainly obtained from bilateral orchiectomy performed in cases of prostatic carcinoma. It is also applicable to specimens acquired through testicular biopsy or through autopsy within 1 h of death.

The tissue is transported immediately after removal in cold to the laboratory in tissue culture medium. (Eagle's basal amino acids and vitamins at double strength in Earle's balanced salt solution adjusted to pH 7.0 with 7.5% $NaHCO_3$, with additives glutamine 2 mM, penicillin 100 units/ml; streptomycin 100 μg/ml; phenol red 7 μg/ml; newborn agammaglobulin bovine serum 15% and USP heparin sodium 20000 units. The tissue is transferred to BSS at 37°C and minced into small pieces. The resultant suspension is transferred to 15 ml centrifuge tubes. On centrifugation for a few minutes, the large particles settle to the bottom. The supernatant, with single cells and small clumps, is transferred to another centrifuge tube. It is centrifuged at 150 g for 4 min and the supernatant BSS rejected. The pellet is resuspended in excess 0.125 M KCl, with heparin added (20000 units/l) and incubated for 1 h at 37°C.

The material is again centrifuged, the excess supernatant KCl rejected and the cells fixed in acetic-methanol (1:3) for 10-15 min. Air-dried squashes are prepared as described earlier. On drying, the slides are stained for 2 h in 1% solution of orcein in 60% acetic acid and mounted with Diaphane (Will Corp).

Dried preparations

Flame dry method

Fresh testicular tissue, obtained by biopsy or orchiectomy, is placed in a watchglass or petri dish and minced finely. About 0.5 ml of the tissue volume is taken and covered with 2 ml of 0.6 hypotonic sodium citrate solution. The tissue is mixed with the solution by inhaling and exhaling through a pipette and allowed to remain for 30 min. The cells are resuspended in the hypotonic solution, 6 ml acetic-ethanol (1:3) is added and the cells fixed in it for 30 min. The mixture is centrifuged at 100 r/min for 5 min. The supernatant is discarded and 6 ml of fresh fixative is added to the cells at the bottom of the tube. The process is repeated two to three times at intervals of 20 min. The cells are then resuspended in 2 ml of fresh fixative, mixed and allowed to stand. The upper layer of dispersed cell suspension is drawn out. One to two drops are pipetted on to a clean, grease-free slide. The slide is held close to a Bunsen burner flame for a few moments till the fixative burns off completely, leaving the slide dry. The slide is then stained by immersion in Giemsa (5-10 min) or any other stain as desired (Sasaki and Makino 1965).

Air drying method

Seminiferous tubules, obtained from fresh biopsy, are placed in a small amount of 2.2% (w/v) sodium citrate solution in a small 5 cm petri dish and agitated gently in the solution to remove the excess fat for 15 *s*. The material is transferred to 3 ml of fresh 2.2% sodium citrate solution in another petri dish. The tissue is held with forceps and the tubules teased out. After 10 min, the total mass is poured into a 4 ml centrifuge tube. The material is centrifuged for 5 *s*. The supernatant is pipetted off to a 4 ml centrifuge tube. The sediment is discarded. The fluid is centrifuged at 500 r/min for 5 min. The supernatant is discarded. The sediment at the bottom is resuspended in a minimal amount of residual supernatant. To the suspension, 3 ml of 1.0% sodium citrate solution is added slowly. The suspension is divided equally between two narrow conical tipped Dreyer tubes and kept for 10 min. The tubes are centrifuged for 5 min at 500 r/min. Supernatant is discarded. The tubes are allowed to stand for one min and the excess fluid that has drained down the sides is pipetted off.

For fixation, absolute ethanol, glacial acetic acid and chloroform are mixed together in the proportion 45:15:1 immediately before use. Two alternative schedules may be adopted for fixation:

(a) In the first alternative, the tube is flicked so that the pellet of cells forms a thick suspension. Two drops of fixative are added, and the tube flicked vigorously. The fixative is added drop by drop till the tube is about three-quarters full. It is centrifuged for 3 min at 500 r/min. The supernatant is discarded and fresh fixative added. Fixative is changed after centrifugation twice more. The period of fixation is 15 min.

(b) In the second alternative, the pellet is not disturbed. Fixative is added carefully down the side of the tube till it is filled. The fixative is immediately pipetted off. The pellet is reeesuspended, fresh fixative is added and allowed to stand for 15 min. The procedure is repeated. The final suspension of cells in the fixative is allowed to stand for 1 to 3 h. A small quantity of the suspension is withdrawn and 2 to 3 drops transferred to a clean slide. The slide is tilted to permit the fluid to spread out to the maximum extent. Air is blown gently on the slide to dry the fluid. The process is repeated with further drops of the suspension. Several successive drops of a dilute suspension may give better preparations than a single drop of more concentrated suspension. The slides can be stored dry or stained immediately in lactic-acetic-orcein or toluidine blue or fluorochromes (Caspersson *et al* 1971).

Study of Chromosomes from Female Germ Cells

Schedule for human chromosomes

Method for the study of first meiotic prophase from fetal tissue

The ovary, obtained from an aborted fetus, is placed in isotonic solution. Its cortical area is dissected out under a dissecting microscope. The tissue is cut into 2 mm pieces and immersed in 10% trypsin solution and incubated at 37°C for 30 min. They are then rinsed in isotonic solution, keeping in it for 10 min each time.

The material is then fixed, squashed and stained according to the method followed for testicular tissue (Priest, 1969).

The nucleus at pachytene shows in a normal female 23 bivalents with chromomeric patterns. The sex chromosome bivalent XX does not condense.

Culture methods for the study of meiosis in oocytes have been developed following similar schedules used in other mammals. A technique, involving the use of hormones, is outlined below (Jagiello, Karnicki and Ryan, 1968).

Patients were admitted to the hospital on the third day of the menstrual cycle. Injections of human pituitary follicle stimulating hormone (FSH) were begun on the fourth day and repeated daily till the ninth day. On the ninth day, human chorionic gonadotrophin (HCG) was given and wedges from each ovary removed next day by biopsy. Ova were harvested in sterile conditions. Some were examined immediately after the schedule described earlier. The others were incubated. After 25 to 29 h incubation, eggs were harvested. Air-dried preparations were made by Tarkowski's method (1967) and stained for 25 min in lacto- aceto-orcein. Ovaries can be fixed in acetic-methanol and sent by post. They can be processed after 24 h.

Schedules for Animals

In mammals, as followed by Evans, Breckson and Ford (1964): Remove testis into 2.2% isotonic sodium citrate solution, swirl. Cut out a portion of the tubule, transfer to 1.125% KCl, mince and tease out the contents. Decant cloudy supernatant and centrifuge at 160 g for 5 min. Discard supernatant, add acetic- methanol (1:3), disperse cells and keep overnight at 4°C. Centrifuge and discard supernatant. Disperse cells in fresh fixative. Place a drop on a pre-cooled clean slide and air dry. Stain with acetic-orcein or lactic-acetic-orcein, dehydrate through ethanol-xylol grades and mount. Variants involve injection of 4 mg/kg of 1% colchicine solution in Hank's medium intraperitoneally in living animals 1-2 h before killing, and use of 0.563% KCl for swelling the cells. In a modified schedule tubules are treated in 0.7% sodium citrate for 20-30 min, 5 ml glacial acetic acid is added, mixed and kept for 30 min and the supernatant is removed after centrifugation. 3ml of 3 M gluconic acid is added to the tissue and treated for

3 h. The acid is then removed and the cells are suspended in acetic-ethanol (1:1). Washing in fixative is done by centrifugation and air-dried smears are prepared.

Hydra

Treat dissected testes successively in 0.001% colchicine solution (2 h), hypotonic sodium citrate (0.37% 15 to 20 min), acetic- methanol (1: 3, 20 to 30 min), squash in 45% acetic acid, air dry and stain in Giemsa (pH 7.2-7.4 for 30 min).

Lepidoptera

Pre-treat testes in hypotonic solution, fix in Carnoy's fluid and prepare both air-dry squash and suspension slides, stain in Giemsa.

Spinous loaches (Cobitus biwae)

Culture cells from gonadal tissue for 7 days in TC 199 supplemented with 20% cell serum. 100 μm/ml penicillin and 100 μg/ml streptomycin, prepare flame-dried slides and stain in Giemsa.

Study of Chromosomes from Somatic Cells

Depending upon the type of tissue available, different methods have been developed and, with variations, have been applied to other vertebrates as well.

Preparations from Bone Marrow Cells

Short-term culture technique

(Ford and Hamerton, 1956, Ford, *et al* 1958; *see* Sharma and Talukder, 1974; Yunis, 1974)

Withdraw 1 ml bone marrow from the individual, disperse in Ringer's solution containing 1:20 000 heparin. Centrifuge at 500- 1000 r/min and discard supernatant. Resuspend in isotonic glucose saline mixed in equal parts with human AB serum (or serum of the subject). Transfer aliquots of 2-4 ml to McCartney bottles, keep at room temperature overnight. Incubate at 37°C for 5 h. Add 0.2-0.4 ml 0.04% isotonic saline solution of colchicine and keep at 37°C for 2 h. Dilute four times by adding 0.37% sodium citrate solution and keep for 10 min. Centrifuge, discard supernatant, fix and stain by acetic-orcein or Giemsa. Modifications of this schedule involve elimination of the culture process. Three of them are described below.

Direct squash preparations

Aspirate 0.5 ml bone marrow and treat in two changes of 2-3 ml colchicine solution (0.85% NaCl solution with 6.6×10^{-3} M phosphate, pH 7, to which is added 0.3g/ml colchicine or colcemid). Keep for 1-2 h at 20-30 °C. Transfer to a watch-glass with a few drops of 2% acetic-orcein N HCl (9:1) mixture, heat gently. Squash in a drop of 2% acetic-orcein or Giemsa. Seal and store .It can be made permanent by dry-ice freezing schedule and mounted in permount (Tjio Whang,1965) . It can be used for leukaemia cells(see Priest, 1969) as well.

Direct air-dried preparations

The preliminary steps are similar for colchicine treatment as the previous schedule. Centrifuge at 400 r/min for 4-5 min. Remove supernatant, add 2-3 ml hypotonic aqueous 1% sodium citrate solution,shake and keep for 30 min. Centrifuge, remove supernatant, add acetic- ethanol (1:3), re-suspend by shaking , keep 2-5 min and repeat this procedure twice. With a pipette, transfer a droplet of the suspension to a slide. Blow on each droplet to assist spreading. Allow to dry. Immerse for 10-20 min in 2% acetic-orcein solution, dehydrate through ethanol-xylol grades and mount in permount. Feulgen reagent, Giemsa stain and crystal violet can be used as alternative stains.

Immediate examination proposes intravenous injection of colchicine in the subject 2 h before removal of bone marrow followed by the usual schedule. It is used extensively in animals.

Medium term culture schedules

Liquid medium technique.

Aspirate out 1 ml bone marrow, suspend in 3 ml culture medium: AB serum 35%; TC 199 medium, 60%; embryonic extract, 5%. Distribute in petri dishes containing cover slips, incubate at 37°C for 24-72h in an atmosphere containing 5% CO_2. Remove the cover slips and treat them in 0.7% aqueous sodium citrate solution, fix and stain following the acetic-orcein schedule.

Solid medium technique

Aspirate 1-2 drops of bone marrow in a heparinised syringe,place on a disc in a Leighton tube covered with a film of chick plasma.Coagulate with one drop of embryonic extract. Add 5 drops of nutrient medium (TC 199,5; AB serum or patient's serum, 5 and embryonic extract, 2). Seal and incubate at 37°C for 3-5 days. Replenish culture after observing under the microscope and disperse; fix and stain as described later under fibroblast culture schedules (Turpin and Lejeune, 1969) .

The bone marrow schedule, when applied to vertebrates other than man, generally includes injection of an antimitotic agent into the animal 1-4 h before killing it. Several modifications are available. An alternative schedule devised by Lee (1969) is given in outline:

Administer a mitotic inhibitor (0.004% vincaleucoblastine) intraperitoneally into the animal at 0.02 ml/g body weight for animals up to 50 g, and 0.01 ml/g for those above 50 g. Kill the animal, and extract the bone marrow into 1% sodium citrate solution. Keep at 37°C for 5-15 min, Suspend the cells by inverting the tube. Decant suspension. Centrifuge twice at 500-700 r/min and discard the supernatant. Fix in 6-8 ml acetic-methanol (1:3) at 4-6 °C for 2-8 h. Re-suspend and remove debris at the bottom. Wash two to three times in fixative by centrifugation. Re-suspend in 0.5-3.0 ml fixative. Place 4- 6 drops on damp, pre-cooled slides and spread by blaze-drying as described later in peripheral blood technique.Any of the usual chromosomal stains can be used. The entire process can also be carried out on a slide.

Peripheral blood culture techniques

These are now almost universally practised and their use has been extended to practically all vertebrates. These methods are based on the finding that under the influence of phytohaemagglutinin(PHA) and certain other more specific antigens, the lymphocyte in the peripheral blood is morphologically changed into a blast-like cell that divides in culture.The first analysis of human karyotype with this method was performed by Hungerford *et al.* (1959). A modification developed by Moorhead *et al.* (1960), combining the air-drying method of Rothfels and Siminovitch (1958), with that of the peripheral blood culture, gave an excellent technique for obtaining well scattered somatic metaphase plates. This technique has two major advantages; first, the ease with which it can be carried out on large numbers of individuals with a high success rate, and secondly, almost all the cells studied in metaphase are in their first division in culture,if suitably controlled. In PHA-stimulated cultures, small lymphocytes increase in size and start to synthesize DNA after 24 h and then divide.

The first step is to prevent coagulation of the blood obtained for the culture. Heparin is found to be ideally suitable and heparinised blood can be stored, without loss of mitotic activity , for 12-24 h in the cold before culturing. An alternative anticoagulant is acid-citrate-dextrose in which successful storage has been carried out for two weeks.

Leucocytes are separated out from red blood cells (RBC) by centrifugation or gravity sedimentation. Centrifugation at a slow speed (25 g) for 10-15 min is required for fresh blood and 5-10 min for blood pre-incubated with PHA (Moorhead *et al*,1960). High-speed centrifugation has also been found to be effective . Gravity sedimentation is a useful tool for separating RBC, the optimum temperature being 25-37°C . More efficient methods are through the application of fibrogen and dextran sedimentation. These chemicals interfere with the quality of staining, but this may be avoided by washing at the time of harvesting. These two latter methods are most efficacious for leucocyte cultures from small animals. The presence of a moderate amount of RBC in the leucocyte suspensions does not appear to interfere with mitosis. Tips *et al.* (1963) devised a blood culture method including the whole blood, thus utilizing the entire WBC in a given volume. Polymorphonuclear leucocytes(PMN) can be removed by storing the WBC suspensions at 5°C for 48 h, when the former degenerate. Alternatively, the culture flask is incubated on its side with the cells exposed to a large glass surface for 30-

60 min and then stored upright. PMN are eliminated due to their property of adhering to glass, while the other cells settle to the bottom for the remaining period of incubation.

Mitosis is initiated in blood culture of normal non-leukaemic individuals by mitogenic agents after a time-lag of 2-3 days. Leukaemic cells, however, divide in culture immediately, without the aid of such chemicals . The principal mitogenic agent used in blood cultures is phytohaemagglutinin (PHA). It is a mucoprotein isolated from seeds of *Phaseolus vulgaris* or *P. communis* by salt extraction.

The basal culture media for leucocyte culture contains mixtures of aminoacids, vitamins and buffered salts, the commoner ones being TC 199, NCTC 109, Parker's, Waymouth's, BME Spinner and Eagle's ME media (see tables). All of them require serum proteins for successful blood cultures in a proportion of 10-40%. Calf serum and both autologous and homologous human sera can be used. Mellman (1965) has advocated adding half the volume of the serum to be used as WBC containing plasma and an equal volume of homologous AB serum. Commercially available fetal calf serum gives better growth. Air-tight bottles are recommended to prevent escape of the CO_2 produced thus increasing the alkalinity, which has an adverse effect on growth. Alternatively,open vessels can be incubated in an atmosphere of 5 % CO_2, or the pH of the medium can be maintained at 7.2 to 7.4 by adding HCl or $NaHCO_3$ daily. Temperature should be kept between 36-37°C to have peak mitotic activity between 60-72 h or strictly at 38°C for optimum activity at 48 h. Bacterial contamination is prevented by the addition of penicillin and streptomycin. Since certain levels of antibiotics like streptomycin and chloramphenicol have been found to be deleterious to mammalian cell cultures, their use can be omitted if sterile precautions are adopted. The major steps in the preparation of metaphase plates for karyotype study from leucocyte culture involve metaphase arrest, hypotonic treatment, fixation, preparation and staining.

For the arrest of mitosis at the metaphase stage, colchicine or its analogue deacetymethyl colchicine (Colcemid, CIBA) is most frequently used. The hypotonic treatment for swelling the cell and dispersing the chromosomes, first devised by Hsu(1952) can be carried out by diluting the balanced salt solution used to wash the harvested cells with distilled water and then incubating the cells in it.

Representative Schedules for Peripheral Blood Leucocyte Culture

Macromethod

Draw in and eject 1 ml solution of 5000 iu heparin/ml in a syringe, thus coating its walls completely with heparin. Draw 20 ml venous blood into the heparinized syringe.

Decant in a test-tube inclined at 45 degrees and permit the RBC to sediment. The blood can also be kept for sedimentation in the syringe itself, with the needle held upward, at room temperature. After 30-50 min, remove supernatant containing exclusively WBC 5 to 10^6/ml. This procedure avoids an alternative method of slow speed centrifugation (200 r/min) for separating the leucocytes. Study an aliquot of the supernatant in a Malassez counting cell to obtain finally a concentration of 1-1.5 × 10^6 cells/ml, in a medium containing 30-35% of the individuals serum, and 65-70% of the basal medium, by adding required amounts of this medium. Add 0.2 ml bacto-phyto-haemaglutinin per 10 ml mixture. Distribute in test tubes, filling them to one- third of their volume. If required, inject a mixture of 5% CO_2 and air before sealing the tube. Incubate at 37°C for 72h. Add, 2 h before harvesting, two drops of isotonic colchicine solution at a concentration of 0.04% per ml of the medium.

Decant into a centrifuge tube, spin at 800 r/min for 5 min, discard supernatant and add hypotonic solution (0.93% aqueous sodium citrate) and treat for 10 min at 37°C . Centrifuge to a button at 800 r/min. Remove supernatant and add fixative (1 : 3) acetic-methanol or acetic-ethanol slowly, suspending the cells . Leave for 30 min and repeat the process using only 2-3 drops of the fixative. Re-suspend cells by pipetting to prevent bubble formation.

Cool several absolutely clean, grease-free slides (or cover slips,) by placing on a paper put on an ice block, or on CO_2 snow, so that they are covered with a fine mist. Place a drop of cell suspension over a cover slip which immediately spreads. Hold over a flame, ignite or place in 60°C cabinet to cause evaporation of the fixative, flattening the preparation . Alternatively, tilt slide on its long edge, touching absorbent paper, and blow directly on slide.Dry in air. Store or stain immediately if necessary.

Modifications generally include the number of centrifugations required and the stains used. Antibiotics can be added if necessary, a proposed incubation medium being: autologous plasma, AB human serum or fetal calf serum, 1 ml; basal medium TC 199, 6 ml; bacto-phytohaemagglutinin, 0.03-0.05 ml; penicillin (100 000 units/ml) and streptomycin (100 mg/ml) solution, 0.02 ml. It can be prepared and stored at 20°C. If in the dried slide, cells are scarce, concentrate the cell suspension, after adding fixative, by centrifugation and removal of part of the fixative . For chromosomes insufficiently spread or shredded, the number of times of suspension in fixative is increased and the flaming schedule is followed for spreading. If, chromosomes are overspread, use a small drop of suspension and do not evaporate quickly. Stain in lactic-acetic-orcein (3 to 24 h) and rinse in 45% acetic acid, or stain in Giemsa (pH 7.2) and air dry; or heat in 0.06 M phosphate buffer (pH 6.8 to 7.2) for 10 min, stain in haematoxylin and mordant in 3% aqueous ferrous ammonium sulphate solution.

Semi micromethods

According to Mellman(1965)

Add 0.1 ml aqueous heparin (1000 units/ml) and 1 ml serum to a 2 ml plastic disposable syringe. Fill to the 2 ml mark with venous blood. Mix the contents. Empty the needle by aspiration. Stand the syringe and allow the RBC to sediment at room temperature, till the clear supernatant is about 1 ml (30-60 min). Bend the needle and inject contents directly in culture vessel without doing WBC count. The vessel contains the complete culture medium described above at room temperature. The later steps are similar to the macromethod.

Adopted by Nowell, Hungerford and Brooks (1958)

Draw 10-20 ml blood in a syringe containing 0.02 ml heparin. Add PHA (M form) to blood (0.2 ml per 10 ml), mix and keep in the cold for 30-45 min. Centrifuge at 25 × g for 10 min at 5°C, remove supernatant and count leucocytes. Plant leuococytes (10^7) in a medium containing: plasma, 3 ml; medium 3 ml; and antibiotics. Maintain a pH of 7.2-7.4 with HCl or $NaHCO_3$. Incubate at 37°C for 72h. The remaining procedure is as above, the hypotonic used being 1 : 5 dilution of serum.

Micromethods

Micromethods were developed by several workers using only a few drops of blood. The major variations involve the type of medium used, its constituents, the relative amounts of PHA and colchicine needed and the periods of treatment and incubation.

Fill centrifuge tubes (capacity 41 ml) with 5 ml human serum, 15 ml TC 199, 4 drops of PHA (mixture of equal parts of phytohaemagglutin Difco M and P, or phytohaemagglutin Wellcome) and 4 drops of Liquemin , (Roche,or equivalent of 5 mg crytalline heparin). Disinfect skin of index finger or thumb carefully. Incise with a vaccinostyle and remove 4-6 drops of blood with a pipette. Transfer directly to the tube containing the medium. Rinse out pipette with medium to suspend the blood. Alternatively, use a few drops of blood drawn from a vein. Seal and incubate at 37°C for 48-72 h. Add 2 ml 0.04% isotonic colchicine solution. Re-suspend and incubate for 2 h. Re-suspend, transfer to a conical centrifuge tube. Centrifuge for 5 min at 800 r/min.

Discard supernatant. Fill up to two -thirds of tube with a mixture of animal serum (1); distilled water (5) and hyaluronidase 2.5 iu/ml mixture. Re-suspend and incubate for 7 min. Centrifuge for 5 min at 800 r/min, keep 2 min, discard supernatant hypotonic and add Carnoy's fixative (acetic-chloroform-ethanol 1 : 3 : 6). Re-suspend and keep for 45 min. Centrifuge at 800 r/min for 5 min, discard supernatant and add acetic-ethanol (1 : 3). It can be sealed and stored in cold. Centrifuge at

800 r/min, discard supernatant and add 5-6 drops of fixative. Re-suspend by pipetting.

Follow next steps as in macromethod. After air-drying, hydrolyse for 7.5 min in N HCl at 60°C, rinse in iced water and stain for 10 min in a solution of 1 part Unna's blue end four parts neutral water. Rinse in water and dry. Pass through xylol or toluene grades and mount in canada balsam or directly in Permount.

A schedule developed by Hungerford (1965) uses a culture medium composed of Eagle's basal amino acids and vitamins, at double strength, in Earle's balanced salt solution (BSS), adjusted to pH 7.0 with 7.5% $NaHCO_3$ supplemented with glutamine 2 mM, penicillin 100 units/ml, streptomycin 100 μg/ml, phenol red 7 μg/ml. Fetal agammaglobulin bovine serum and phytohaemagglutinin M (Difco) are added to make 15 and 2% respectively, of the final volume and also 20000 USP units of heparin sodium per litre of medium. After usual schedule of inoculation, incubation and treatment with colchicine, the cells are separated by centrifugation and the medium replaced with 0.075 M KCl, 16 USP units/ml heparin sodium, and incubated for 10 min. Subsequently, KCl is removed and the material fixed in two changes of acetic-methanol (1 : 3), air-dried and stained in 1% orcein in 60% acetic acid. It has the advantage of using a very small quantity of blood.

Preparations from fibroblast and other cultures

Cultures

Though relatively time-consuming, fibroblast culture methods can be effectively used for checking unusual karyotypes observed in the blood cells, determining mosaicism by studying chromosomes of different tissues, and for other experimental work where cytogenetic studies are correlated with biochemical, virological or other studies. In general, any tissue aseptically removed, can give usable cultures. The different techniques have evolved essentially as a result of the manner in which the cells are transferred: *en bloc* in the state in which they are (explant technique, Turpin and Lejeune, 1969), or after trypsinization (Hamden and Brunton, 1965).

For chromosome preparations, attempts are made to obtain partially synchronous divisions by changing the medium or subculture. The culture is allowed to become acidic and the pH restored to 7.4 by adding $NaHCO_3$ and embryo extract, resulting in a large number of mitosis after 16 h. In most techniques, however, colchicine is added to the culture to obtain metaphase configurations. The next stage is the use of a hypotonic solution, ranging from distilled water to dilute sodium citrate, or diluted salt solution or serum. For obtaining suitable plates, two methods are available. Either the cells grown on cover slips and processed while attached to them, or the cells are suspended through trypsin digestion and then processed by any of the methods available for peripheral blood culture. The second method gives more satisfactory results. In fixation, acetic-ethanol or acetic-methanol (1 : 3) is commonly employed. Fixation can be carried out without breaking up the pellet, or by suspending the cells prior to fixation. The cells can be spread by the air-drying schedule used for blood cells. For better spreading, they can be transferred to 75% acetic acid immediately before drying. Squash preparations are not very successful. Giemsa, Feulgen and Unna's blue are the frequently used stains. The length of time from culturing to chromosome preparations may range between 1 and 3 weeks, depending on the method and the material used.

Constituents used in the techniques

Cockerel plasma (CP) can be stored at 4°C for several weeks in wax-or silicone-coated glassware. It is prepared by drawing 20 ml blood from the wing vein of a young bird, centrifuging at 1500 r/min for 30 min at 4°C and storing in cold.

Trypsin (0.25%) is prepared by dissolving 2.5 g Bactotrypsin 1-300 in a few ml Hank's solution without Ca and Mg and adjusting the pH to 8.0 with NaOH. The solution is made up to 100 ml with Hank's solution and later diluted with it 10 times before use.

Sodium bicarbonate (1.4%) contains $NaHCO_3$ 3.5 g; phenol red (0.2%) 2.5 ml, and neutral water 247.5 ml. 5 ml is added to 200 ml Hank's solution before use.

Colchicine stock solution contains 0.5% colchicine in neutral water. Working solution of 0.005% colchicine is prepared in BSS. 0.5 ml this solution is added to each 10 ml culture to have a final concentration of 0.0025%.

Antibiotics Kanamycin, mycostatin, neomycin, penicillin and streptomycin are all dissolved in neutral water to obtain stock solutions and later diluted with culture medium for the required concentrations.

Stain : 8 ml Giemsa stock solution in 192 ml distilled water.

Chick embryo extract (CEE) can be obtained commercially. For large quantities, embryos from hen's eggs, incubated for 10 days, are collected and homogenized by forcing through a syringe. Hank's BSS medium is added (1.25 vol), the mixture centrifuged at 2500 r/min for 30 min at 4°C. The supernatant is decanted, mixed with 2 mg hyaluronidase/100 ml extract, incubated at 37°C for 1 h, ultracentrifuged at 25000 r/min for 1 h and filtered through No. 03 porosity Selas candle filter under pressure of 8-10 lb/in^2.

Serum: AB serum can be obtained by allowing a pint of human AB venous blood to coagulate at 4°C, drawing out the serum with a syringe, centrifuging at 1500 r/min for 30 min and filtering through a No. 03 porosity Selas candle filter.

Growth medium used in almost all cultures, contains Eagle's medium with antibiotics, 70 parts; human AB serum 20 parts and CEE 10 parts.

Trypsin-digestion Culture Method

Biopsy : Clean skin with spirit, inject local anaesthetic, make an incision with scalpel and cut off V-shaped skin (2 mm). Place in about 4 ml Eagle's medium. Cover the wound.

Primary culture : Transfer tissue to glass dish and cut into small pieces. Place equal amounts of CEE and CP in two petri dishes. Transfer one bit of tissue to a third petri dish and suck out the excess medium. Mix one drop each of CEE and CP and draw them and the piece of tissue into the pipette. Transfer to a culture flask spreading out the CEE/CP mixture in a thin layer. Treat other bits of tissue similarly so that a culture flask with 4 cm diameter has five pieces of tissue. Allow the plasma to clot for a few min. Add 5 ml growth medium. Add 5% CO_2 in air to flask and close with siliconed stopper and incubate at 37°C.

Sub-culture for First sub-culture, remove growth medium, add 10 ml 0.25% pre-warmed trypsin solution, incubate at 37°C for 15 min, transfer to centrifuge tube and centrifuge at 500 r/min for 5-10 min. Discard supernatant and re-suspend cells in 1 ml growth medium. Transfer to fresh culture flask containing 9 ml fresh growth medium, seal and incubate at 37°C. For subsequent sub-cultures, use trypsin for 5 min only and after suspending cells in 1 ml growth medium, use only 0.1 ml for each sub-culture, to give a 1 in 10 dilution of cells. For maintenance, check cultures every two days for growth, sterility and pH, and replace growth medium. Adjust the pH with 5% CO_2 or $NaHCO_3$ solution (1.4%) when required.

Processing : To a healthy culture, add 0.5 ml 0.005% pre- warmed colchicine solution, incubate for 2-4 h, prepare cell suspension with trypsin as given for first sub-culture. Centrifuge at 500 r/min for 10 min, discard supernatant and resuspend cells in pre-warmed Hank's BSS. Centrifuge at 500 r/min for 10 min, discard supernatant, resuspend cells in pre-warmed 0.95% sodium citrate solution and keep 20 min in 37°C. Again centrifuge at 500 r/min for 10 min, discard supernatant, re- suspend cells in small quantity of fluid left, add acetic-ethanol (1:3) drop by drop with agitation. Keep in excess of fixative from 30 min to overnight.

Preparation of slides : Centrifuge at 500 r/min for 10 min, discard supernatant, add fixative and again centrifuge at 500 r/min for 10 min. Discard supernatant and resuspend in a few drops of 75% acetic acid. Place a few drops on a wet, pre-cooled slide, heat over a flame to dry, cool and stain in 2% acetic- orcein for 2-3 h at 37°C. Dehydrate in cellosolve, treat in euparal essence for 2 min and mount in euparal (Harnden and Brunton, 1965).

Plasma clot culture method

Biopsy : For surgical cases, remove under general anaesthesia, a tissue about 4 x 4 x 4 mm in measurement. Wrap in a sterile square of gauze, place in a wide-necked flask 4 cm in diameter. Add 5 ml of sterile physiological saline, seal the flask and store for 24-36h at room temperature, if necessary. For skin biopsy, after thorough cleansing with soap and sterile water, and sterilization in ethanol (twice) and ether, pinch the skin between the jaws of a 'Coprostase' clamp, keeping a piece protruding above it. Anaesthetize locally with ether, remove with a scalpel a piece 3-4 mm long, place in a sterile square of gauze and treat as for surgical biopsy. Disinfect and close the suture with band aid.

Explant : Wash the tissue in physiological saline and cut into pieces 1-2 mm square. Place a sterilised cover slip in the distal depression of a Leighton tube. Spread a drop of CP over the coverslip. Place the bits of tissue (2-3 per cover slip) on the plasma. Add a drop of CEE to coagulate the plasma and fix the tissues to the glass. Close the tube with a rubber stopper, incubate for several hours or overnight at 37°C. Add culture medium containing per tube: human AB, five drops; Hanks' solution with 200 μm/ml penicillin, 50 μg/ml streptomycin and 5 μg/ml chloramphenicol, five drops, and embryonic extract, two drops. Incubate.

Sub-culture : After development of a crown of fibroblasts around the explants in 4-6 days, transfer the explants to other tubes following the method described under explant. Replace the medium in the tubes with cover slips with fresh medium. Incubate.

Processing : After 36 h incubation, add three drops of CEE to the medium in each tube. Incubate again for 16 h to obtain a large number of mitosis due to change in the substrate. Place the coverslip with a curved tip pipette, with the cells uppermost, in a mixture of sterile human or mammalian serum, 1 part; neutral water 5 parts and hyaluronidase 2.5 per ml (used to hasten absorption) of mixture. Incubate at 37°C for 35 min. Transfer the cover slips to Carnoy's fixative (chloroform-acetic-ethanol, 3:1:6) and treat for 45 min.

Preparation of slides : Take out the cover slip and dry in open air. Hydrolyse in N HCl for 7.5 min at 60°C. Rinse in neutral water, stain in Unna's blue for 10 min (1 part Unna's blue solution, 4 parts neutral water). Rinse in neutral water, dry for 5-10 min in open air, pass through toluene and mount in canada balsam.

Liquid medium method

This has been applied to surgical biopsies, mainly of tumours.

Cut into small bits, wash three times in Dulbecco PBS solution (with calcium). Transfer to Erlenmeyer flask, add trypsin solution in PBS buffer (25 mg/l), seal and agitate in a magnetic shaker at 37°C from 10-20 min. Replace trypsin solution every 10-15 min with continued agitation, up to 1-4 h according to the tumour studied. Transfer each change of trypsin solution to a centrifuge tube in cold. Centrifuge for 5 min at 800 r/min. Wash three times in a solution of casein hydrolysate or a culture medium. Resuspend sediment in 1-2 ml casein hydrolysate or the culture medium. Place the culture in test tubes with cover slips containing: casein hydrolysate or synthetic culture medium (like 199), 2 parts and human AB serum 1 part to have a concentration of 5×10^5 cells/ml. Incubate at 37°C after adjusting pH. Replace the medium every three days. Cell multiplication depends upon the nature of the tumour. Processing and preparation of the slides are similar to the schedule given for peripheral blood cultures.

Several alternative schedules are in practice, combining or altering different stages in the representative ones described here. For metaphase arrest in cultures grown on cover slips, they can be treated directly with colchicine in culture medium (0.1 g/ml) for 18-24 h or colchicine can be added to the culture medium to give a concentration of 0.0025%. A saturated solution of Abopon in 0.2 M phosphate buffer at pH 7.0 can be used as a suitable mounting medium after acetic-orcein staining and rinsing in water.

Application in other vertebrates

The schedules, with suitable variants, have been employed for the study of tissues from different parts of the body of different animals, as well as human subjects. Chromosomes have been studied successfully in cell cultures from kidney tissue of primates, from skin cultures of lemurine lemurs, from shank skin of *Ardea cinerea*, from embryonic tissue, from feather pulp culture of the ibis; from rabbits derived from eggs fertilised *in vivo* and from gill culture. In an agar-fixation procedure, trypsinized cells from the culture are suspended in Hanks medium and then transferred to an agar surface containing 1.5% agar and 0.5% sucrose for 10-15 min. An agar square is cut out, processed and stained as usual. Both change of medium, and addition of colchicine, were adopted to obtain a large number of metaphase plates in opossum, followed by double staining in 2% acetic-orecein and Harris's haematoxylin. For feather pulp, inject 0.05% colcemid (IP) at 0.8 ml/kg body weight, incubate for 1 h, pluck out feathers, transfer pulp to 0.45% (w/v) sodium citrate solution for 15-20 min, fix in 50% acetic acid for 30 min, squash.

Sex Chromatin Studies

The sex chromatin, as seen in the interphase nuclei, has been used in the detection of errors of sex development, to determine the number of X chromosomes in the complement of an individual, and in studies requiring the indication of the sex chromosome complex of a particular tissue. The different techniques are based on the fact that the X chromosome may show the property of heteropycnosis in interphase nuclei and may form a distinctive chromatin mass or chromocentre. It is a female characteristic in all mammals, except the opossum, which shows such masses in the cell nuclei in both sexes. The presence of a female-specific chromocentre was first demonstrated in the nerve cells of the cat by Barr and Bertram (1949), leading to the term 'Barr body'. A long series of publications has established sex chromatin analysis as an effective diagnostic tool for sex chromosome studies, according to the rule that the maximum number of sex chromatin masses is one less than the number of X chromosomes present in the individual studied.

The proportion of nuclei with demonstrable sex chromatin masses varies with the type of preparation and the tissue, ranging from less than 60% in buccal smears to almost 100% in thick sections of nervous tissue. The sex chromatin mass shows a Feulgen-positive reaction due to its DNA content and stains with chromosomal dyes like cresyl violet, fuchsin, gallocyanin, haematoxylin, thionine and particularly with orcein. It has an affinity for methyl green in pyronin methyl green staining and persists after mild acid or ribonuclease treatment. It may be studied from buccal smears, sections of tissues and peripheral blood neutrophiles.

The *buccal smear* technique for studying sex chromatin mass from buccal mucosa is the simplest one and is most widely used for clinical studies of sex determination. A simple schedule is given below:

Label clean slides with the subject's reference number and side of the body. Draw the edge of a metal spatula firmly over the buccal mucosa. Discard the material; scrape the mucosa gently a second time to obtain healthy epithelial cells from a deeper layer. Spread over a small area of an albuminized slide, using separate slides for smears from right and left sides. Do not spread too thinly. Immerse immediately in fixative (95% ethanol) for 15-30 min, treat successively in absolute ethanol for 3 min, and 2 min in a 0.2% solution of Parlodion in ethanol-ether mixture (1:1) to attach the cells firmly. Dry in air for 15 *s*, pass through 70% ethanol (5 min), two changes of distilled water (5 min each) and stain. Alternative fixation can be carried out in ethanol-ether mixture (1:1), immediately after smearing, for periods ranging from 12 h to two weeks, followed by gradual hydration.

Staining may be done in 1% aqueous cresyl violet solution or acetic-orcein solution. Barr (1965) suggests staining 5-10 min in working solution of carbol fuchsin followed by differentiation for 1 min in 95% and absolute ethanol successively. (Stock solution of carbol fuchsin: 3 g basic fuchsin in 100 ml of 70% ethanol. Working solution: stock solution 10 ml; 5% carbolic acid in distilled water, 90 ml; glacial acetic acid, 10 ml; 37% formaldehyde, 10 ml. Keep for 24 h before use.) Double staining, with Biebrich scarlet as chromatin stain and Fast green as counterstain, has been suggested. Dehydration and clearing are carried out as for other squash schedules.

Sex *chromatin studies from tissues* have a much more limited application, being restricted mainly to tissues obtained during operation or post mortem. A recommended fixative is: 37% formalin, 20 parts; 95% ethanol, 35 parts; glacial acetic acid, 10 parts, and distilled water, 30 parts. After 24 h fixation, tissues are transferred to 70% ethanol. Paraffin blocks are prepared after dehydration through ethanol and xylol grades, sections 5 μm thick are cut. Staining is carried out in Harris's haematoxylin, counterstained with eosin, or in other chromosomal stains like Feulgen, thionin or gallocyanin. Sex chromatin can also be studied from a monolayer of cells growing *in vitro*, or skin biopsies or leucocyte cultures, by omitting treatment with hypotonic solution during processing. Prenatal sex chromatin analysis is useful in detecting male conceptuses with sex-linked recessive hereditary disorders when pedigree data are available. It can be carried out by concentrating cells from fluid samples by centrifugation, fixing on a slide and staining by thionin or Feulgen.

Studies from peripheral blood neutrophiles: A drumstick-shaped nuclear appendage, which stains with Wright's, Giemsa or haematoxylin schedules, is observed in occasional neutrophile leucocytes in normal females, but never in the male, and is thought to contain sex chromatin. The number of neutrophiles per drumstick varies in different females. They can be studied from relatively thick blood smears, drawn on cover slips or slides, and followed by staining. Both buccal smear and neutrophile methods should be used simultaneously for the detection of sex chromosome mosaics.

B. CHROMOSOME STUDY FROM MALIGNANT TISSUES

Cancer represents an unchecked, malignant form of rapid growth, perpetuated through several cell generations and probably originating from several causes, both internal and external, including transformation by viruses. Irrespective of its mode of origin, cancerous tissue is always characterized by distinct cytological features. Mitotic instability and chromosomal unbalance, associated with lack of differentiation, are features of cancerous cells. However, no cancer is unquestionably characterized by specific chromosomal changes, except chronic myeloid leukaemia, where a structural alteration in chromosome G(Ph') has been found to be a common feature. Several, as yet unconfirmed, reports of other such associations are also available.

The development of techniques for the study of cancer chromosomes and their behaviour was principally stimulated by the discovery that certain human abnormalities or diseases, such as mongoloid idiocy, can be correlated with certain chromosomal irregularities, and since then various techniques have been developed for the study of cancer chromosomes from tissues extracted from individuals, or after growing them in culture.

Data on cancer cytology was mostly nebulous until the discovery of the *ascites tumours* and a method for their observation. The single cells, or groups of cells, suspended in the body fluid are convenient for cytological observation, and the methods of study have undergone so much improvement that solid tumour tissue can now ultimately be converted into ascites tumours for facilitating cytological study. Techniques have also been devised to study the comparative cytology of cancerous and normal cells under culture conditions and their behaviour following cell fusion *in situ*.

Though cancer is the principal unsolved medical problem in human beings, the latter cannot, for obvious reasons, be utilized for detailed experimental purposes. In addition to blood and bone marrow, only biopsies for studying the chromosomes can be taken from patients suffering from cancer in various organs. In carrying out extensive experiments, using different chemical agents and testing the effect on the development and nuclear behaviour of cancerous tissue under laboratory conditions, inbred strains of white rat, mouse, guinea pig and rabbit are mainly employed.In addition to the development of spontaneous cancers, different chemical agents, termed *carcinogens*, can be used for their artificial induction. The different types of carcinogens available in the market belong to the coal-tar derivatives; principally, the hydrocarbons, azodyes and also certain mustard derivatives, steroids and other compounds like aerosols, alkaloids, nitrosamines, sulphonamides, etc.

The different methods by which a carcinogen can be administered into tissue to induce cancerous growth in animals are: (a) feeding by gavaging a specific dose in the food; (b) intraperitoneal injection; (c) subcutaneous injection and (d) rubbing into the skin. These are applied with different solvents like water, olive oil, sesame oil, lard, glycerine, benzene, acetone, in specific doses for a prolonged period, at regular intervals. For example, tumorous can be induced with intraperitoneal injections of 3,4 benzpyrene and methyl cholanthrene in rats within 3-6 months. For inducing multicentric hepatoma, n-nitrosomorpholine can be administered to inbred rats in drinking water (20 ml daily for six days), the dosages being 6 mg% solution for up to 34 weeks, 12 mg% up to 12 weeks, and 20 mg% up to 7 weeks. In established mouse fibroblast cell cultures, to study the effect of carcinogens, methyl cholanthrene at 1μg/ml of medium can be administered for a prolonged period (one to several weeks). The behaviour of loss of contact inhibition can be detected even after 2-3 days exposure to either methyl cholanthrene or to 3,4- benzpyrene. When the treated culture is injected into C_3H mice, maximum tumour production was obtained with 6-day treated cultures. The two terms *carcinoma* and *sarcoma* are applied, depending on the type of origin. The latter type represents tumorous growth initially originating from the mesoderm, whereas all other types are included within the category of carcinoma. Cancerous growth involving blood cells in the body fluid are known as *ascites* tumours. Specific names are given to growths in the different organs, based on the organ affected and detailed cytological studies have been carried out on almost all kinds of human cancer. After induction of cancer in the body of the subject, cells are maintained in culture and chromosomes can be studied in them through the method outlined here for the study of mammalian chromosomes.

Study of Cancer Tissue in Culture

The rapid advance of research on cancerous materials has been responsible for the development of several techniques for culturing malignant cells *in vitro*. To secure growth *in vitro*, methods for the culture of *tissue explants, monolayers* as well as *cell suspensions* have been devised. Most of these schedules have been discussed in detail earlier in the section on mammalian and human chromosomes and tissue culture. Only the schedules specially modified for cancerous cells are described here.

Explant culture

Most methods for cancer tissue explant culture are identical with those described for the study of mammalian chromosomes and tissue culture. Mammalian serum is one of the major constituents of the medium but plasma clot consisting of 50% plasma in balanced salt solution and 50% embryo extract in serum provide the most congenial medium for tissue growth.

Suspension cultures from cell effusions can be obtained most successfully from ascites tumour cells. De Bruyn (1965), De Bruyn and Hampe(1961) effectively applied the *monolayer technique* for culturing ascites cells and cell effusions obtained from human pleural and peritoneal fluids. In monolayer culture, cancer cells do not exhibit contact inhibition,often manifested in closely adhering normal cells.

Of all the techniques evolved for cancer tissue culture, processing of cell suspensions from solid tissue is the most elaborate one. As with normal tissues, disaggregation of cells poses the principal problem, for which mechanical, chemical and enzymic methods have been adopted as described earlier for mammalian tissues.

Of the *chemical* methods, the most effective is the use of versene or EDTA which may or may not be supplemented with homogenization. Among the different enzymes tried for dispersion, trypsin is undoubtedly the most useful, both for HeLa cells and other malignant tissues. Trypsin, with a small amount of deoxyribo-nuclease, has been applied successfully in some cases but this treatment is limited by the possibility of digestion of DNA. Elastase and collagenase have also been employed, the latter especially in the adult organs. Fibrinolysin was used for disaggregating tissues from cancerous human colon.

Both natural extracts, like serum and embryo extract, and synthetic media are in use. All laboratory adapted cell strains in culture grow profusely in Waymouth's (1956) simple medium M.B. 752/1. The optimum growth is generally seen at pH between 7.0 and 7.8, cell disintegration occurring at over 8.0. Lymphocytes survive for long periods in hypotonic media. A buffering system with bicarbonate, employing an atmosphere of 5% CO_2 in air, or with Tris-HCl in air, atmosphere or containing 2% carbon dioxide, gives good growth and pH control. *Suspension cultures* have been found to be very suitable for malignant cells. For such cultures, continued rotation of the drum containing the culture tubes, steady shaking, silicone-coating of the inner wall of the container to prevent adhesion, continuous stirring and automatic replacement of the medium through cryostat are all applied as in normal tissues(Paul,1959; Bjorklund, Bjorklund and Paulsson, 1961). After continued and repeated sub- culturing,*cell lines* can be secured containing a colony of cells showing a rapid rate of proliferation.

Several schedules for isolated *single cells* have been evolved but in general, single cell cultures of HeLa cells are easier to prepare than cultures of *fibroblast* cells. Methods for obtaining clones from a single cell originally derived from suspensions are also identical with those described for other animal tissues.

The potential of *hybridizing mammalian cells in culture* for the suppression of malignancy has been elucidated in the work of Harris *et al* (1969).The technique for fusing somatic cells in culture has been described in detail in another chapter.

Excellent work has been carried out on the virus-induced malignancy in animals. Under culture conditions, oncogenic viruses cause changes in the host cells resulting in tumorous growth of the latter, the process being known as *transformation.* This behaviour is similar to that of neoplastic or malignant growth induced by different carcinogens. Both RNA and DNA viruses can cause transformation (Klein,1966).

Schedules for Preparation of Tissue

Direct biopsies are taken from patients suffering from cancer of particular organs. The tissue is directly observed in acetic- orcein, acetic-carmine or acetic-dahlia after staining for a few min. If necessary the fluid may be slightly warmed to facilitate staining and squashing. To secure well-scattered chromosomes, pre-treatment in aqueous coumarin solution for a few minutes prior to acetic-carmine staining, yields excellent results.

Ascites tumours

The peritoneal or pleural fluid can be taken out with the help of a hypodermic syringe or pipette and a drop of this can be stained directly on the slide. The cells may be treated in a hypotonic salt solution(preferably 1.12% sodium citrate) before staining. To obtain a concentrated solution the fluid is centrifuged. Makino(1957) observed that pre-treatment in a drop of water on the slide for 20-30 min prior to staining with acetic-dahlia yielded well-scattered chromosomes. Propionic-sudan black B is also an effective stain. For fixing, squashing and staining the fluid directly in acetic-orcein, the material is pressed between specially prepared siliconed surfaces of slide and cover slip.

Direct observation of bone marrow cells, as described for mammalian and human chromosomes from normal cells can be followed.

Mitotic division of living cells in pleural fluid

Full the depression of a sterile grooved, or depression, slide with liquid paraffin above the level of its edge. Cut open a tumour-bearing rat. Draw out a small amount of tumour ascites fluid from the peritoneal cavity with a pipette. Place the fluid on a sterile dry cover slip without any medium. Invert the cover slip over the depression so that the ascites fluid lies in the liquid paraffin. Apply slight pressure and remove excess fluid with a blotting paper. Seal and observe under a phase contrast, or light microscope.

Study from peripheral blood

The different techniques for studying chromosomes from peripheral blood cells, as described for mammalian and human chromosomes, can be effectively used for cancer chromosomes as well.

Schedules

Cloning of HeLa cells

The dilution technique

Materials:

A strain of HeLa cells growing on a medium of pH 7.4;

A balanced salt solution (BSS) without Ca,Mg,or PO_4; 0.05% trypsin dissolved in BSS; medium containing 20% human or calf serum and the remaining synthetic nutrient medium; a stainless steel or glass cylinder of approximately 5mm diameter and 10mm height; 60mm petri dishes; CO_2 incubator; pasteur pipettes.

Prepare a healthy culture of HeLa cells on a clean medium at a pH of 7.4. Drain off the nutrient medium from the container. Add a quantity of BSS and drain again. Add 0.05% trypsin dissolved in the BSS to the culture and incubate at 30°C for 5-10 min. Distribute nutrient medium, containing 20% serum and the rest of the synthetic medium, into petri dishes and test tubes, putting 5 ml into the former and 4.5 ml into the latter. Keep the petri dishes in a CO_2 incubator. Suspend the cells in the medium by shaking the container gently. Aspirate the suspension in and out of a pipette several times. After all cells have been suspended, add to the suspension half its quantity of growth medium to stop the action of trypsin. With a haemocytometer, determine the number of cells in a known volume of suspension. Add 0.5 ml of the suspension to 4.5 ml of the medium in the first one of the series of test-tubes. Shake well and add 0.5 ml of the fluid from this test tube to 4.5 ml of the medium in the next one, and so on along the set of test tubes until, on counting, a suspension yields 1000-2000 cells/ml. Pipette out 0.5 ml of the final suspension on to one of the petridishes containing the medium, to give 100 cells per dish. Treat the other dishes similarly and keep them in the CO_2 incubator at 37°C for 1 week. For counting the colonies, drain off the medium and stain.

For further isolation, outline on the outer glass surface of the petridish, with a glass pencil, the location of a particular colony. Coat the bottom edge of a glass cylinder with silicon grease. Drain out the medium and place the glass cylinder on the petri dish so that it encloses the colony to be isolated and is attached to the petri dish by the silicon grease. Pour a few drops of BSS into the cylinder, drain and add a few more drops. Drain out this solution, washing the culture within the cylinder. Add a few drops of trypsin solution and incubate at 37°C for a few minutes. Disperse the cell suspension thoroughly by aspirating in and out with a pipette. Draw out the suspension and inoculate a fresh nutrient in another vessel for sub-culture.

To Study the Behaviour of Chromosomes and Cell Division in Different Sub-cultures obtained from the Primary Tumours, the following method may be applied.

Ascitic Fluid

May be taken from an affected rat and injected into the peritoneum of a normal rat. Oksala (1956) inoculated ascitic fluid, suspended in buffered physiological saline (1:9), in a dosage of 0.2 ml per mouse. Single cell subculturing from ascitic fluid is applicable to cases in which the chromosomal behaviour of a cancer cell colony derived from a single cell of a heterogeneous population has to be studied. The inherent heterogeneity in the chromosome behaviour of every cancer cell has been substantiated from such studies.

Tumour ascitic fluid with tumour cells is first removed from rats. These are then diluted in the diluting fluid in a proportion of 1:20 000, the diluting fluid being prepared by injecting 15-20 ml of physiological saline with a sterilized syringe into the peritoneum of rats weighing about 100-120 g. With the same syringe, the peritoneal fluid is taken out from the treated rats after 15 min and centrifuged at 3000 r/min for several min. The supernatant can serve as the diluting fluid provided it is cell free on microscopical examination. With a sterilized pipette, a drop of the diluting fluid containing tumour cells is taken on a clean cover slip. A microscopic droplet containing a single cell is sucked into the micropipette. The droplet is ejected into the peritoneum of a normal rat. Cytological observation of the sub-culture following the method discussed above can be carried out after a few days.

Sub-culturing of solid tumours

Solid extracted tumours can be mashed in a homogenizer or tissue press and the tissue suspension can be inoculated subcutaneously or intraperitoneally into normal animals. The remaining steps are similar to earlier schedules.

Ascites developed from solid neoplasms

This technique is very important in the study of cancer cytology, because solid tissues, which are otherwise difficult to observe, can conveniently be converted into free tumour cells suspended in the body fluid. The steps are:

Intraperitoneal inoculation of sarcoma or lymphoma cells in mice is performed. Week-old subcutaneous implants of the subsequent tumour are taken and ground in a tissue press and a suspension in 0.85% NaCl solution is prepared. The number of viable tumour cells in the mashed suspension is first determined by separating the tissue fragments and cell clumps from the suspension by sedimentation and decantation; a suspension sample is then placed in a haemocytometer previously coated with 0.02% solution of neutral red in ethanol, and cells coloured with neutral red are considered dead. This step is necessary to find out the amount of dilution required to obtain the requisite number of viable tumour cells in 0.1 mm^3. Mitotic activity can be counted with acetic-orcein staining. The suspension, after proper dilution of 1:10 or 1:20, is then inoculated into another mouse intraperitoneally. Growth characteristics, cell division, etc. can be studied from the peritoneal exudate after a few days.

The chromosomes can be identified following the schedules described under Banding Pattern Analysis and In situ Hybridization.

Conclusions

Lately, the aspect of cancer research involving chromosomes has also been greatly influenced by the discovery of proto-oncogenes in chromosomes which are homologous to some viral sequences. Varied lines of studies have established the presence of cellular genes with potential for oncogenic activity. *In situ* hybridization has emerged as a powerful tool in identifying cellular oncogenes which are the homologues of transforming genes of RNA tumour viruses. The malignancy induced by a genotoxic virus may involve activation of the cellular proto-oncogenes. The virus may undergo integration, comparable to the action of transposons. The integration may even lead to mutations of specific oncogenes as in Burkitt's lymphoma or chronic myelocytic leukaemia. On the other hand, the mechanism of interaction of genotoxic compounds with the oncogenes is yet to be fully resolved.

The consistent association of specific chromosome translocations with human lymphomas and leukaemias has given a new dimension to cancer research. Such translocations which can be induced by different agents involve breakage points where cellular oncogenes responsible for carcinoma have been located. The translocations may lead to alterations or activation of oncogenes. The discovery that induced segmental rearrangement of chromosome, which normally leads to position effect, may contribute to malignancy, has opened up a novel approach to genotoxicity. This rearrangement can be accelerated through mobile sequences capable of insertion at different lici. In all these aspects of research, *in situ* hybridization with oncogenic probes has yielded valuable data in relation to carcinogenesis.

Nutrients for Animal Tissue Culture (see White 1954, 1959; Paul, 1959)

Balanced salt solutions (g/1000 ml)

The ingredients are all dissolved successively in distilled water. $NaHCO_3$ and phenol red are dissolved separately in water, filtered through a Selas candle and saturated with CO_2, bubbling through a plugged and sterile pipette. Before use, the two solutions are mixed.

Solution	NaCl	KCl	$CaCl_2$	$MgSO_4 7H_2O$	$MgCl_2 6H_2O$	$NaH_2PO_4 H_2O$
Locke (1895)	9.00	0.42	0.24	-	-	-
Ringer (1886)	9.00	0.42	0.25	-	-	-
Tyrode (1910)	8.00	0.20	0.20	-	0.16	0.05
Glucosol	8.00	0.20	0.20	-	0.10	0.05
Gey and Gey (1936)	7.00	0.37	0.17	0.07	0.21	-
Simms (1941)	8.00	0.20	0.147	-	0.20	-
Earle (1954)	6.70	0.40	0.20	0.10	-	0.125
Gey (1945)	8.00	0.375	0.275	-	0.21	-
Hanks (1946)	8,00	0.40	0.14	0.10	0.10	-
BSS (Hanks, 1955)						
Holtfreter (1929) (Amphibia and Pisces)	3.50	0.05	0.10	-	-	-
Ringer A (Amphibia)	6.50	0.14	0.12	0.20	-	-
Modified Locke (Insects)	9.00	0.42	0.25	-	-	-
Carlson (Grasshoppers)	7.00	0.20	0.02	-	0.10	0.20
White (1949)	14.00	0.75	-	0.55	$Ca(NO_3)_2. H_2O$	0.42
(dilute 20 times)					$Fe(NO_3)_3, 9H_2O$	0.011

Solution	$Na_2HPO_4 2H_2O$	KH_2PO_4	Glucose	Phenol red	$NaHCO_3$	Gas phase
Locke (1895)	-	-	-	-	0.20	Air
Ringer (1886)	-	-	-	-	-	Air
Tyrode (1910)	-	-	1.00	-	1.00	Air
Glucosol	-	-	1.00	-	-	Air
Gey and Gey (1936)	0.15	0.03	1.00	-	2.27	5% CO_2 in air
Simms (1941)	0.21	-	1.00	0.05	1.00	2% CO_2 in air
Earle (1954)	-	-	1.00	0.05	2.20	5% CO_2 in air
Gey (1945)	0.15	0.025	2.00	-	0.25	Air
Hanks BSS (1946)	0.06	0.06	1.00	0.02	0.35	Air
Holtfreter (1929)						
(Amphibia and Pisces)		-	-	-	0.20	-
Ringer A (Amphibia)	-	-	-	-	-	-
Modified Locke (Insects)	-	-	2.50	-	0.20	
Carlson (Grasshoppers)	-	-	0.80	-	0.05	-
White (1949)	0.58	0.104	17.00	0.01	2.20	Water 600 ml
(dilute 20 times)	-	-	-	-	-	Sat. with CO_2

Some standard nutrient solutions for animal tissue culture

Earle's medium (1955)

Synthetic (mg/l)

	mg		mg
Penicillin	0.50	Pantothenic acid	1.0
l-Arginine	17.4	Pyridoxal	1.0
l-Cystine	6.0	Riboflavine	0.1
l-Histidine	3.2	Thiamine	1.0
l-Isoleucine	26.2	Inositol	1.0
l-Leucine	13.1	Biotin	1.0
l-Lysine	18.2	Folic acid	1.0
l-Methionine	7.5	Glucose	2000.0
l-Phenylalanine	8.3	NaCl	8000.0
l-Threonine	11.9	KCl	400.0
l-Tryptophane	2.0	$CaCl_2$	140.0
L-Tyrosine	18.0	$MgSO_4.7H_2O$	100.0
l-Valine	11.7	$MgCl_2.6H_2O$	100.0
l-Glutamine	146.0	$Na_2HPO_4.2H_2O$	60.0
Choline	1.0	KH_2PO_4	60.0
Phenol red	20.0	$NaHCO_3$	350.0
Nicotinic acid	1.0		

Parker and Healy's medium (1955)

Amino acids	mg		mg
l-Arginine	70.0	l-Leucine	120.0
l-Histidine	20.0	l-Isoleucine	40.0
l-Lysine	70.0	l-Valine	50.0
l-Tyrosine	40.0	l-Glutamic acid	150.0
l-Tryptophane	20.0	l-Aspartic acid	60.0
l-Phenylalanine	50.0	l-Alanine	50.0
l-Cystine	20.0	l-Proline	40.0
l-Methionine	30.0	l-Hydroxyproline	10.0
l-Serine	50.0	Glycine	50.0
l-Threonine	60.0	l-Cysteine	260.0

Synthetic (mg/l)

Vitamins			mg
Pyridoxine	0.025	*p* Aminobenzoic acid	0.05
Pyridoxal	0.025	Vitamin A	0.10
Biotin	0.01	Ascorbic acid	50.00
Folic acid	0.01	Calciferol	0.10
Choline	0.50	Tocopherol phosphate	0.01
Inositol	0.05	Menadione	0.01

Parker and Healy's medium (contd.)

Coenzymes			
95% DBN	7.0	**Liquid sources**	
80% TPN	1.0	Tween 80 (oleic acid)	5.0
75% COA	2.5	Cholesterol	0.2
88% TPP	1.0		
60% FAD	1.0		
90% UTP	1.0		

Nucleic acid derivatives	**Inorganic salts**		
Adenine deoxyriboside	10.0	NaCl	6800.00
Guanine deoxyriboside	10.0	KCl	400.0
Cytosine deoxyriboside	10.0	$CaCl_2$	200.0
5-Methylcytidine	0.1	$MgSO_4$, $7H_2O$	200.0
Thymidine	10.0	NaH_2PO_4, H_2O	140.0
		$NaHCO_3$	2200.0
Miscellaneous		$Fe(NO_3)_2$	0.1
Sodium acetate	50.0	**Antibiotics**	
d-Glucuronic acid	3.6	Sodium penicillin G (just before use)	1.0
1-Glutamine	100.0	Dihydrostreptomycin sulphate	100.0
d-Glucose	1000.0	*n*-Butyl 4-hydroxybenzoate	0.2
Phenol red	20.0	No organic supplement is needed	
Ethyl alcohol	16.0		

White nutrient medium (White, 1954)

It contains 70 ml of sterile water and 15 ml of the mixture of White's inorganic salt solution, $Fe(NO_3)_3$ solution and sugar buffer given in table. To this are added in succession 5 ml each of amino acids, stock, AC stock, B stock and B_{12} stock to prepare the final nutrient with pH 7.4 and a pale red colour. The ingredients are:

Inorganic salt solution, $Fe(NO_3)_3$ solution and sugar buffer as described in Table.

Amino acid stock, containing (mg/80 ml of water)

l-Lysine-HCl	312 mg	*dl*-Isoleucine	208 mg
dl-Methionine	260 mg	*dl*-Phenyalanine	100 mg
dl-Threonine	260 mg	l-Leucine	312 mg
dl-Valine	260 mg	l-Tryptophane	80 mg
l-Arginine-HCl	156 mg	l-Glutamic acid	280 mg
l-Histidine-HCl	52 mg	l-Aspartic acid	120 mg
l-Proline	100 mg	l-Cystine	30 mg
Glycine	200 mg	0.01% phenol red	10 ml

AC stock, containing:

Group A	Carotene	10 mg
	Vitamin A	10 mg
	Ethyl alcohol	100 ml
Group B	Ascorbic acid	10 mg
	Glutathione	20 mg
	Cysteine	20 mg
	Water	100 ml Filter

Mix 2 ml of A, 10 ml of B, 2 ml of C and 86 ml of water and maintain at pH 7.4 by adding 0.1 N NaOH.

B Stock containing:

Group B		*Group FA*	
Thiamine HCl	10 mg	Folic acid	10 mg
Riboflavin	10 mg	$NaHCO3$	10 mg
Ca-pantothenate	10 mg	Water	100 ml
d-Biotin	10 mg		
Pyridoxin HCl	10 mg	Sterilise and filter.	
Nicotinic acid	10 mg		
Inositol	10 mg		
l-Alanine	10 mg		
Choline	100 mg		
Water	100 ml		

B_{12} stock containing 0.015 per cent aqueous vitamin B_{12} solution, sterilised.

CHAPTER 8

SOMATIC CELL FUSION IN ANIMALS

Cell fusion was first reported by Barski, Sorieul and Cornefort (1960) on the basis of their observation that two different cell lines of mouse, when grown together, result in the production of a nucleus containing chromosomes of both parents. A virus SV5 was found to be the main agent responsible for fusion .

The lipoprotein envelope of the Sendai virus is essential for fusion and the inherent infectivity of nucleic acid is lost by ultraviolet treatment (*see* Harris, 1979). Since the discovery of Sendai virus as a fusion inducing agent, several other viruses as well as chemicals have been shown to induce or promote somatic hybridization.

SCHEDULE FOR SOMATIC FUSION BETWEEN CELL LINES DERIVED FROM DIFFERENT MAMMALIAN SPECIES

A. Use of Sendai Virus

Take two cell types in quantity, as suspensions of single cells, namely HeLa cells from suspension cultures and Ehrlich ascites tumour cells from the peritoneal cavity of Swiss mice. (The virus selected in this case was a strain of Sendai virus supplied by Dr. H. G. Pareira of the National Institute for Medical Research, Mill Hill). This member of the para-influenza I group of myxoviruses was chosen since one strain of these viruses (HVJ) was found to induce rapid fusion in suspensions of Ehrlich ascites cells *in vitro* (after Harris 1970).

Inactivation of Virus

(1) Expose 1 ml of the concentrated Sendai virus suspension for 3 min (in a watchglass) to ultraviolet light on the surface of the membrane (from a Philips 15W18 in germicidal tube, with an intensity of 3000 ergs/ml/*s*). Mix the suspension by pipetting after each min. To test the infectivity of the irradiated virus, incubate pieces of chorioallantoic membrane with it in TC 199 medium in a tray. After 3 min of ultraviolet treatment, a drastic reduction is seen in the ability of the virus to multiply in the membrane, but its capacity to induce cell fusion *in vitro* remains intact.

(2) Inactivation can be carried out by β-propiolactone as well. Prepare a 10% solution of β-propiolactone and dilute in 1% saline bicarbonate (1.68 g $NaHCO_3$ and 0.5 ml phenol red in 100 ml isotonic saline). Treat the virus suspension (4°C) with β-propiolactone (9:1), shake the mixture for 10 min in ice bath, incubate at 37°C for 2 h with stirring at every 10 min and incubate at 4°C overnight for hydrolysis of β- propiolactone.

Cell Fusion with Sendai Virus

Centrifuge and separate out HeLa cells from a suspension culture. Resuspend in Hank's solution to a concentration of 2×10^7 cells/ml. Draw out Ehrlich ascites cells from the peritoneal cavity, wash by centrifugation in Hank's solution and resuspend in it at a concentration of 2×10^7 cells/ml.

Draw out 0.5 ml of each cell suspension with a pipette and transfer to a chilled inverted T-tube. Add 1 ml of the virus, and dilute, if required, with Hank's solution. The cells clump together and the size of the clumps depends on the quantity of the virus added. Keep the tube with the mixture at 4°C for 15 min and then shake in a water bath at 37°C for 20 min at the rate of 100 oscillations/min. The cells in the clumps undergo various degrees of fusion during this period.

Transfer 1 ml cell suspension of 5 ml culture medium, with the help of a pipette, into a 6 cm diameter petri dish containing 15 coverslips, each 1 cm in diameter. The culture medium contains: 20% calf serum and 1% tryptose, both in TC 199 with 100 units ml penicillin and 100 µg/ml streptomycin. Incubate the petri dishes at 37°C in a gas mixture of 5% CO_2 in air. Transfer the coverslips to fresh medium after one day and again after four days. Multinucleate cells are observed adhering to the cover slips 4 h after introducing the suspension into the petri dishes. Within 24h most of them flatten out on the glass. Cell counts show that about 10% of the original single cells in suspension adhere as multinucleate cells to the cover slip. Each multinucleate cell contains about two to 20 nuclei of two morphological types.

In addition to Sendai virus, cell fusion can be induced by several other viruses including arboviruses all of which have the common property of having envelopes.

B. Virus Induced Fusion Mediated by Lectin

As virus induced cell fusion is associated with virus cell surface interactions, plant lectins, which often bind strongly with cell membrane have been explored in this connection. A number of plant lectins are known to inhibit fusion by viruses. But phytohaemagglutinin (PHA), a plant lectin obtained from red kidney bean (***Phaseolus vulgaris***), ***accelerates*** fusion to a significant extent. This technique is described below:

Take egg grown cultures of Sendai virus or (NDV) Newcastle disease virus and inactivate in β-propiolactone as mentioned in earlier schedules. Take thymidine kinase deficient (CI-IID) mouse cells (Dubbs and Kit, 1964) and hypoxanthine guanine phosphoribosyltransferase deficient (WI-18Va2) human cells (Weiss, Ephrussi and Scaletta, 1968) in equal numbers (5×10^6) of each and suspend in 1.0 ml serum free Eagle's basal medium with PHA (100 μg/ml). Adjust the pH to 8.00. Incubate for 90 min at 37°C. Add 2000 HAU-inactivated Sendai virus or 2000 (EID_{50}) egg infectious disease of NDV and incubate for 1 h at 37°C. Inoculate mixed cell suspensions on cover slips and incubate for 16 h at 37°C. Fix and stain as usual with Giemsa or test hybridity by culturing in HAT medium.

C. Use of Lipid Vesicles in Cell Fusion

In order to eliminate the toxic properties of lysolecithin, the undesirable effects of Sendai virus, like interference in the release of transforming virus, alteration in the surface properties and metabolic pathways and chromosomal aberrations, lipid vesicles prepared from phospholipids have been utilized to fuse mammalian cells of different lines. A brief outline of the technique for securing interspecific mammalian cell hybrids is described.

Prepare unilamellar (small vesicles with single lipid bilayer 25-50 nm) vesicles in the following way:

Disperse 10-20 μM of lipid phosphate in 2-4 ml aqueous buffer with 100 mM NaCl, 2 mM N-tris methyl-2-aminoethane- sulphonic acid (TES) and 0.1 mM EDTA, pH 7.4 and shake for 10 min at 37°C (above the transition temperature of the lipid) in Vortex mixer. Sonicate with nitrogen for 1 h (for natural phospholipids, 24-26°C is suitable). Equilibrate the vesicles at 24-26°C after sonication for 1 h. Dilute the vesicles in phosphate-buffered saline for use to secure 10^2-10^8 vesicles per cell for fusion; (1 μM of sonicate contains 2×10^{14} vesicles).

Take two different cell types, thymidine kinase (TK) deficient Cl-1D mouse cells and hypoxanthine guanine phosphoribosyltransferase-deficient (HGPRT) WI-18Va2 human cells and suspend equal number of cells (5×10^6) of each in 1.0 ml serum free Eagle's basal medium (pH 8.0). Incubate with vesicles (10^4 per cell) for 2 h at 37°C. Take a drop of the mixed cells and inoculate on cover slip, incubate for 16-24 h at 37°C, fix in methanol and stain in Giemsa to observe fused cells. Inoculate the rest of the cell mixture in 2.0 ml HAT medium (pH 8.0) in 35 mm plastic petri dishes (1×10^6 cells per dish) and incubate as usual. Replace the medium by fresh medium after 1,3,6,9 and 12 days. After 14 days, fix and stain the cells following the usual procedure or follow starch gel electrophoresis to study enzyme activity of hybrid cells as expressed in HAT selective medium.

D. Use of Lysolecithin in Cell Fusion

In view of the fact that several lipolytic agents affect the cell membrane, lysolecithin, a phospholipid, was used to induce cell fusion. Several lypophilic molecules have later been shown to have the same capacity.

Mix H^3 thymidine labelled 5×10^6 CV-1 (African green monkey kidney continuous line) with unlabelled 5×10^6 F5-1 cells (hamster line transformed by SV40) in 1 ml Hank's solution. Centrifuge at 24-26°C for 5 min at 160-180 × g. Treat the pellets (0.5 to 1×10^7) with 0.1 ml of LL solution in BSA for 1 min at pH 8.0 and shake the tubes to secure maximum exposure of cells to LL by detaching the pellets from the bottom. In order to neutralize the action of LL, add 1 ml MEM containing 30% fetal calf serum (FCS) inactivated for 30 min at 56°C. Centrifuge the samples at 160 × g for 5-10 min at 24-26°C, discard the supernatant, add 1 ml MEM on the pellets. Incubate at 37°C for 15-20 min. Re-suspend the pellets in MEM containing 10% FCS. Seed the culture (one million cells or so) in petri dishes or flasks. After 18-22 h of seeding, for observation, wash the cells three times in the medium, fix in methanol for 5 min and stain in Giemsa.

E. Use of Polyethylene Glycol in Cell Fusion

Polyethylene glycol $[HOH_2C(CH_2OCH_2)_nCH_2OH]$ normally used for inducing cell fusion in plants has been effective in mammalian systems as well.

For Monolayers

For fusion experiments, mouse fibroblast (3T34E) a bromodeoxyuridine resistant line and rat glial cells (RG6A-TgA), an azaguanine resistant line, may be chosen.

In 60 mm Falcon tissue culture dishes, take Dulbecco's modified Eagle's medium supplemented with 10% fetal calf serum (E-10FCS), and inoculate with 2.5×10^6 3T3 and 2.5×10^5 RG6 cells. Incubate for 24 h at 37°C. Remove the medium by aspirating first, and then tilting the dish to drain off the residual medium and aspirating again. Add PEG solution (1000 at 50%) at 26-28°C or room temperature sufficiently to cover the cells and keep as such without disturbing for 1 min. Remove quickly the PEG by aspirating and wash the dishes thrice with Eagle's medium (Dulbecco's modification). Cover the cells with 5 ml E-10FCS medium and incubate for 24 h at 37°C. Trypsinize the cells and harvest. Plate the cells in E-10 FCS medium containing hypoxanthine, aminopterin and thymidine (E-10 HAT). Addition of these compounds dose not allow the parental cell lines to grow because of enzyme deficiencies associated with drug resistance, but the hybrid lines can grow and as such can be selected. Renew this medium every four to five days. After seven days, fix the dishes containing the cells in methanol and stain with Giemsa and count colonies.

The PEG concentration (1000) at 50% seems to give the optimum result as below it, there is less fusion, and increased concentration results in decreased viability of cells. Normally the frequency of hybridization is maximum at pH 6 and minimum at pH 9 but the effect of change is rather small. PEG 1000 at 50% is chosen because of high rate of fusion and its least sensitivity to dilution effects.

For Suspensions

For cell lines in suspension, for one partner, Chinese hamster (Wg 3H) or mouse (RAG), and for the other, human peripheral leukocytes (WBC) or mouse peritoneal macrophages may be chosen.

In order to secure cells of Wg 3H or RAG, cell lines can be grown in Eagle's medium (Dulbecco's modification) supplemented with 5% fetal calf serum (E-5 FCS). For plasma suspension of WBC, sedimentation for 1-2 h in a mixture of equal volumes of heparinised whole blood and dextran solution (3 g of dextran of molecular weight 100 000-200 000, in 100 ml of saline), is to be followed by washing thrice with Hank's balanced solution before use.

Mix 5×10^6 WBC and 5×10^6 Wg3H cells in a conical centrifuge tube. Centrifuge the mixture and remove the supernatant by aspirating, with 0.05 ml still remaining in the tube. Mix 1 ml PEG by pipetting and keep for 1 min. Add 9 ml of Eagle's medium, centrifuge and again remove the supernatant by aspirating. Suspend the cells again in 5 ml E-5 FCS medium and divide the suspension in the same medium in five 60 mm falcon culture dishes. After 24 h at 37°C follow the same procedure as above and after 14 days, fix the dishes containing cells with methanol, stain with Giemsa and count colonies.

In order to secure optimum yield of hybrid colonies, absolute removal of PEG is needed after use.

F. Use of Microcells for Cell Fusion

As mentioned already in mammalian somatic cell hybrids, chromosome elimination of the human cells is one of the common phenomena from fused cells with mouse or Chinese hamster. This has been extensively utilized in the mapping of genes. As the elimination of chromosomes is a rather slow process and occurs in an erratic way, several authors have tried to use deficient cells with single or a few chromosomes as one of the partners in fusion, so that genes can be conveniently mapped. The use of irradiated cell which has already suffered chromosome damage and loss has been advocated as well.

One of the effective methods applied involves the use of subdiploid cell fragments termed '*microcells*' containing small genetic material as one of the partners in fusion. Such microcells have a small deficient nucleus with surrounding cytoplasm and wall. These are ultimately produced from micronucleate cells often induced by colcemid as well as other microtubular poisons. Cytochalasin B and centrifugation can also be used for the production of microcells prior to enucleation needed for cell fusion from monolayers as well as from cell suspension.

From Monolayers

Remove cultured growing cells to small round (25 mm diameter) plastic discs punched from tissue culture dishes and allow to grow for 12-24 h. Add colcemid or vinblastine at a final concentration of 1-3 μg/ml and incubate for 72h. After micronucleation, remove the medium and place the cells facing downwards in centrifuge tubes containing 4 ml of phosphate buffered saline (PBS). Stabilise the position of the plastic disc in the tube with plastic plugging.. Centrifuge at 10 000 r/min for 10 min in a centrifuge at 37°C to remove loose cells. Transfer the discs to fresh centrifuge tubes containing PBS with 10% calf serum and 10 μg cytochalasin B per ml. Centrifuge at 14 000 r/min for 20 min at 37°C. Re-suspend the pellet, pool and centrifuge at 400 r/min for 5 min. Remove the supernatant, wash and re-suspend the pellet in Eagle's MEM without serum for fusion experiment.

From Cells in Suspension

With use of cytochalasin, discontinuous Ficoll gradient has been employed for separation of nucleated and enucleated cell fragments. This method is quite rapid as it allows enucleation of almost 6×10^7 cells in a 10 ml gradient.

The production of microcells can also be secured through cold treatment. The method involves initially the mitotic arrest by N_2O of synchronized cells at 5 atm followed by treatment for 9 h at 5°C. The cells are then incubated at 37°C for 2 h to induce abnormal chromosome segregation and cytokinesis resulting in bunches of microcells which can be separated by shearing. The microcells can also be separated through Ficoll gradient or 5μm nucleopore filter.

FUSION BETWEEN PLANT AND ANIMAL CELLS

In order to secure fusion between plant and animal cells, both polyethylene glycol and Sendai virus have been employed. The technique employed is as follows:

Prepare hen erythrocytes and suspend in modified Eagle's basal salt solution buffered at pH 7.4 with 20 mM HEPES buffer (6×10^8 cells/ml). Prepare yeast protoplasts with the use of helicase from exponentially growing cultures of *Saccharomyces cerevisieae* strain NCYCO263B. Centrifuge the protoplasts at1000 × g for 15 min in Ficoll (400) gradient dissoloved in osmotic stabiliser (0.7 M sorbitol; 0.01 M citric acid, pH 6.5).

Prepare the gradient in the following way:

(a) Prepare a layer of 8% w/v Ficoll (8 vol).
(b) Over (a) pour another layer of 5% w/v Ficoll (8 vol).
(c) Layer osmotic stabiliser on (b) above.

Re-suspend the sedimented protoplasts in the osmotic stabiliser (6×10^8 protoplasts/ml). Centrifuge the protoplast preparation for 3 min at 1000 × g and remove the supernatant. Add erythrocyte preparation on the protoplast pellet (1 : 50 or 1 : 250) and centrifuge for 5 min at 800 × g and remove the supernatant. Resuspend the preparation and mix with 1 ml 40% w/v PEG (molecular weight 6000) in the HEPES buffered Eagle's medium, pH 7.4. Incubate at 37°C for 15 min which will allow the formation of aggregates. Dilute 1 ml of the incubate with 5 ml of buffered Eagle's salt solution at 37°C. Centrifuge at 800 × g for 5 min and resuspend 1 ml solution at 37°C. With continued incubation, note the swelling of erythrocytes and observe cell fusion under light microscope. For electron microscopy glutaraldehyde fixation may be employed.

CHAPTER 9

EFFECTS OF EXTERNAL AGENTS AND MONITERING FOR ENVIRONMENTAL TOXICANTS

SECTION A: EFFECTS OF PHYSICAL AND CHEMICAL AGENTS ON CHROMOSOMES

The discovery of x-ray-induced mutation in *Drosophila* by Müller in 1927, followed by Stadler in Maize in 1928, provided the necessary impetus for research on the effects of outside agents on chromosomes. The result of this enthusiasm led to the discovery of the mutagenic property of chemicals, by Oehlkers (1943) and Auerbach and Robson (1946), followed by a host of others. In the meantime, the polyploidizing action of colchicine and its effective application on plants had been disclosed through the works of Blakeslee and Avery (1937) and their collaborators.

The standardization of a method for a systematic attempt to explore the properties of different chemical agents was first made by Levan (1949) and his collaborators. The technique applied by them is known as the Allium test, in which the experimental materials consist of bulbs of *Allium cepa*, the common onion.

The chemicals to be tested are prepared in solution and kept in wide-mouthed jars. A bulb with roots intact is then placed over the mouth of the jar so that its roots dip in the solution—the jars may be covered with black paper to allow healthy growth of the roots. After treatment for a desired period, the roots can be excised, fixed, stained and observed directly or kept for recovery in water or nutrient solution under similar conditions before fixation, staining and observation.

The above method is applicable only to root meristems, however, and for meiotic cells or pollen grains the entire inflorescence is generally treated by dipping the stalk in water. But the mode of treatment with these materials may vary and, if necessary, the anthers may be dipped in the fluid directly. For animal materials, the usual technique of application is through feeding or injection, and for post-treatment cultures, the organ can be dissected out or cultured in artificial medium, or the animals can be reared in cages with natural feeding.

For physical agents such as x-rays, ultraviolet rays, both plant and animal materials are placed in front of the source and the required dosage is applied, the subsequent treatment being the same as that with chemical agents. Radioactive precursors can be administered orally by gavaging or through injection.

Chemical Agents

Metaphase Arrest

In the majority of plant and animal materials, the nuclear division within the different cells is not synchronous, with the result that the meristematic or the dividing zone represents a heterogeneous mass of cells, in which the nuclei are at different stages of division. The difficulty of such a medium is twofold: (a) metaphase stages—the best nuclear phase for chromosome analysis—cannot be obtained in high frequency, and (b) with regard to the analysis of an effect, the exact stage affected cannot be ascertained. A large number of cells, showing induced synchrony in division, is therefore necessary.

The most suitable chemical for the purpose of securing a large number of metaphase plates is colchicine as it causes metaphase arrest by inhibiting the operation of the spindle mechanism. To secure metaphase arrest, the organs can be treated directly with aqueous colchicine solution for a required period, being added to the medium in tissue culture or injected into tissue or added to the food. Before harvesting the cells in leucocyte cultures, the addition of colchicine or colcemid is essential for securing a large number of metaphase plates. The concentration, needed to secure this effect may vary from a very dilute concentration, 0.01%, to even 2% and the period of treatment from 10 min to 16 h in some amoeba. In *Allium cepa*, spindle in root tips can be arrested by just $1\frac{1}{2}$ h treatment in 0.2% colchicine solution, but before fixation and observation, a thorough washing in water for at least 15-20 min is necessary to remove any superficial deposits of this alkaloid. In leucocyte cultures, application of colchicine 10-20 h before harvesting is desirable. The characteristic appearance of metaphase stages, showing clear separate segments, is otherwise known as colcincine-mitosis or *c*-mitosis. In addition to colchicine, a number of other compounds such as gammexane, chloral hydrate, acenaphthene, actidione, etc., are all employed for metaphase arrest.Colchicine is significantly more effective than any of the others.

Polyploidy

The importance of polyploidy in agricultural and horticultural practices is well known. The properties of arresting metaphase and the induction of polyploidy are inter-related. Polyploidizing chemicals like colchicine inhibit the formation of the spindle within the two poles and confine the chromosomes within one nucleus though their division remains unhampered. Levan (1949) has classified colchicine action under narcosis, as the narcotic action allows the tissue to recover as soon as the influence of the chemical is removed. Evidently polyploid cells, which are formed by colchicine action, divide normally and give rise to polyploid shoots.

Protocol for use of colchicine:

On seeds and young seedlings

The seeds may be immersed for 2-48 h in concentrations of colchicine solution varying from 0.02-0.1% before sowing. Just- germinating seeds can be treated with similar concentrations of colchicine solution for 12-48 h with the plumules dipped in the solution, or the entire germinating seedling can be immersed completely in solution.

On mature seedlings

(i) Colchicine is added in the form of soaked cotton plugs on the growing shoot, the period of treatment varying from 2-4 h, and the range of concentrations used being the same as the previous schedule. Cotton plugs, placed over the growing tip, should be moistened at regular intevals by adding drops of colchicine solution with a brush, and after the treatment, the plug should be removed and the tip washed by brushing with water. The same method can be followed for treating young inflorescences.

(ii) In the form of a paste mixed with lanolin or with glycerine. This method has been found to be effective where the growing point lies within the plumules, as in monocotyledonous plants.

On pollen grains and tissues in culture

Colchicine may be added in the agar medium meant for pollen tube growth (2 ml of 0.2% colchicine in 8 ml of agar medium or in culture medium for other tissues). Other chemicals such as chloral hydrate, gammexane and acenaphthene, which cause metaphase arrest, can also be applied for the induction of polyploidy. The success or failure to induce polyploidy can be determined as follows:

(i) The chromosome number of young shoots, leaf tips and root tips can be counted, following acetic-orcein or Feulgen squash.

(ii) Polyploidy is associated with an increase in stomatal size and decrease in stomatal frequency per unit area of the leaf, so the lower epidermis of the mature leaf of the polyploid can be peeled off and mounted in 50% glycerin solution. Stomatal size and frequency per unit area can be noted and the result can then be compared with that of diploids obtained following a similar procedure adopted for a control diploid plant. Post-treatment with x-rays of colchicine-treated plants results in better survival of polyploids as x-rays are more effective against diploid cells, causing them to be eliminated in selection.

Chromosome Fragmentation and Other Effects

The capacity of inducing chromosome breakage is a property of several chemical agents. Fragmentation followed by translocation of some fragments may bring about a new patterning of chromosome segments resulting in heritable phenotypic difference. The chromosome breaking property of chemicals has an important bearing on the chemotherapy of cancer.

The study of chromosome fragmentation by chemicals has a special significance in bringing out the differential nature of chromosome segments. Several chemical agents such as 8—ethoxy caffeine induce chromosome breaks at certain specific loci. This differential break can be taken as an index of the different chemical nature of susceptible segments from the rest of the chromosome parts.

The modes of action of different chemical agents causing chromosome breaks vary. Some of them affect sulphydryl groups of proteins whereas others act through their influence on hydrogen bonds of nucleic acids. Guanidine cross linkages are held to be involved with mustard compounds. Some agents may affect the oxidation-reduction system within the nucleus. The final upset of the nucleic acid metabolism ultimately results in hazards in protein reduplication causing chromosomes to break at different loci. The effects of different chemical agents and their modes of action have been dealt with in detail by several workers including our group. The effects can be classified into:

i) *clastogenic* — showing chromosomal aberrations;
ii) *turbagenic* — affecting the spindle;
iii) *carcinogenic* — capable of inducing cancer;
iv) *mutagenic* — able to induce mutations;
v) *mitogenic/mitostatic* — affecting the frequency of cell division.

These effects overlap and may be shown by the same chemicals under different conditions and using different concentrations and /or period of exposure. Several plant products, such as pigments, alkaloids, certain flavonoids, etc influence chromosome breakage. Certain antibiotics and extracts of bacterial culture have been shown to affect DNA replication, transcription and/or protein synthesis and thus initiate clastogenic effects. On the other hand, several plant products have been shown to reduce the frequency of chromosomal aberrations induced by known clastogens. Well known are chlorophyllin, a sodium potassium salt of chlorophyll, and crude plant extracts containing high amounts of vitamin C, vitamin B or its precursors, sulphur groups, anti- oxidants and fibres as shown by our group.

The mode of action of antibiotics like mitomycin C, actinomycin D, streptomycin, puromycin and chloramphenicol has been thoroughly investigated.

Schedules for Inducing Chromosome Breaks in Plants

1. For Random Breakage

Place a healthy bulb of *Allium cepa*, with root tip intact, on top of a jar containing 0.005 m/l solution of pyrogallol. Keep the jar in a temperature of 25-30°C. After 4 h, cut a few roots, fix in acetic-ethanol mixture (1:2) for 30 min, and follow the usual method of orcein squashing or Feulgen staining for root tips. Mount in 1%

acetic-orcein solution or 45% acetic acid and count the number of random fragments in metaphase and anaphase. Continue the treatment of roots in pyrogallol solution up to 24 h and observe at regular intervals to study the increase or decrease in the frequency of fragments.

2. For Localised Breakage

Place germinated seeds of *Vicia faba* on a sieve fixed over a jar containing 0.075 M 8-ethoxycaffeine solution, in such a way that the roots passing through the sieve remain dipped in the solution. Continue the treatment for 6 h at 10°C. Allow the roots to recover in tap water at 20°C for 24 and 48 h. Cut the root tips and follow the same schedule for observation as for random breakage.In the study of the effect of chemical agents on chromosomes, the technique employed for observation should be taken special care of, due to the fact that chromosome breaks may result during slightly prolonged heating with acetic-orcein-HCl mixture (Sharma and Roy, 1956), which is an essential step in the procedure for orcein squashing. Chromosome breaks have also been observed through water treatment by Sharma and Sen (1954).

3. For Demonstration of Orcein Breakage

Treat excised root tips of onion in 0.002 M solution of 8-oxyquinoline for 2 h at 16—18°C. Fix in acetic- ethanol (1:2) for 30 min. Heat the root tips gently over a flame in a mixture of 2% acetic-orcein and normal hydrochloric acid mixed in the proportion of 9:1 for 30 *s*. After a few minutes, mount and squash in 1% acetic-orcein and observe. Fragments can be observed in metaphase and anaphase stages, but if the heating is prolonged for a few seconds more, the frequency of fragments shows an increase. High frequency of fragments can also be obtained if acetic-ethanol fixation is omitted. In view of the above results, it is always desirable to keep a check on the period of heating, not exceeding 10 *s* in the procedure for orcein staining.

4. For the Demonstration of Chromosome Breakage Induced by Water Treatment

Treat young healthy roots of *Crinum asiaticum* in tap water at 30°C for 2 h.

Follow the usual procedure of fixation and orcein staining. Fragments can be detected both in metaphase and anaphase.

In the study of the effect of chemical agents on chromosomes, control experiment should be set up with water treatment alone in cases where water is used as the solvent of the chemical agent.

Division in Differentiated Nuclei

Division in adult nuclei, when otherwise the nuclei have ceased to undergo apparent division, can also be induced with the aid of chemical agents. The importance of this line of investigation was realised after the demonstration of the endopolypoid constitution of differentiated nuclei by Huskins and his group. They claimed that the adult nuclei, though apparently non- dividing, undergo endomitotic re-duplication of the chromonemata, thus exerting control over the process of differentiation. Huskins and Steinitz (1948), with the aid of a special technique using hormones, induced division in the adult nuclei, thus permitting the chromosomes to complete the nuclear cycle; during this division their polytenic and polyploid constitution was revealed. Afterwards other chemicals were worked out having the same property. Sharma and Mookerjea (1954) demonstrated that of the nucleic acid, the sugar constituent alone can induce division, and claimed that the polytenic condition of the adult nuclei is due to deficiency in nucleic acid. Torrey (1961) used kinetin (6-furfuryl-amino-purine) for the induction of division in endomitotic plant cells. A considerable amount of work has been carried out in our own laboratory on the induction of division in adult nuclei. With the aid of 2,4-dichlorophenoxyacetate indolylacetic acid, division could be induced even in the vascular zone (Sen, 1970) where both diploid and polyploid nuclei have been recorded. The occurrence of diploid chromosome numbers in adult nuclei in the differentiated zone can be explained on the basis of the fact that transcription of RNA responsible for gene action does not depend on replication of DNA, and one strand of DNA is active in this respect. But the occurrence of polyteny may possibly suggest that there is a limit for continued transcription of a strand after which replication is necessary for the production of a fresh strand to be used in transcription (Sharma, 1978).

The chromosomal control of differentiation has been well illustrated in the puffing pattern in salivary gland chromosomes of Diptera at different phases of development. Excellent reviews on this aspect of differentiation have been published.

Protocol for the induction of division in differentiated nuclei

Take cutting or seedlings of *Rhoeo discolor* and remove all the roots. Grow in jars containing the following culture solution:

	mg/l		*mg/l*
$Ca(NO_3)_2$, $4H_2O$	95	Mn	0.5
NH_4NO_3	129	Cl	1.9
$MgSO_4$, $7H_2O$	180	B	0.5
KH_2PO_4	133.5	Cu	0.02
K_2HPO_4, $3H_2O$	7.35	Zn	0.01
$Fe_2(C_6H_4O_6)$, H_2O	5.0		

Use distilled water and adjust the pH to 5.7. Fix the cuttings in such a way that a portion of the stem dips in water. Cover the jar with black paper and keep at 22-24°C. Keep the materials in the jar until freshly generated roots, 3-10 cm long, are obtained. Transfer the materials to a similar jar containing 50- 100 ppm indole 3-acetic acid and keep for 24-72 h. Re-transfer the materials to fresh culture and keep for a maximum of 72 h. Take out the materials after every 24 h. Remove about 4 mm from the root tip, which is meristematic, then cut 1-2 mm. Follow the usual procedure for fixation in acetic-ethanol and orcein squash technique for root tips. Polyploid cells can be seen in metaphase in the differentiated region. Control preparations should be set up allowing the roots to grow in culture solution, in which no division will be found in the similar zone of the root tip.

Somatic Reduction

The occurrence of reductional separation of chromosomes in tissues, other than the gonadal ones, was demonstrated on slides prepared by Huskins and his associates and by Kodani. By treatment with sodium nucleate, reductional separation of chromosomes can be initiated in root-tip cells. Somatic reduction, through treatment with sodium salts of ribose nucleic acid, evidently suggests that the balance of the nucleic acid within the cell is at least one of the principal controlling factors of mitosis and metosis.

Protocol for the induction of Somatic Reduction in Plants

Take a healthy young bulb of *Allium cepa* and let it grow in a jar containing tap water until a fresh crop of roots germinates. When the roots are just 2-3 mm long, fit the bulb on top of a jar containing 0.1-0.2% sodium nucleate solution in water. Sodium nucleate supplied by Schwartz Laboratories, New York (S.N. 4509) yields good results. Place the bulbs in such a way that the roots dip into the solution. Treat the bulbs for 6-12 h.

Cut the tip portion of the root (meristematic region) and follow the usual method of fixation and Feulgen or orcein squashing.

Reductional separation of chromosomes can be observed in a few % of the cells. Each chromosome unit with two chromatids intact moves as it is to either of the poles, so that the distribution of 16 chromosomes is equal between the two poles, i.e. eight at each. For inducing somatic reduction through temperature treatments onion bulbs may be grown in tap water at 5- 6°C for 5-64 days. At every seven or eight day interval, healthy young root tips may be fixed, stained and observed after the usual procedure. In all cases of reductional grouping in somatic cells, it is preferable to confirm the observation from materials sectioned from paraffin blocks.

Physical Agents

X-rays are employed for effectively altering chromosome structure, both at microscopic and ultramicroscopic levels. In addition, other types of radiations are used, such as α-rays, β- rays, fast neutrons, ultraviolet and infrared rays. The biological action of these agents depends on their ionising and non-ionising properties, and as such they are classified under two categories, ionising and non-ionising rays.

Ionizing radiations dissipate energy during their passage through matter, by ejecting electrons from the atom through which they pass. The ionised atom losing the negative balance, becomes positively charged and is known as an ion. The result of ionisation is a chemical change of the molecule concerned. Whenever a binding electron between the two molecules is affected, serious after-effects ensue.

Non-ionizing radiations, such as ultraviolet rays, infrared rays, etc., cause dissipation of energy within the tisssue by molecular excitation, the principal biological action being attributed to the absorption of energy by particular cellular constituents, the most important one being nucleic acid.

The production of gene mutation is one of the most important uses of the physical agents. Fragmentation of chromosomes is another outstanding effect. Breakage may involve both the chromatids of the chromosome or a single chromatid, depending on the stage in which the nucleus has received the radiation. When the chromosome is already split, either one or both chromatids may break, depending on the path of the rays, the breaks being either immediate or delayed. The subsequent effects of chromosome breakage are translocation, deletion, inversion, rejoining as well as stickiness, pyknosis and polyploidy, and can all be studied in treated materials (see table 1). The importance of irradiation, either through x-rays or other agents, in working out the time of chromosome reproduction as well as the structure of chromosomes is well-known. Results of radiation breaks are good indices of timing of chromosome reproduction. Treatment, followed by a study of the successive cycles and the scoring of chromosome or half-chromatid aberrations, provides significant clues in this direction and aids the understanding of the effects occurring at G_1, S and G_2 phases.

Methods have been devised for the application of different physical agents under different conditions on the chromosomes, but before undertaking any work on the effects of physical agents on chromosomes certain factors, which control their sensitivity should be considered. The principal factors affecting radiosensitivity include both internal ones like genotypes: chromosome type, size and number; stage of cell division, level of ploidy and age of the tissue and external ones like moisture, temperature, presence of oxygen and other atmospheric gases, dosage, storage, presence of other chemicals and ionization density of the rays used.

The comparative action of different rays and chemicals has also been dealt with by different authors (*see* Ghosh and Sharma, 1968). The importance of nucleic acid and protein in the manifestation of chromosome breaks has also been studied (Sharma and Sharma, 1960).

The two theories regarding the mode of action of x-rays on chromosomes are the direct hit or target theory and the chemical theory. The *target theory*, first postulated by Lea, suggests that direct hit by an electron may cause a gene to mutate, bringing about a chemical change with a different phenotypic expression or a chromosome break. The chemical theory holds that the x-ray effect on chromosomes is indirect and is principally conveyed through the cytoplasm. Some evidence, including the similar effects of radiomimetic chemicals, oxygen, etc., points to the plausibility of this theory. The dissociation of water molecules into H and OH ions by x-ray and the later formation of HO_2 or H_2O_2 are considered to be the main effects of x-rays, these chemicals ultimately bringing about chromosome breakage. X-rays have also been thought to affect the hydrogen-bonding of nucleic acids or —SH groups of proteins. With ultraviolet rays, it is suggested that the linkage between adenine and thymine undergoes a breakage and two adjacent thymines of a single strand of the nucleic acid undergo union, forming a thymine dimer, which causes difficulty in replication and chromosome breakage. If the organisms are brought to light, there is photoreactivation of the repair enzyme complex and the frequency of the effects decreases. Research on this aspect, carried out in this laboratory, has shown that histone acts as repressor for the repair complex and the digestion of histone accelerates the ultraviolet-induced breaks.

Schedules for Treatment with Different Types of Physical Agents

X-rays

Apparatus consists in the simplest form of an x-ray tube with attached rectifier, transformer, control and dosimeter.

Application on plant materials

Place dry *seeds*, in a single row, in a petri dish. Expose to x-rays at 30 000R. Germinate in moist sawdust. Remove root tips when about 1 cm long, treat in acetic-ethanol mixture (1:3) for 30 min and squash following orcein or Feulgen schedule. If soaked in water before irradiation, the effect is usually greater and may become drastic.

Seedlings and bulbs with root tips: Place young seedlings on a petri dish with their radicles pointing in the same direction. For bulbs, take healthy ones with a tuft of healthy roots about 2 cm long and place them with the root tips facing the source. Expose to x-rays for the desired period. Transfer them to sawdust or to nutrient medium. For immediate effect, remove root tips at regular intervals of an hour, fix for 30 min in acetic-ethanol (1:3), followed by staining by Feulgen or orcein squash methods. For prolonged effects, study the root tips at intervals of 24 h.

Inflorescences

Grow the plants in flower-pots. Expose the young inflorescences to x-rays by bending them to face the source. For immediate effect, select a flower bud of suitable size and observe the pollen mother cells after smearing in acetic-carmine so-

Table 1: Effects of radiation resulting in chromosome breakage in some animal tissue

Radiation	Material used	Effective dosage
Beta rays	*Drosophila*, larva	^{32}P in food at 1 mg per fly approx. 1000 nm approx. 72, 144 and 216 h preceded and followed by 2000R x-rays
Near infrared	*Drosophila*, sperm	1000 nm approx, before and after 3000R x-rays
Neutrons	*Drosophila*, whole fly, male	Different dosages
X-rays	*Drosophila*, whole fly,	1000-5000R
	Drosophila, egg	500R, 105R
	Drosophila, embryo	5000R
	Locusta, testes	500R approx
	Chortophaga, embryo	4-8000R

Table 2: Chromosome breakage induced by radiation in some plants

Radiation	Material	Dosage	Radiation	Material	Dosage
Alpha rays	*Tradescantia*, pollen tube	Different dosages	X-ray	*Hyacinthus*, root tip	150-1000R
	pollen grain	Inflorescence treated in radon solution		*Lilium*, p.m.c.	150R
	Vicia faba, root tips	Treated in radon solution 7-8 units		*Scilla*, endosperm	50R
Beta rays	*Tradescantia*, pollen grain	^{32}P plaque, externally		*Tradescantia*, p.m.c.	18R and 360R
	Tradescantia, anthers	(a) ^{14}C from ammonium carbonate at 0.9-8.2 μCi/ml for 4-8 days		anther	150R
		(b) ^{32}P from sodium hydrogen phosphate at 1-10 μCi/ml for 1-9 days		pollen grain early	(a) 360R (b) 30-300R
		(c) ^{3}H-thymidine 1 μCi/ml for 8-56 h		pollen grain late	180R, 250R
Gamma Rays	*Tradescantia*, pollen grain	(a) 5/min for 100-2000 min (b) ^{40}Co 100-400R		*Trillium*, root tip	5-45R
		(c) ^{40}Co : 1.1-1.3 MeV in air and nitrogen		p.m.c.	50R
Near infra-red	*Tradescantia*, pollen grain	(a) 1000 nm approx for 3 h before and after 107R x-rays		pollen grain	45-375R
		(b) 1000R approx. for 3h before and after 90 and 350R x-rays		*Uvularia*, p.m.c.	90R
	Tradescantia, meiosis	Different dosages		*Vicia*, root tip	50-200R
Neutrons	*Tradescantia*, pollen grains			*Zea mays*, pollen grain	800-1500R
Ultraviolet	*Tradescantia*, pollen tube	(a) 254 nm at 2 x 10^{-3} ergs/mm^2 for 60 s (b) 253.7 nm before and after x-rays			
	Zea mays, pollen grain	253.7 nm at 546.000 ergs/cm^2			

Table 3: Types of Chromosomal Alterations

Term	*Definition*
Chromatid gap	An achromatic region in one chromatid, the size of which is equal to or smaller than the width of the chromatid
Chromatid break	An achromatic region in one chromatid, larger than the width of the chromatid. It may either be aligned or unaligned
Chromosome gap	Same as (i) only in both chromatids
Chromosome break	Same as (i) only in both chromatids
Chromatid deletion	Deleted material at the end of one chromatid
Fragment	A single chromatid without an evident centromere
Acentric fragment	Two aligned (parallel) chromatids without an evident centromere
Translocation	Obvious transfer of material between two or more chromosomes
Triaradial	An abnormal arrangement of paired chromatids resulting in a triradial configuration.
Quadriradial	An abnormal arrangement of paired chromatids resulting in a four-armed configuration
Pulverised chromosome	A spread containing one fragmented or pulverised chromosome
Pulverised chromosomes	A spread containing two or more fragmented or pulverised chromosomes, but with some intact chromosomes still remaining
Pulverised cell	A cell in which all the chromosomes are totally fragmented
Complex rearragement	A abnormal translocation figure which involves many chromosomes and is the result of several breaks and mispaired chromatids
Ring	A chromosome which is a result of telomeric deletions at both ends of the chromosome and the subsequent joining of the ends of the two chromosome arms
Minute (min)	A small chromosome which contains centromere and does not belong to the karyotype
Polyploid or endo reduplication	A cell in which the chromosome number is an even multiple of the haploid number (n) and is greater than $2n$
Hyperdiploid	A cell in which the chromosome number is greater than $2n + 1$ but is not an even multiple of n
Dicentric	A chromosome containing two centromeres

lution. Allow the inflorescence to develop and observe meiotic stages at intervals of 24 h. Endosperm and pollen grain mitosis can be studied in a similar manner.

In tissue culture, different agents can be tried on tissues growing in culture.

Application on animal materials

For small animals

Place the animals to be irradiated inside a tube and expose to the x-rays. Transfer them to their normal condition. At regular intervals, dissect out the gonads in a drop of Ringer's solution, stain with acetic-carmine and observe. Other tissues can be exposed directly.

Ultraviolet rays

The apparatus consists, in simpler forms, of a mercury lamp or a quartz mercury arc in conjunction with a monochromator. The method of treating the material, both plant and animal, is similar to that followed with x-ray.

Infrared rays

The apparatus generally consists of special types of tungsten lamps used for rapid drying. In some later models, arcs, for the production of both ultraviolet and infrared rays, have been built within one source with separate controls. The unwanted rays are screened off with suitable filters. The method of application is similar to that of x-ray.

Most metallic compounds are effective mitotic poisons. When administered to higher organisms most of them are clastogenic, the effects being S-dependent. In general, the clastogenicity is directly proportional to the increase in atomic weight, electropositivity and solubility of the cations in water and lipids within each vertical group of the Periodic table. The effects are modified by interactions between metals and the addition of external agents.

Treatment of Tissue with Both Physical and Chemical Agents

It may sometimes be necessary to treat the tissue with both physical and chemical agents, either singly or in a combined form. Radiation effects and the consequent cytological irregularities have often been found to be much minimized in the absence of oxygen or in the presence of certain chemicals. As the effects of radiation on the organism as a whole are mostly deleterious, such chemicals are referred to as 'protective chemicals', against radiation damage. Their study is considered as one of the most important aspects of radiobiology since it opened up possibilities of affording protection against the destructive effects of radiation to human beings.

Protective chemicals can be applied before, after, as well as during, the time of radiation, the procedure varying according to the type of chemical and the nature of the radiation. From a cytological aspect, protection is afforded against chromosome breakage and inhibition of cell division.

It is very necessary to have protective chemicals with selective action, otherwise, along with other parts of the body, the radiation treated area may also develop radio-resistance, and so nullify the purpose of radiation treatment. Another purpose for which both the agents are often applied to the tissue is to secure a high number of metaphase stages for treatment with x- rays.

Schedule of Treatment

Place young denuded bulbs of *Allium cepa* in jars containing tap water and allow the roots to germinate till they are 1-3 cm long. Place the bulbs horizontally in large glass petri dishes filled with a solution of the chemical whose protective action is to be studied. For this purpose 4 × 10^{-3} M sodium thiosulphate may be used. Treat for 30 min in this solution. Irradiate the bulbs in the same solution with an x-ray dose of approx. 150R at an output of 30- 40R/min. The x-ray machine may be operated at 200 kV peak and 15 mA. After irradiation, keep the bulbs in the fluid for another 10 min. Transfer the bulbs to beakers containing tap water where they can be kept for 4-5 days for immediate observation and for observation of the root tips at regular intervals. Cut healthy young root tips, treat in 0.5% colchicine solution for 1 h to secure a large number of metaphase stages necessary for the study of chromosomal interchanges and deletions. Fix in acetic-ethanol (1:3) and follow the usual procedure for orcein or Feulgen staining of root tips and observe the interchanges and deletions at the metaphase stages. To measure the protection afforded by the chemical, set up a control experiment in which all the above steps are followed, substituting distilled water in place of the chemical used. The difference in the frequency of interchange and deletions between the two sets will give a measure of the protection afforded by the chemical. For animal and plant tissues in culture, the tissue is irradiated *in vitro* and treated with chemicals added to the medium.

Synchronization of Division through Physical and Chemical Agents

In a logarithmically growing cell population, methods of analysis present some problems due to non-synchrony of cell division. Cells in certain cases show natural synchronous division, such as, eggs at the time of cleavage, or p.m.c.s in the anthers of *Lilium* or antheridial filaments in *Chara*. Induced synchrony becomes essential for accurate analysis of several aspects of the cell cycle. The principle involved in induced synchrony is to allow all the cells to start DNA replication simultaneously. The different schedules for inducing synchrony fall under two categories: physical and chemical. Of the physical methods, shock treatment like chilling at 4°C for 1 h and subsequent removal to 37°C has been found to be successful in HeLa cell cultures but limited in other cases. Even irradiation with x-rays has been utilized for synchronization.

The chemical methods appear to be more promising in inducing synchronization as specially noted in the case of mammalian cells. With this object in view, methods have been devised to cause thymine deficiencies, thus blocking DNA replication. After treatment, the cells in G_1 remain in the same phase while those in the later phases complete their cycle and are blocked prior to the succeeding S phase. Cells in the S phase, on the other hand, are heterogeneous in the sense that a few per cent of the cells are at the beginning, some at the middle, and some at the end of the S phase. They are all blocked in their respective positions. As soon as the synthetic activity starts after the addition of thymidine, the cells lying at the different stages of the S phase resume the activity from their respective points and become non-synchronous, while all other cells start from the *S* phase and show synchrony.

In monolayer cultures of HeLa cells, interphase cells are not attached to the surface as the mitotic cells. As such, the intermitotic cells can be separated out from the mitotic ones and synchrony can be induced in the latter. An antibiotic which also checks DNA replication is mitomycin C.This compound permits RNA synthesis to continue, checking only the synthesis of DNA. Cells treated with mitomycin C show the two polynucleotide strands being linked together. Because of the cross linkage, replication, which required separation of the strands, is obstructed. The alkylating action of the ethylene amine group of mitomycin C is responsible for linking it to a DNA strand.

Similar inhibiting effects on replication have been obtained with fluorodeoxyridine (FUDR). It is also widely applied. Hydroxyurea has been used as a DNA synthesis-inhibiting agent as well. Several other chemical agents have so far been used for achieving synchronisation in cultures. In all cases, the principle in general is to block synthesis of DNA or arrest metabolic activity of mitosis. Other chemicals used are thymidine, colcemid, a combination of cytosine arabinoside and colcemid and vinblastine. The induced deficiency of certain metabolites such as leucine, isoleucine as well as glutamine may also lead to synchrony. There are some serious limitations in the use of chemical agents for induction of synchrony since chromosome damage and toxicity are often caused by these chemicals.

An effective technique utilises hypotonic solution in inducing synchrony by hypertonic arrest in metaphase (SHAM). Ooka (1976) described a method utilizing selection of postmitotic cells through combination of techniques for detaching cell monolayers and gravity sedimentation on a column of culture medium.

The schedule is outlined below:

Grow mammalian cells in suspension (2×10^5 cells/ml in suitable medium Eagle MEM). Seed suspension culture cells in Roux bottles (8×10^6 cells per bottle). Remove the old medium after 48 h and add 30-50 ml of fresh medium. Shake horizontally. Recover the detached cells (20- 30% by centrifugation at 500 r/min for 10 min. Resuspend in 3 ml of serum free medium. Carefully deposit with pipette 70×10^6 cells suspension in serum free medium on the top of a sterile glass column (2.6×19 cm) containing 90 ml of medium. Incubate and allow to settle at 37°C for 40 min. Pipette out the upper third of the cells in medium (30 ml), mix with 170 ml of medium and incubate at 37°C to obtain synchronised cell populations.

SECTION B: METHODS FOR MONITERING CYTOGENETIC EFFECTS OF ENVIRONMENTAL TOXICANTS

The exposure of the human sustem to a wide variety of agents— physical, chemical and biological—following increasing release of hazardous toxicants in the environment, has considerably enhanced the importance of the study of genetic toxicology. A large number of protocols has been developed for short term assays, both *in vitro* and *in vivo*, of the toxicity of different agents acting either on the genetic architecture, that is, on DNA skeleton, or as promoters or inhibitors of genotoxic effects. Over the past decade, these researches have become oriented towards precise risk assessment and risk benefit analysis as applied to human populations. The test systems designed to detect point mutations and chromosome anomalies cover a wide range, *in vivo* and *in vitro* (see tables 4, 5, 6, 7). The number of short-term assays exceeds 74 and almost every major laboratory group has its modified schedules. As a part of E.P.A. Genetox Assessment Panel program, 2622 were listed. For correct assessment, both single test systems with multiple endpoints and mixed batteries of tests should be employed.

A large number of compounds have been tested to detect their potential for inducing aneuploidy or alterations in chromosome number, initially developed as an ancillary to clastogenesis. This aspect is gaining importance following the observation that aneuploidy has been associated with stillbirth, abortions and certain human syndromes. Genotoxic effect as reflected in carcinogenicity is the other facet of the problem. The relationship between mutagenic and carcinogenic effects is, however, not very simple; particularly since several compounds have been shown to behave as a carcinogen not necessarily through a direct action on DNA, but through their crucial role on growth and development of transformed cells.

Specific chromosome translocations have been located in human lymphomas and leukemias. Such translocations, which can be induced by different agents, involve breakage points where cellular oncogenes responsible for carcinoma have been located. The translocation may lead to alterations or activation of oncogenes. The discovery of induced segmental rearrangement of chromosome which otherwise leads to position effect, contributing to malignancy, has opened up a novel approach to genotoxicity. This rearrangement can be accelerated through mobile sequences capable of insertion at different loci, which may be modified through exposure to genotoxic compunds. All these advances emphasize the need for understanding the relationship between functional segments of chromosomes and genotoxicity.

Several human recessive disorders have been associated both with increased spontaneous frequencies of chromosomal aberrations and/or sister chromatid exchanges and with one or more classes of mutagens. Such sensitivity to specific mutagens may be employed in the cytological detection of such syndromes. For example, Fanconi's anemia shows both increased frequency of spontaneous chromosomal aberrations and enhanced susceptibility to X-rays and chemicals like mitomycin C and acetaldehydes.

The chemical carcinogens and genotoxicants represent a varied spectrum. Their biological activities are strikingly different, from the highly reactive compounds able to alkylate macromolecules and cause mutations in many organisms to the hormonally reactive substances which cannot cause such aberrations. The activity may vary by a factor of at least 10 at the different ends of the spectrum.

Two major categories of synthetic mutagens that human beings are exposed to in higher limits are the halogenated hydrocarbons and the alkylating agents due to their application in large amounts in a number of forms. Almost all the commercial halogenated hydrocarbons exhibit toxicity, particularly the lower molecular weight C_1-C_4 chlorinated alkanes and alkenes. They are extensively used as solvents, fumigants, aerosol propellants, etc. Another very important group of industrial chemicals of recognised toxicity includes the chlorinated aliphatic saturated and unsaturated hydrocarbons (methyl chloride, chloroform, carbon tetrachloride, methylene chloride, ethylene dichloride, 1, 1, 2- trichloroethane, methyl chloroform, trichloroethylene, perchloroethylene, 1, 1, 2, 2-tetrachloroethane, epichlorohydrin, alkylchloride, BCME, Freons F-11 and F-12 and halothene). In the indoor environment, a number of chlorinated hydrocarbons exist as pollutants, administered as ingredients of aerosols. They may even be present in drinking water, either through disinfection by chlorination or as effluents.

The different groups of alkylating agents, which have been identified as genotoxicants and/or carcinogens, are; (i) epoxides, (ii) lactones; (iii) aziridines; (iv) aliphatic sulphuric acid esters; (v) cyclic aliphatic sulphuric acid esters; (vi) alkyl fluorosulphuric acid esters; (vii) sultones; (viii) diazoalkanes; (ix) aryl dialkyl-triazenes; (x) phosphoric acid esters; (xi) aldehydes; (xii) alkane halides; (xiii) alkyl halides; (xiv) haloethers and (xv) haloalkanols. Of these, considering the populations at risk and the total volume produced, the most toxic ones are the epoxides. Some compounds of this group, of identified mutagenic potential, are ethylene oxide, propylene oxide, butylene oxide, epichlorohydrin, glycidol, glycidaldehyde, diglycidyl resorcinol ether and glycidyl ethers. A relationship between the size of the chemical structure and mutagenicity has been suggested for aromatic amines.

Various techniques have been developed for assessment of the environmental status, the hazards to which an organism is exposed and the extent to which such studies can lead to remedial measures. Such techniques are based on the principle of the relationship between internal levels of exposure and the effects caused thereby and may be utilized for both indication and monitoring of environmental pollutants. The tests principally relate to an analysis of toxicity, mutagenicity and carcinogenicity of the different types of pollutants found in the environment.

The principal endpoints for monitoring are:

(i) mutational events, involving DNA, and

(ii) chromosomal aberrations, involving breaks and interchanges at different levels. These changes may occur in the somatic cells or in the germ cells. The methods include both *in vivo* and *in vitro* systems (tables 4, 5, 6, 7).

Some Common Test Systems Used (table 4)

Yeast and fungi

Strains of yeast have been utilized *in vitro* as indicators of reverse mutations, gene conversion, mitotic recombination, aneuploidy and other forms of chromosome damage. Growing cells of different species of *Saccharomyces* produce the enzymes necessary to activate many promutagens which would not be able to induce mutations *in vitro* without S9 mix. Mutation studies have also been carried out with species of *Neurospora* and *Aspergillus*. These systems have a relatively limited scope.

Table 4: Assays to Identify Chromosomal Effects (after Gene-Tox Assessment Panel)

Description of assay	Biological class
Saccharomyces cerevisiae—aneuploidy studies	LE
Aspergillus—aneuploidy studies	LE
Neurospora crassa—aneupoloidy studies	LE
Plant chromosome studies—grouping of tests with Allium, Hordeum, Tradescantia, and Vicia	HE
Drospohila melanogaster aneuploidy studies—whole sex chromosome loss	HE
Drosophila melanogaster aneuploidy studies—sex chromosome gain	HE
Drosophila melanogaster aneuploidy studies—sex chromosome loss	HE
Drosophila melanogaster heritable (reciprocal) translocation test	HE
Mammalian cytogenetics—all in vitro cell culture studies, nonhuman studies	HE
Mammaliann cytogenetics—all in vivo bone marrow studies, nonhuman studies	HE
Mammalian cytogenetics—all in vivo lymphocyte or leukocyte studies, nonhuman	HE
Mammalian cytogenetics—all male germ cell studies	HE
Mammalian cytogenetics—in vivo oocyte or early embryo studies	HE
Micronucleus test—mammalian polychromatic erythrocyte assay, all species	HE
Micronucleus test—plant	HE
Micronucleus test—in vitro mammalian cell culture studies, nonhuman	HE
Micronucleus test—in vitro human lymphocyte studies	HE
Dominant lethal test—rodents	HE
Heritable translocation test—rodents	HE
Mammalian cytogenetics—in vitro cell culture studies, human	H
Mammallian cytogenetics—in vitro lymphocyte or leukocyte studies, human	H
Mammalian cytogenetics—in vivo bone marrow studies, human	H
Mammalian cytogenetics—in vivo lymphocyte or leukocyte studies, human	H

LE, lower eukaryote : HE, higher eukaryote; H, human

Table 5: Assays evaluated in phase I of the Gene-tox program

Gene mutation	*Chromosomal Aberrations*
Bacteria	Mammalian cytogenetics
Salmonella typhimurium	Plant cytogenetics
*Escherichia coli*WP2 and WP2 *urr* A	Sister-chromatid exchange
	Yeast
Yeast	Fungi
Fungi	Drosophila
Neurospora crassa	Dominant lethal assay
Aspergillus nidulans	Micronucleus test
	Heritable translocation test in mice
Plants *in vivo* and in culture	*DNA Damage and Repair*
Mammalian cells in culture	Repair-proficient and repair-deficient bacteria
Chinese hamster lung (V79)	Unscheduled DNA synthesis
Chinese hamster ovary (CHO)	Unscheduled DNA synthesis
Mouse lymphoma L5178Y	DNA repair
	Oncogenic transformation
Drosophila sex-linked recessive lethal test	Cell strains
	Cell lines
Mouse spot test	Viral infected cell lines
Mouse specific-locus test	*Ancillary assays*
	Host-mediated assay
	Body-fluid analysis
	Sperm morphology

Yeast-Agar Overlay Protocol (Parry 1985):

Suspend stationary-phase cells of an appropriate tester strain in saline. Take the test extract up to 4% and mix with yeast minimal medium kept at 45°C with requisite number of yeast cells. 35×10^2 or 35×10^3 per ml are adequate for the detection of histidine - independent and tryptophan - independent prototrophs respectively. If needed, an exogenous S9 mix may be added. Supplement the minimal media with 20 µg/ml of tryptophan and 0.1 µg/ml histidine for detection of respective prototrophs as used. This allows at least 3 cell divisions during incubation along with the tissue extract. Keep the plates in the dark for 9 days at 28°C, count the frequency of prototrophic colonies.

Plants

Higher plants are used extensively for monitoring genotoxicity of chemicals due to the complexity of their capacity in some cases to activate promutagens to mutagens and the ease of their use. The more commonly used ones are different species of *Vicia* and *Allium, Tradescantia, Hordeum* and maize. For the localization of chromosomal alterations, karyotypes with marker chromosomes are required. Pretreatment of root tips with other chemicals prior to exposure alters the susceptibility in some cases. Gene mutation studies have been carried out on *Arabidopsis,*

Table 6: Relative reliability of some test systems as applied to man (after Anderson, D. and Ramel C., 1979, In : Genetic damage in man caused by environmental agents, Academic Press, New York, 50).

Test system	Time to run tests	Relative ease of detecting	
		Gene mutations	Chromosomal aberrations
Microorganisms with metabolic activation			
Salmonella typhimurium	2 to 3 days	Excellent	—
Escherichia coli	2 to 3 days	Excellent	—
Yeasts	3 to 5 days	Good	Unknown
Neurospora crassa	1 to 3 weeks	Very good	Good
Cultured mammalian cells with metabolic activation	2 to 3 weeks	Excellent to fair	Unknown
Host-mediated assay with			
Microorganisms	2 to 7 days	Good	—
Mammalian cells	2 to 5 weeks	Unknown	Good
Body fluid analysis	2 days	Excellent	—
Plant			
Vicia faba	3 to 8 days	—	Relevance to
Tradescantia paludosa	2 to 5 weeks	Good potential	man not clear
Allium cepa	One to 30 days	—	
Insects			
Drosophila melanogaster	2 to 7 weeks	Good to very good	—
Mammals			
Dominant lethal mutations	2 to 4 month	—	Unknown
Translocation	5 to 7 months	—	Potentially very good
Blood or bone marrow cytogenetics	1 to 5 weeks		Potentially very good
Specific locus mutation	2 to 3 months	Unknown	—

Table 7: Some Mammalian Cell lines used for studying Chromosomal Aberrations

Human lymphoid cells — B35M, B 411 - 4, CCRF-CEM, HLCL, P3J, RAJI, RPMI-1788, SNN 1029

Human fibroblast cels — Hum eue, Hum fibro, JA, JHV-1, MRC5, WI-38

Human cells, other types— CA', FL, HeLa, Hum amn, VVP-1.

Chinese hamster cells— B14F28, B241, CH fibro,CH1 - L, CHL, CHMP/E, CHO, CHO-A7, CHO-AT3-2, CHO-CL-10, CHO-CL-H, CHO-KI, CHO-KI-A, CHO-K1-BH4, Don, Don D6, HY, V79, V79-4, V79-CL-10, V79-CL-15, V79-E,

Syrian hamster cells — A(T1) CL-3, SH-fibro.

Rat cells—MCT1, Rat fibro, Rat lymph, RL1, RL4.

Mouse cells— C3H10T1/2, FM3A, L cells, L 929, L 5178Y, M10, Mus fibro, Q31

Cells from other species— DH/SV40, MA61, Munt fibro, PTK 1.

Glycine, Hordeum, Tradescantia and *Zea*. Of these, *Zea mays* has been found to be a very good bioindicator because of the convenience of detecting mutation in the specific gene locus in identified chromosomes.

Maize

The two tests which are widely employed involve chromosome 9.

Waxy locus pollen test: The gene is located on the short arm of chromosome 9 and is responsible for the synthesis of carbohydrate amylose. The starch in individuals carrying the dominant allele *Wx* is a mixture of amylose and amylopectin, whereas amylose is absent in the double recessive individuals. Iodine test colors amylose blue-black. Its absence gives a red or tan color. Since the pollen grain is haploid, containing either *Wx* or *wx*, the color can be used as monitor.

Take an early - synthetic type of maize and allow it to grow till anthesis in an area containing the mutagenic pollutants. Harvest each tassel from double recessive sx-C/wx-C and label. Dehydrate to 70% ethanol for 48 hours. Shake the selected tassels in pure ethanol and remove a few unopened flowers. Dissect the anthers from the floret and place in a stainless steel cup of microhomogenizer. Prepare: Stock A - Iodine solution with 25 ml of water, 100 ml of KI and 95 mg of Iodine. Stock B - gelatin solution with 25 ml of hot water and 15 g of gelatin. Mix A and B in equal proportion. Take 0.6 ml of mixture and add in the cup. Add a drop of Tween-80. Homogenise the minced anthers for 30 *s*. Filter the homogenate through cheese cloth on a microscope slide and put the cover slip on the suspension. When the suspension solidifies, examine the slide under the microscope. The black grains would indicate the revertants and the percentage of mutation can be worked out.

Yg-2 locus on short arm of chromosome 9 in maize induces pale yellow green and slowly maturing leaves in homozygous recessive plants (Yg-2/yg-2). The dominant Yg-2 causes dark green coloration. In the heterozygous state (Yg-2/yg-2) leaves are a normal green in color. In a heterozygous individual, mutation of the dominant allele leads to deficiency in the production of chlorophyll in the leaf sector. If a leaf primordial cell is involved, yellow green sectors develop in the green leaf.

Choose heterozygous Yg-2/yg-2 seeds from Early-Early synthetic type. Expose the seeds under aeration to specific environmental pollutants by immersion. Germinate the seeds and grow the plant in growth chambers at 28°C till the emergence of 4 to 5 leaves. Dissect the mature leaves with ligule and observe under the ultraviolet light.

Micronucleus test: In Early-Early synthetic type of maize, the presence of micronuclei in root tips has been taken to indicate chromosome aberrations produced as a result of environmental toxicant. Micronucleus test has also been utilized in other plants, e.g. in root-tips of *Vicia faba* following exposure to radiation. Micronuclei are taken to indicate non-disjunction, resulting in whole chromosomes or acentric fragments being left out of the spindle.

Collect young tips from germinating seeds of plants grown in a specific environment which is to be monitored. Fix for 24 h in ethanol - acetic acid mixture 3:1. Rinse in demineralized water. Hydrolyse in 1 N HCl at 60°C for 7.5 min.

Rinse the root tips again and place in Feulgen solution for 1 h. Treat the root tips with pectinase solution for softening and clearing. Squash the root tips in 45% acetic acid. Count the number of interphase cells under microscope. At least 1000 interphase cells per slide should be scored. It is a convenient and rapid assay method for monitoring acute or chronic exposure and shows direct correlation with the degree of pollution.

Tradescantia

Staminal hair method: A change in color in high mutagen sensitive staminal hairs has been utilized to monitor mutagens in ***Tradescantia***. The mutational events change the color from blue to pink, which is observed in the stamen under the microscope after dissection.

The microspore has been used to measure the extent of radiation as absorbed by the plant when exposed. In general, the microspores of ***Tradescantia*** are excellent materials for scoring the effects of pollutants as shown by the formation of chromosomal aberrations during mitosis of pollen grains.

Pick up pollen with a brush from anthers of plants growing in polluted environments. Heat 12 g of lactone and 1.5 g of agar with 100 ml distilled water till agar dissolves. Add 0.01 g of colchicine when the temperature cools down to 80-60°C. Keep at 60-70°C. Coat slides with a thin layer of egg white. Dip in the medium at 60°C until it warms up. Withdraw the slide, drain off the medium, and wipe the back with a piece of clean cloth. Dust a thin layer of pollen on the medium after it has set. Immediately place the slide in a moist growing box which is a horizontal glass staining dish lined with damp blotting paper on the two sides and top, at 20-25°C. Observe at intervals till the optimum period for maximum division is found. Add a drop of 1% acetic-carmine solution to the pollen tube, squash under a coverslip and observe.

Barley, Vicia faba and species of ***Arabidopsis*** offer good systems for environmental monitoring. In the latter, due to the short life cycle, the effect of the pollutant can be measured in all phases of growth.

Allium cepa, Allium sativum

The ***Allium*** test, initially worked out by Levan is one of the most convenient methods of monitoring genotoxicity. The chromosomes of ***Allium***, being quite large, allow a detailed analysis. The bulbs can be easily handled and roots with meristems grow profusely at regular intervals. Such rapid successive growth enables an analysis of both long and short term effects. The method involves the culturing of healthy bulbs on the top of small jars containing the solution to be tested for genotoxicity with the growing roots immersed in the solution. The roots are cut at regular intervals to study the effects of different periods of treatment at different concentrations.

In order to study the recovery from the toxic effects, if any, the bulbs with roots after treatment are transferred to Knop's nutrient medium for observation at successive periods.

The effects normally are classified as subnarcotic, narcotic and lethal. The subnarcotic effects include chromosomal alterations and consequently lethality. Certain vital processes are switched off but enough of the basic processes are maintained so that the tissue can recover when the external influence is removed. On the other hand, certain compounds are highly toxic, leading to pycnosity, liquefaction of chromatin and subsequent lethality.

Place healthy bulbs of *Allium cepa* at the mouth of jar containing the chemical to be tested. Keep the jar at room temperature. Excise young healthy root tips and fix in acetic- ethanol mixture 1:2 for 30 min. Transfer to 45% acetic acid for 15 min. Prepare a mixture of 2% acetic-orcein in 45% acetic acid and N HCl mixture in the proportion of 9:1. Slightly warm the root tips on flame for 4 to 5*s* for staining, separation of middle lamella and clearing. Keep in the stain mixture for 1 h. Mount in 1% acetic-orcein or 45% acetic acid. Observe the number of chromosome fragments in metaphase and anaphase stages. The frequency of breaks gives an indication of the extent of chromosome damages caused.

Animals

A wide range of animals, from Drosophila to fish and rodents like mouse, rabbit and chinese hamster, has been utilized for monitoring of agents present in the environment. Both *in vivo* and *in vitro* systems have been employed. A major addition is peripheral blood lymphocyte culture from human systems. The latter is extensively used, partly due to the lower cost and partly to the relative ease of extrapolation of the results obtained from human leukocytes *in vitro* to human subjects *in vivo*. The usual endpoints observed in mammalian test system *in culture* are: (i) chromosomal aberrations in metaphase and anaphase: (ii) micronuclei in interphase; (iii) sister chromatid exchanges in metaphase and (iv) unscheduled DNA synthesis as measured by microauto-radiography. These endpoints, particularly the first three, are common to *in vivo* test system as well.

1. ***The fruitfly Drosophila melanogaster*** is a very good indicator for *in vivo* testing of genetic damages from the chromosome to gene level. In addition to the wellknown dominant lethal test, recessive sexlinked lethal (SLRL) and Muller 5 tests, which identify single gene mutations, chromosomal changes can also be utilized for monitering. Non disjunction or maldistribution of chromosomes may be utilized to identify changes at chromosomal level. Survival of the aneuploid offspring is the endpoint recorded. The test therefore is limited to the X-chromosome or the small fourth chromosome. In order to distinguish the offspring resulting from non-disjunction, one set of chromosomes must carry markers.

2. ***Fishes*** have been utilized for monitoring of the aqueous environment and chromosome damages caused. The damage is principally measured through chromosome breaks, bridges and other aberrations. The organs utilized are the gill cells as well as kidney and intestine. Laboratory fish such as *Umbra pygmaea* (mudminow) can be exposed for different periods in various aqueous environments. Short-term bone marrow culture technique has also been employed.

Select healthy fishes exposed to specific environments under analysis. Approximately 7 h prior to sacrifice, inject fishes with 25 μg colcemid or colchicine and return to well-aerated aquaria. Sacrifice by decapitation. Dissect out gills, kidneys, intestines and place in 10 times its volume of 0.4% KCl hypotonic solution for 30 min. Remove tissue and place in 3:1 ethanol - or methanol/acetic acid fixative for 30 min. Remove tissue, touch them briefly to absorbent paper. Immediately place them in the depression of a concave slide. Add 2-3 drops of 50% acetic acid. Mince the tissue with forceps to form cell suspension and return unsuspended tissue to fixative. Withdraw 30 μl of suspension with micropipette or microhematocrit capillary tube and expel it into a clean glass slide heated to 40-60°C on a slide warmer. After 5 *s*, withdraw the suspension from the heated slide back into the capillary tube or pipette leaving behind a ring of dried cellular material 1-2 cm in diameter. Repeat, producing 2 or 3 rings per slide. Stain slides in 4% Giemsa in 0.01 M phosphate buffer (pH 7) for 10 min. Rinse slides in distilled water, air dry, and place in xylene for at least 10 min. Mount with suitable mounting media using glass coverslips.

3. ***Mammals***: Almost all rodents have been used as test systems. Of these, the mouse has been most extensively employed in monitoring the mutagenic potentials of hazardous chemicals and wastes. Toxicity can be measured in terms of the production of lethals or of chromosome aberrations *in vivo*. Embryonic lethality can be judged by scoring of dead implants in rodents. Chromosome aberrations can also be studied following culture techniques, either in the bone marrow or in leukocytes.

Dominant lethal in mouse is a quick *in vivo* test representing a mutation either in the sperm or the egg, which is lethal to the zygote produced. Most dominant lethals can be related to chromosomal aberrations and they reduce the litter size in different degrees.

Heritable translocation test in mouse has been used to identify partial or full sterility amongst the first generation offspring of mutagen-treated animals. Heterozygosity for a reciprocal translocation leads to partial sterility. This test also gives an estimate of the transmissibility of induced chromosomal exchanges.

4. Dominant lethal test - Protocol:

Take ten mice after exposure to toxicant. Mate each male separately with 7 sets of virgin females, one set each week for consecutive 7 weeks. Determine the successful mating by the appearance of vaginal plug in the mating females. Sacrifice two pregnant females on the 16th day of gestation. Expose the viscera and clear the two horns of the uterus. Determine the number of live-implants by the conspicuous swelling at the place of implant and the dead implants by the scar mark. Keep records separately for 7 weeks. Sacrifice one or two mating females at the end of the weaning period after parturition, each week. Expose and remove carefully the uterus and put it in a petri dish containing normal saline. Slit open each horn and examine the wall of the lumen under dissecting microscope and count the number of brownish colour scar surrounded by significant amount of vascular tissue representing living-implantation and the light brown or whitish scar mark with less vascular surrounding, representing dead implantation. Combine the data of different mothers of a week and find out the number of living and dead implants per mother each week. Use the following formula to determine the post implantation

$$\text{lethality } \% = 1 = \frac{\text{no. of live–implants/total imp. exp. gr.} \times 100}{\text{no. of live implant/total imp. in control gr.}}$$

and Mutagenic index

$$= \frac{\text{Number of dead implants} \times 100}{\text{Total number of implants}}$$

In vivo chromosomal alterations induced by external agents can be scanned in all dividing cells.

A routine test involves the *bone marrow,* following a short term colchicine pretreatment—hypotonic—air drying schedule. The aberrations scored include breaks of different types—chromatid and chromosome—depending on the phase (G_1 or G_2) of the cell treated and the nature of the clastogen—whether dependent or DNA synthetic (S) phase or not. Translocation may also be seen. Breaks not included in the daughter nuclei later form micronuclei. Such micronuclei persist in the polychromatic erythroblasts formed from bone marrow and can be scored easily to give the frequency of breaks induced. In liver micronucleus assay, treated animals are partially hepatectomized and the hepatocytes from the regenerating liver are analyzed for the presence of micronuclei. Metaphase and micronucleus analyses may also be combined in the same animal.

In vivo bone marrow test

Expose the animal to the toxicant by gavaging or by injection or inhalation. Inject intraperitoneally 0.4% colchicine solution at a dosage of 1 ml/100 g body weight into the anmal. After one and half h, sacrifice the animal by cervical dislocation. Cut open abdomen. Remove the femurs. Wash out the marrow into a small phial with warm 1% aqueous solution of sodium citrate, using a hypodermic syringe with a fine needle. Gently aspirate the marrow in and out of the syringe till it breaks up into a fine suspension. Keep the phial with the suspension in a water bath at 37°C for 20 min. Filter by centrifuging the suspension through nylon bolting cloth in a bacterial inflitration tube to obtain a fine clean suspension without any debris. Fix in chilled acetic-methanol (1:3) mixture for 30 min to 2 h. Centrifuge, wash twice in fixative and resuspend in small amount of fresh fixative. Draw out the suspension by means of a Pasteur pipette with rubber bulb. Tilt a chilled greasefree clean slide at an angle. Allow 2 or 3 drops of the suspension to fall on upper end of slide. Allow the drops to spread and dry by blowing or under a fan. Stain the slides in diluted Giemsa stain (4 ml Giemsa stock and 20 ml each of 0.1 M Na_2HPO_4 and KH_2PO_4 solution) for 10 to 15 min, stain in distilled water, dry and mount with DPX under a coverslip. To observe aberrations induced at all phases, observation should be made 6, 12, 18 and 24 hours after exposure.

In vitro cytogenetic tests: The popularity of the *in vitro* system stems principally from the ease and low cost of performing these tests. The most frequently used systems for *in vitro* screening for mutagens are established Chinese hamster cells or stimulated human peripheral lymphocytes. Agents which induce DNA double strand breaks, lead to chromosome breaks in G_2, chromatid breaks in G_1 and both types in S phase of the cell cycle. Other mutagens induce only chromatid breaks irrespective of the cell cycle stage treated.

A list of the more common mammalian cells lines is given in the table 7.

Human leucocyte culture test (see also Appendix A)

Draw 2-10 ml blood aseptically from the vein of exposed individuals by sterile disposable syringe and collect in a vial containing 0.1 ml heparin (5-10,000 IU/ml) sterile.

Sediment leucocytes by gravity at room temperature (22-38°C) for 2 h. Add 0.5 ml of plasma separated out at top of vial to 5 ml tissue culture medium. The culture medium is prepared by mixing 210 ml of RPMI media with 40 ml of inactivated AB+ human serum and filtering 10 ml of reconstituted phytohaemagglutinin (PHA, M from Difco) is added at a rate of 6 μg/ml and mixed. 5 ml of this medium is stored in each sterile culture tube. Volume of plasma depends on the white blood cell count. Add chemical to be tested to culture medium. Incubate the bottles at 37°C for 70 h in dark.

Add colchicine to 0.4 μg/ml final volume, 0.04 μg/ml or 0.1 μg/ml of colcemid or velben may be added. Shake vials mildly.

Harvest after 2 h. Centrifuge tubes at 1000rpm for 10 min. Discard supernatant. Add to pellet 0.075 KCl hypotonic, gently flush. Again incubate for 8-10 min and recentrifuge for 10 min at 1000rpm. Remove supernatant. Add fixative slowly to the pellet and leave undisturbed for 30 min at 4°C. Shake slowly. Keep for 15 min. Recentrifuge for 10 min. Add fresh fixative to pellet and mix with a long siliconized pasteur pipette. Keep for 10 min to overnight at 4°C. Recentrifuge and add enough fixative to the deposit to give a suspension. Drop on cold, clean wet glass slides drops of suspension from a height of 6 cm. Air dry by blowing or under a fan. Stain with diluted Giemsa stain as given for *in vivo*.

Sister Chromatid Exchange (SCE) in Animals and Plants as Indicators of Toxicity

Sister chromatid exchanges are cytogenetic processes related to DNA damage and repair. It is an effective cytogenetic parameter for monitoring the damage. These are due to unrepaired lesions present at the S phase followed by chromosome replication and can be induced by different compounds capable of alkylating DNA. The SCE for visualization requires at least two replications in the presence of Bromodeoxyuridine (Brdu). This leads to DNA of one chromatid of metaphase chromosome to become bifilarly substituted by Brdu, the other remaining singly substituted. Following the use of fluorochromes such as Hoechst 33258 and observation under fluorescent microscopy or Giemsa staining, the contrasting color of the two chromatids can be clearly noted.

In vivo Sister Chromatid Exchange in Mice:

Implant 5-BrdU tablet (50 mg each) to each etherised mice from specific environment, subcutaneously into the neck. Inject colchicine (5 mg/kg) intraperitoneally 21-22 h after subcutaneous implantation of BrdU tablet. Sacrifice the animal by cervical dislocation 2 h after injection of colchicine. Prepare bone marrow chromosomes by usual procedure. Stain the slides by fluorescence - plus-Giemsa technique. Stain 4-5 days old slides in 5 mg/ml 33258 Hoechst dissolved in NaCl/KCl solution for 10 min. Rinse and mount with M/15 Sörensen's phosphate buffer (pH 6.8). Irradiate the slides with a 254 nm UV mineralogic lamp for 30 min. Incubate the slides in 2 × SSC (0.3 M NaCl, 0.03 M trisodium citrate) solution for 90 min at 59°C immediately after irradiation. Wash the slides thoroughly with distilled water and stain with 7% Giemsa in phosphate buffer for 20 min (after Talukder 1985).

In vitro SCE Test in Mouse:

Grow cells of mouse from affected environment in modified Eagles MEM supplemented with 10% fetal calf serum. Alternatively, add test chemical to medium. Incubate the cells is dark and grow in a medium containing 10^{-5} M BrdU for 18-20 h or 30-32 h (one or two cycles of DNA replication). Add colcemid (0.1 µg/ml) to culture. Harvest after 2 h by trypsinization. Prepare usual air dry smear. Stain with 1 µg/ml of 4-6 diamidino-2 phenylindole (DAPI) in phosphate buffered saline (pH 7.0) for 10 min. Rinse in deionised water. Mount in 0.2 sodium phosphate (pH 11.00). Observe chromosome fluoroscence with Zeiss microscope with HBO 200 w/4 Hg lamp dark field illumination, a BG 12 exciter filter and on

470 nm barrier filter. Chromosome aberrations and micronuclei formation can be scored (Lin and Alfi 1976).

In vitro SCE Test in Chinese Hamster

Trypsinise exponentially growing exposed Chinese Hamster cells (v-79). Plate 1 ml of a suspension containing 10^5 cells/ml on object glasses placed in sterile 10 cm petri dishes. Keep for 2 h. Add 9 ml of culture medium (Dulbecco's MEM) supplemented with 10% newborn calf serum (Gibco), penicillin (Gist-Brocades 200 units/ml) and streptomycin (Gist-Brocades 100 µg/ml). Keep for 24 h. Add to the plates 0.1, 0.2, 0.3 ml of dimelthylsulphoxide. After 2 h exposure, replace the culture medium by a medium containing Bromodeoxyridine (Brdu 10 µM).

Continue Brdu exposure for 24 h as the cell cycle is 12 h. Treat the cells following the procedure of Parry and Wolff (1974) with slight modifications.

Observe the sister chromatid differential staining. Score at least 50 metaphase plates per experiment per concentration.

In vitro SCE in Human System

In the usual method for leucocyte culture of exposed individuals, BrdU should be added 24 h after initiation and reincubated for additional 48 h. The culture should be grown in dark. Two hours of colcemid treatment (0.5 µg/ml) before harvesting for usual air dried preparation is needed.

In vivo SCE in Plant Test System

Treat lateral roots of exposed plants with to aqueous solution containing 100 µM-5-BrdU, 0.01 µm 5-FdUrd and 5 µM (Urd) for 22 h. Transfer to aqueoous solution containing 100 µM thymidine (dThd) and 5 µM Urd for 21 h. Treat in 0.05% colchicine for 3 h. Fix overnight in acetic-methanol cold (1:3) in dark at 20°C. Rinse in 0.01 M citric acid - sodium citrate buffer (pH 4.7). Incubate for 75 min at 27°C with 0.5% pectinase dissolved in same buffer. Squash in 45% acetic acid. Coat with a mixture of 10:1 gelatin and chrome alum. Remove cover slip by dry ice and bring the preparation to water through descending ethanol grades.

Incubate in moist chamber for 60 min at 27°C with RNAse (1 mg RNase in 10 ml 0.5 × SSC) 200µl; cover. Rinse in 0.5 × SSC, stain 20 min in H33258 (1 mg dissolved in 100 ml ethanol). Add 0.1 ml of this solution to 200 ml 0.5 × SSC. Rinse and mount in 0.5 × SSC. Store for four days at 4°C. Incubate for 60 min at 55°C in 0.5 × SSC. Rinse in 0.017 M phosphate buffer, pH 6.8. Stain for 6-7 min in 3% Giemsa (R66) solution in same buffer. Rinse in phosphate buffer, then in water, air dry, pass through xylene and mount in Canada Balsam.

Appendix A

General Laboratory Procedure for Human Cytogenetic Studies

Three ml of blood are drawn from each participant by venipuncture and gently mixed with heparin. If blood is to be shipped, 0.7 ml acid:citrate:dextrose (ACD) or appropriate amount of another pre-tested anticoagulant should be added per 3ml of blood. The blood samples are coded at the time of collection and not decoded until the chromosome analysis is complete.

Whole blood cultures are established by placing 0.4 ml of the blood-ACD mixture in 5 ml of medium in appropriate culture vessels. The culture medium should contain Minimal Essential Medium (or other suitable medium, except those low in folate levels or those that will induce low folate levels), fetal calf serum (10-15%), an antibiotic, heparin, phytohemagglutinin as a mitogen, and a standardized concentration of bromodeoxyuridine. The BrdUrd is used to identify cells for scoring aberrations that have undergone only one DNA synthesis period. Seven cultures are established per person.

One of the cultures for the aberration analysis is grown without any BrdUrd. Colcemid is added to three of the cultures (two with and the one without BrdUrd) at 45 h and they are harvested at 48 h. These would be used for the aberration analysis. Two cultures with BrdUrd receive Colcemid at 69 h and are harvested at 72 h. These would be used for the SCE analysis. Two cultures with BrdUrd, to be harvested at 96 h, are reserves. This protocol will ensure that an appropriate number of first and second division cells will be obtained. Cells are harvested for chromosome preparations by standard procedures. Basically, the cells are concentrated by centrifugation and resuspended in KCl (0.075 M) for 8 min at room temperature. At the end of this time the cells are again concentrated and resuspended in absolute methanol:glacial acetic acid (3:1). This fixation procedure is repeated three times. The fixed and washed cells are suspended in fresh fixative and dropped onto clean microscope slides. Usually 5 slides are prepared from each culture and the unused cells are stored in 2 ml vials at -32°C. The slides are allowed to dry for a period of two days and are then stained by a suitable method to provide sister chromatid differentiation.

The priority for scoring the aberration or SCE endpoints is dependent upon the type of exposure. In the case of exposures to radiation or radiomimetic agents, chromosomal aberrations should be scored first. In the case of chemical exposured, SCEs should be scored first. Since all cultures have been previously coded, scoring of aberrations is done blind. Cells from the 45-48h harvest are analyzed first and, if the necessary number of first division cells is obtained from these cultures, the first division cells from the 69-72 h harvest are not analyzed for aberratons. Only cells with 46 centromeres are scored unless the constitutional karyotype for the individual is different. Aberrations are classified by standard criteria into chromosome and chromatid types. Each of these two major subdivisions is classified into exchanges and deletions. Gaps (defined as discontinuities along the chromosome that are less than the width of a chromatid and are not displaced) are tabulated but are not included as aberrations in the final calculation of data.

For the SCE analysis a minimum of 50 cells should be scored whenever possible. Data should be reported on a per cell basis, i.e., for each cell, data can be collected and analyzed on a per chromosome basis but the SCE frequency should be normalized to 46 chromosomes per cell. Cells with less than 43 chromosomes should be identified and accepted or rejected under low magnification, and scored under high magnification. Appropriate statistical tests should be applied to both the SCE and aberration data to verify internal consistency among the scorers and to identify the presence of abnormally high individuals that may invalidate standard statistical methods.

CHAPTER 10

ISOLATION AND EXTRACTION OF NUCLEI, CHROMOSOMES AND COMPONENTS

A. ISOLATION OF NUCLEI

The primary requisite for the extraction of nuclei is to isolate them in normal condition in adequate amounts and to keep the chromatin in an undamaged state. The procedure followed may be *direct* or *indirect*. In the former, the nucleus is isolated directly from the cytoplasm with the aid of a micromanipulator and observed under the microscope. In the indirect mass isolation technique, the general stages include the disintegration of the cytoplasm through mechanical or chemical means, keeping the nucleus intact, followed by filtration through mesh or cheesecloth, differential centrifugation in suitable liquids and sedimentation. The general specific gravity of the liquid lies between 1.35 and 1.45 and centrifugation at 1000-6000 r/min is needed.

A) For Animal Cells

From S3 Hela cells

Grow monolayers in F-10 medium supplemented with 10% calf serum, penicillin (50 units/ml) and streptomycin (50 μg/ml). Add 5 ml 0.25% trypsin in saline D-2 at pH 7.0-7.4. Incubate for 5-10 min at 35-37°C and suspend in an additional quantity of 5 ml saline D_2. Centrifuge at 1000 × g at 0-2°C for 5 min and remove supernatant. Wash thrice in 50 ml 0.154 MKC1 and centrifuge at 1000xg for 5 min. For lysing the cells and isolating the nuclei, resuspend the cells (0.1 ml) in 4 ml 0.1% (v/v) Triton ×100 isolation medium (Rohm and Haas, Philadelphia) containing 0.001% (w/v) spermidine phosphate trihydrate (spermine) dissolved in redistilled water. Agitate continuously to rupture cell membrane. After complete lysis, add 0.25 M sucrose with 0.001% spermine to a final volume of 50 ml and centrifuge at 700 × g at 0-2°C for 20 min. Resuspend the sediment in 0.25 M sucrose with spermine and centrifuge. Repeat twice. Saline D_2 with spermine is more satisfactory than sucrose.

From Mammalian Organs and Other Animal Systems

Immerse the dissected organ from physiological saline solution, in a cold mixture of 0.3 M sucrose -3 mM $CaCl_2$ solution. Remove connective tissues. The remaining procedure is carried out at 4°C.

Weigh 5 g of tissue and mince. Homogenise in a Teflon pestle tissue grinder with 7.5 ml 0.5 M sucrose - 3mM $CaCl_2$. Add 20 ml 0.3 M sucrose and 3 mM $CaCl_2$ and filter through fine cloth. Adjust to 42 ml with same solutions. Centrifuge 40 ml of the sample at 1000xg for 9 min. Suspend the pellet in 20 ml of 0.3 M sucrose, 3 mM $CaCl_2$ and 0.5% Nonidet P40 to preserve the nuclei and grind continuously.

Prepare a layer of 10 ml suspension in a centrifuge tube on 10 ml 1.8 M sucrose - 1 mM $CaCl_2$. Centrifuge at 40000xg for 15 min. Suspend the pellet in 0.7 ml of 1.0 M sucrose -3 mM $CaCl_2$ and add initially 1.0 ml of 0.5 M sucrose - 3 mM $CaCl_2$, followed by 1.0 ml of 0.3 M sucrose and 3 mM $CaCl_2$ through continuous stirring for 15 min.

Make the solution up to 30 ml with 0.3 M sucrose and 3 mM $CaCl_2$. Centrifuge for 10 min at 1000xg. Resuspend the pellet in 10 ml 0.3 M sucrose and 3 mM $CaCl_2$, which forms the nuclear fraction. Precipitate with 2 vol of 95% ethanol. Store at –20°C.

In an alternative method the mammalian tissue is initially homogenized in a 3-5 vol solution of 0.32 M sucrose; 5 mM $MgCl_2$ — 0.2 mM $CaCl_2$ in a Potter-Elvehjem homogenizer (200 μg clearance, 50 ml capacity), Teflon pestle being mechanically operated. The homogenate is filtered through eight layers of gauze for removal of connective tissue, diluted four times with the same solution and then centrifuged at 1000xg for 10 min. For further removal of the contaminants, the pellet is suspended again in 5 vol of 2.2 M sucrose -5 mM $MgCl_2$ -0.2 mM $CaCl_2$, and after a few strokes in homogenizer, diluted further five times with the same solution and centrifuged at 15000xg for 1 h. The pellet finally contains purified nuclei. The method allows nearly 80% recovery of the nuclei. For pachytene nuclei separation of Syrian hamster multilayered discontinuous sucrose gradient method can be adopted.

Dipteran ovaries are suspended in a buffer mixture (0.1 M NaCl, 0.0019 M KCl, 0.001 M $MgCl_2$, 0.01 M Tris, 0.4% (w/v) Triton X100 of pH 7.2), slowly homogenized with a pipette, and suspension filtered through a gauze (69μm). The process is repeated several times with the residue. The total filtrate is layered on a two step Ficoll (polymer of sucrose) gradient (5 and 20%) and kept at 4°C. By 15 h, the follicle cell nuclei form a small band between the buffer and 5% gradient, and nurse cell nuclei accumulate at the interface or top of 20% Ficoll. The bands are separated with pipette and pellets of nuclei can be obtained by slow centrifuging (50xg for 4 min) at 4°C.

For polytene nuclei of Drosophila, several methods have been proposed including even mass isolation as they provide ideal materials for the study of gene action. The procedure is carried out at 0-4°C. Two sample schedules are given here.

(i) Isolation of nuclei in aqueous medium after Royd (1975 a):

Isolate salivary glands in 1-5 ml isolation buffer consisting of 0.11 M NaCl, 0.002 M KCl, 0.01 M Tris (hydroxymethyl) aminomethane, 0.0025 M $MgCl_2$ (pH 7.2 at 20°C), with 0.001% spermidine. Reduce the suspension into small masses by aspirating through a pipette. Add 0.2% Triton X 100 and continue gentle pipetting for 5 min or more to secure 95% of free nuclei without rupturing the wall. Filter through a nylon screen (53 μm opening), saturated with isolation buffer. Add 9 vol of isolation buffer through the tube and pipette slowly to free the nuclei. Centrifuge at 40 x g for 5 min at 4°C. The pellet consists of nearly 95% polytene nuclei.

For further purification, follow discontinuous sucrose gradient separation. Suspend the pellet in isolation buffer containing 1.67 M sucrose (15 ml/g of gland). Prepare a layer of 2 ml isolation buffer, 2.3 M sucrose under 10 ml nuclear suspension in the tubes. Slightly mix the interface and centrifuge at 40000xg for 42 min at 4°C (SW 40 rotor-Beckman a2-65B). Suspend the pellet by heavy agitation. Scrape off the nuclei from the bottom of the tube.

(ii) In a modified method for mass isolation of polytene nuclei of *Chironomus* salivary glands, the larvae are dried on filter paper, frozen immediately in liquid propane and stored either at −70°C or liquid nitrogen. Two to three g of larvae are broken in 2-4 mm pieces and filtered through 1000 μm mesh nylon grid. The filtrate is treated with pre-warmed (55%) SNKE (30% sucrose w/v in 100 mM NaCl, 5 mM KCl, 0.5 M EDTA, 10 mM Tris maleate, pH 6.3 per g of frozen tissue), and immediately cooled. The filtrate solution is introduced by a plastic syringe (30 ml) with an eccentric outlet into a specially prepared glass tube with six regularly spaced capillary constrictions. After 5-10 min in the glass tube, agitating at intervals, two-third of the supernatant is removed and the residue mixed with 5 vol cold NKE. The glands are collected after dissection and kept in ice-cold NKME, and centrifuged (2500xg) for 10 min.

The pellets are re-suspended in 1% digitoxin solution. Then 500 glands/ml in NKMC (80 mM NaCl, 5 mM KCl, 2.5 mM $MgCl_2$, 10 mM Tris maleate pH 6.3) are ruptured by passage through glass capillary (0.4 mm diameter). Contaminants are removed and nuclei pelletted again (2800 x g) for 5 min.

(iii) Isolation of nuclei in non-aqueous medium is based on the principle of replacing the Ringer with liquid nitrogen and freezing (Busch 1967).

Homogenize slightly the gland in Ringer, and discard the supernatant. Pour liquid nitrogen in the tube containing the pellet, against a dry ice acetone bath. Lyophilize and suspend frozen tissue in dried petroleum ether. Centrifuge for 10 min at 500 x g, pipette off the supernatant and evaporate the rest. Resuspend the tissue in equal vol of dry cold benzene and carbon tetrachloride. Centrifuge for 15 min at 12000xg. Dry the pellet and later hydrate the nuclei if necessary.

For tumour and cultured cells and for tissues from old animals, a slightly stronger method is adopted. For separation of nuclei from cytoplasm, low clearance homogenizers (50-77 μm) are necessary. Nuclear swelling is obtained through the use of hypotonic medium and suitable detergents like Triton x100 NP40, Tween 40, Tween 80, sodium deoxycholate; high pH and the use of citric acid.

For transformed cell lines the method used is as follows:

Take 300 falcon flasks (250 ml) or 25 roller bottles (0.59 gallon) containing at least 3 x 10^9 cells in culture. Pour off the original medium. Scrape the cells and collect in approx 1.0 1 of reticulocyte standard buffer (RSB) (0.01 M NaCl, 0.01 M Tris, pH 7.4 and 1.5 mM $MgCl_2$). Centrifuge at 1000xg for 10 min. Resuspend the cells in 150 ml of 1:3 dilution of RSB and keep for 5-10 min to secure optimum swelling without bursting. Transfer to a Potter homogeniser (77-100 μm clearance, 50 ml capacity) and completely disrupt the cells. Centrifuge the homogenate at 1000xg for 5 min. Wash the pellet of crude nuclei and unbroken cells with 150 ml of 1.3 RSB and centrifuge again. Resuspend the pellet in 300 ml of 2.2 M sucrose-1.5 mM $MgCl_2$ by slight mechanical stroking in the Potter homogenizer (200 μm clearance). Centrifuge the suspension at 15000xg for 1 h in an angle centrifuge. The pellet at the bottom of the tube contains almost purified nuclei with slight cytoplasmic tags. The recovery of nuclei is nearly 60-70%.

Isolation of Mitotic Apparatus:

The different steps principally involve synchronization of mitotic cells, stabilization of the mitotic apparatus, release through cell lysis, and fractionation of the suspension for separation of the released mitotic apparatus. Hexylene glycol (HG) is used to stabilise the spindle, 2-*n*-morpholino sulphonic acid for pre-lysis washing off the medium at a higher pH and calcium for checking spindle shrinkage and securing chromosome contrast.

B) For Plant Tissues

Wash fresh, young roots of leaves in chilled water, remove the midribs of leaves, weigh 20 g and remove to 4°C chamber for next stages.

Transfer the tissue to a Waring blender containing 120 ml extraction buffer (0.25 M sucrose, 20 mM Tris HCl, pH 7.8; 10 mM NaCl, 1 mM $MgCl_2$, 2.5% Ficoll and 5% dextran 40) and continue blending for 15 *s* (at Variac Setting 50 volts).

Moisten a flannelette of two layers with buffer. Filter homogenate through it and adjust the final volume to 120 ml. Centrifuge for 10 min at 2500xg. Separate the pellet and suspend in 10 ml extraction buffer. Slowly layer 5 ml of the suspension in a Corex tube (30 ml) on the upper surface of discontinuous sucrose gradient solution (7 ml 60%, 7 ml 50% and 7 ml 25% sucrose). Centrifuge for 30 min at 8000 r/min. The pellet contains the nuclei only. Suspend it in 4 ml extraction buffer. It can be precipitated with 2 vol 95% ethanol and stored at -20°C.

The method is also applicable to algae and protozoa.

B. ISOLATION OF CHROMOSOMES

Chromosome isolation often becomes necessary for identification of functional segments and mapping of gene loci. Individual chromosome isolation through micromanipulation is well suited for preparation of chromosome specific DNA library through cloning and amplification. Isolated chromosomes can be subjected to scanning electron microscopy as well.

For chromosome isolation, metaphase cell populations treated with colcemid are cooled at 4°C in fresh medium to inactivate trypsin, dissolve mitotic apparatus and remove residual colcemid. It is also necessary to incubate the pellet in cold buffer after centrifugation at 37°C for 10-15 min to allow the cells to be broken easily and to equilibriate the chromosomes with the buffer.

Hexylene glycol in the buffer prevents instability and disintegration. Calcium concentration is raised as otherwise there is dissolution of chromosomes. Deft handling is needed in syringing through a 22 gauge needle for rupturing the cell membrane. For different cell lines, the number of times the solution is to be passed through the needle and the force necessary to secure the desired result have to be standardised. All steps are continuously checked through phase contrast microscopy and the temperature should not be allowed to drop below 37°C before cell breakage. After the chromosomes are liberated, the remaining process is carried out at 4°C. Isolated chromosomes can be stored in a stable condition without disintegration at 4°C for several months. Detailed schedules are given at the end of this chapter.

Separation of DNA and Protein after Chromosome Isolation

After chromosome isolation, the initial step for the analysis of components is the separation of DNA and protein, and then histone and nonhistone moieties. The methods earlier tried for separation of the protein components involved (a) use of dilute mineral acids, (b) quantitation with polyacrylamide gel electrophoresis, and quantitation after initially fractionating chromatin proteins through chromatography on Bio Rex-70; different media, like hydroxyapatite; Sephadex and cellulose. Detailed representative schedules are given at the end of the chapter.

To eliminate limitations in these techniques, Sonnebichler *et al* (1977) suggested a method for analysis of chromosomal proteins after complete separation of DNA and protein from chromatin through high speed centrifugation with salt and urea.

The outline of this method is as follows:

To separate DNA from protein

Wash isolated chromosomes once with 0.024M EDTA, 0.075M NaCl (pH 7.0) and twice with 0.15M NaCl. Add 2M NaCl, 5M urea and 0.01M $NaHSO_4$ (for checking protein degradation), transfer the chromosomes to centrifuge tubes, and centrifuge in an angle rotor at 100 000xg for 35 h. The DNA forms a pellet and protein remains in the supernatant. Subject the pellet after separation to the same procedure to free it completely from protein. This pellet is now pure DNA. (For small amount of nucleoprotein, separation can be achieved by density gradient centrifugation with 2 M CsCl and 5 M urea for 70 h at 100 000xg).

Dialyse the supernatant for 3 h, reducing the salt concentration to 0.6M and precipitate protein with 6 vol of acetone. Wash the protein with pure acetone. Dry the purified protein in vacuum.

For separation of protein components, dissolve proteins in 1 mg/100 ml 1% acetic acid, 0.01m β- mercaptoethanol, and 8M urea. For electrophoresis use refrigerated teflon surface in a moist chamber.

Prepare 0.6 M ammonium borate buffer with 6M urea, 0.01M EDTA, 0.01 M mercaptoethanol (pH 10). The presence of strong urea checks histone-non-histone overlapping during electrophoresis. Adjust the buffer with 25% ammonia.

Equilibrate cellogel strips (4x17 cm) with buffer, put in the chamber, neatly blot with filter paper to prevent formation of air bubbles. Apply 1 mm of protein in 1 to 2 µl.

Use electric power of 60 v/cm for 1 h. Histones migrate towards the cathode and nonhistones move towards the anode or remain stationary. Slight nonhistones may move towards the cathode, but for their slow rate can easily be differentiated.

Stain cellogel strips with 0.5% amido black in 45% methanol, 45% water and 10% glacial acetic acid. Remove excess dye with successive washing with solvent for 15 min. For quantitation, cut the coloured areas from blank cellogel strips, dissolve in glacial acetic acid. Measure the intensities at 630 nm on the basis of the standards prepared with known quantities of protein on cellogel strips.

This method does not allow any loss of protein and the analysis gives an accurate assessment of protein composition. With low resolution electrophoresis for short duration, a series of samples can be analysed within a short period.

For separation of chromosome proteins after chromosome isolation by electrophoresis Wray (1977) used the same method as described for Chinese hamster and HeLa tissue cell line isolation. The cells are broken in a pressure vessel and chromosomes are purified through sucrose buffer gradient centrifugation at 1500 x g for 30 min. The isolated chromosomes can also be studied for ultrastructure analysis in electron microscope following critical point drying (see Chapter on electron microscopy).

For analysis of chromosome proteins, the method using polyacrylamide gel is as follows:

Dissolve chromosomes in 3% SDS (sodium dodecyl sulphate), 0.062M Tris (pH 6.8) at 100°C for 10 min and observe in spectrophotometer. Analyse chromosome protein in 9% polyacrylamide gel using Tris-glycine-buffered SDS system. Stain the gels with 0.05% Coomassie blue in methanol-acetic acid-water (40:7.5:52.5). Destain by diffusion in 7.5% acetic acid. Analyse the protein bands in spectrophotometer. For photography, develop in D-11 (Kodak) for 5 min at 20°C. Scan in a scanner.

Isolation and Separation of Heterochromatin and Euchromatin of Chromosomes

The disruption of nuclei is best achieved through removal of outer membrane. The method involves repeated suspension of nuclei in 0.01 M Tris buffer (pH 7.1), containing divalent calcium and magnesium, through stirring and centrifuging at 500 x g for 5 min. The quantity of buffer used is 100 ml/ml of nuclei. The treatment slightly differs with different tissues depending on the type and concentration of cations used in the initial homogenizing medium and Tris buffer. The suspension is to be prepared at least thrice.

If only calcium or both calcium and magnesium are used, repeated suspensions with heavy stirring for 30 *s* in each are necessary for removal of the outer membrane. It is always desirable to use in buffer the same cations originally used in homogenizing medium.

In order to disrupt the nuclei, and secure chromosomes and chromosome fragments for further separation through density gradient centrifugation, two methods are in vogue - sonication and passage through a French pressure cell. In both, nuclei treated with Tris buffer are suspended on 0.25 M sucrose (25 vol) and stirred with Potter-Elvehjem pestle for 30 *s* in plastic tube (clearance 1 mm, 20%). To secure nuclear swelling, the optical density at 420 nm is adjusted to 1-3 with 0.25 M sucrose and the mixture stirred gently for 20 min in cold. Fibrous materials are removed by passing through two layers of flannelette. The nuclei are periodically observed to detect optimal swelling which is indicated by a typical spherical shape, hyaline nature and size almost twice that of the original one.

In sonication method, a sonifier is used such as Branson sonifier operating at 7 to 11A, generating sonic waves at 20 kc/s. The sample is kept at 0-4°C. Nuclear suspensions in 15 to 18 ml aliquots are sonicated for 5 *s* bursts and the sonicate is periodically examined for disruption of all nuclei.

French pressure cell method is employed to disrupt nuclei by processing through a pressure cell. The technique in outline is as follows, the entire procedure being carried out at 2-4°C.

Homogenise the tissue in 10 vol of 0.25 M sucrose, with 5 mM NaCl and filter through several layers of cheesecloth. Stir the filtrate gently wih a magnetic stirrer. Take 40 ml of aliquot in a French pressure cell and apply pressure through the hydraulic press up to 7000 psi. Slowly open the needle valve of the cell for the homogenate to come out, without allowing the pressure to reach below 5000 psi. Take 20 ml of the pressate in 30 ml tubes (Spinco 25.1 rotor), follow a two layer system with different concentrations of sucrose and separate the nucleoli as pellet at 25000 r/min. Pipette out the upper layers and separate chromatin through high speed centrifugation.

For separation of euchromatin and heterochromatin, the chromatin suspension is filtered through two layers of flannelette and subjected to differential centrifugation or separation of chromatin fractions of different densities, referred to as heterochromatin, intermediate chromatin and euchromatin. As the compaction and amount of heterochromatin are variable from species to species, the degree of centrifugation also differs. Different velocities, namely 500xg, 1000xg, 4000xg, 6000xg, 12000xg, 20000xg and 78000xg per 30 min may be used and sediments collected for each fraction. The euchromatin part of very low density is obtained after the final centrifugation from the supernatant by making it 0.15 M with NaCl and precipitating with 2 vol of cold ethanol. The preparations can be periodically fixed with acetic-methanol and stained with Wright's stain.

For a precise estimate of the amount of heterochromatin present in the fraction, it is preferable to extract the DNA from the fraction and its satellite or highly repeated content through CsCl or Cs_2SO_4 -Ag^+ density gradient centrifugation. The distribution of satellite DNA in various fractions does not follow a set pattern and is species specific.

For isolation of different components of chromatin,

A method was devised in which isolation and sonication of mouse liver nuclei are carried out in a medium containing 2.2 M sucrose, 1 mM Tris (pH 7.5), 25 mM KCl, 0.9 mM $CaCl_2$ and 0.14 mM spermidine. After sonication and differential centrifugation, two heavy fractions (1000 x g, 3500 x g) containing nucleoli, nucleolar associated chromatin and extranucleolar heterochromatin, one intermediate fraction (78000 x g) containing nuclear membrane and other particles and a light fraction (105000 x g) with euchromatin fibrils only, can be separated. This method thus allows a separation of euchromatin and heterochromatin. Detailed representative schedules for isolation of DNA are given at the end of the chapter.

C. REPRESENTATIVE SCHEDULES FOR ISOLATION OF CHROMOSOMES AND DNA

A) Isolation of Chromosomes in Animals

Isolation of Mammalian Chromosomes at Alkaline pH

Add to exponentially growing fibroblast culture of Chinese hamster, 0.06 μg/ml colcemid, and keep for 3 h. Follow differential trypsinization to keep only metaphase cells. Centrifuge at 1000 r/min for 2 min and discard the supernatant. Inactivate trypsin by suspending in fresh medium. Keep at 4°C for 20 min for dissolving the spindle. Recentrifuge at 1000 r/min for 2 min and discard the supernatant. At 4°C, wash thoroughly in freshly prepared isolation buffer (pH 10.5), composed of 1.0 M HG/2×10^{-3} M $CaCl_2$/1×10^{-3}M CAPS (cyclohexylamino propane sulphonic acid). Adjust pH before addition of HG. Centrifuge at 2000 r/min for 3 min and discard the supernatant. Suspend at 4°C in buffer. Incubate at 37°C for 10 min in water bath. Follow gentle syringing by a 22 gauge needle and observe under phase contrast microscope. Centrifuge for 5-10 min at 3000 r/min. Discard the supernatant. The pellet contains isolated chromosomes (Wray 1973).

This method allows separation of short and long chromosomes through fractionation on a 20 ml sucrose buffer gradient (8-40%) and centrifugation for 30 min with a Sorvall HB-4 rotor at 2000 r/min. Several fractions can be separated from the top containing large and small chromosomes respectively. One of the difficulties often faced after centrifugation is the adherence of chromosomes on the walls and their separation as clumps. In order to avoid clumping, Wray (1977) developed a method using metrizamide-a-tri-iodinated benzamido derivative of glucose: (3-acetamido- metrizamide 5-N-methylacetamido-2,4,6 tri - iodobenzamido)-2-deoxy- 2-glucose). Metrizamide is used for layering on a preferred gradient, gradually allowing it to rise till the buoyant density is reached.

Treat the isolated mass for 5 min with dimethyl dichlorosilane in CCl_4(1%) and bovine serum albumin in 15 ml centrifuge tubes. Dry and treat in silicone at 140-160°F for 24 h. Prepare chromosome suspension in 1 ml of chromosome isolation buffer containing 0.75 M metrizamide and siliclad, ethanol, isoamyl alcohol and 5 mM disodium salt (NDA: 2-naphthyl-6-8-disulphonic acid). Centrifuge for 10 min at 10,000 r/min using HB-4 rotor (Sorvall). Fractionate the gradients on the top in 0.25 ml aliquots.

Isolation of Mammalian Metaphase Chromosomes by Extraction from Cell Culture at Neutral pH

Grow suspension cultures of HeLa strain S3 cells, mouse strain L- cells, Chinese hamster cells, strain V-79-379A from the lung of a female and a strain of Syrian hamster cells transformed by SV-40 virus, in Eagle's medium supplemented with non-essential amino acids and 5% fetal calf serum.

Add to these logarithmically growing cultures, vinblastine sulphate (Eli Lilly and Co.) to obtain a final concentration of 0.01 µg/ml for HeLa and Chinese hamster cells, and 0.5 µg ml for Syrian hamster and L-cells, incubate at 37°C for 15 h and 8 h, respectively, for the two former strains and 11 h for the two latter strains.

The selection of the optimum incubation period should be checked against the formation of micronuclei, because once formed, they are very difficult to separate from the chromosomes and the chromosome yield is thus lessened (Maio and Schildkraut 1969).

For hypotonic treatment and homogenization, harvest in each experiment for extraction, 4-12 1 of cell culture containing 2×10^9 to 6×10^9 cells. Centrifuge at 500 × g for 15 min at 10°C and wash the sedimented cells twice in Earle's BSS. Resuspend in a mixture (TM) containing 0.0001 M of each of the chlorides of calcium, magnesium and zinc in 0.02 M Tris, maintained at pH 7.0, in a proportion of cell suspension to medium (1:10) and keep for 20 min. Add 5% filtered saponin solution to make a final concentration of 0.05% and keep for 5 min. Transfer aliquots of the suspension to a 40 ml capacity Dounce homogenizer with pestle, and break up the cells to release the chromosomes.

To prepare crude chromosome suspension, add twice the volume of TM, containing 0.05% saponin, to the homogenate and transfer to centrifuge tubes. Fill about 3 cm of each tube and centrifuge for 5 min at 120 g. Decant the suspension, containing the chromosomes, and store. Resuspend the sediment in each tube by pipetting in TM solution to the original vol. Centrifuge and store the supernatant to extract the chromosomes left in the sediment. Repeat the process to extract most of the chromosomes; observing through a phase contrast microscope.

To obtain a purified extract of chromosomes, centrifuge the collected suspension in an anglehead centrifuge at 2500×g for 10 min. Wash the sediment in only 0.02 M Tris containing 0.1% saponin (pH 7.0), because the divalent metallic salts in the TM solution prevent the liberation of chromosomes from the debris.

Resuspend the sediment in 2.2 M sucrose in 0.02 M Tris (pH 7.0) and 0.1% saponin. Layer the suspension in cellulose nitrate in ultracentrifuge tubes over 10 ml of the dense sucrose solution (sp.gr.1.28). Stir with a glass rod to mix the two solutions, leaving about 1 cm of the sucrose solution undisturbed at the botton of each tube. Centrifuge at 50,000 × g for 1 h. Discard supernatant. The sediment contains the chromosomes (sp. gr. approx. 1.35) plus the few nuclei that are left as contaminant.

Resuspend the sediment with purified chromosomes in TM and observe under phase contrast. If the chromosomes are still mixed with impurities, repeat the process of washing in Tris-saponin buffer and centrifuging in dense sucrose. Remove the sucrose by suspending the chromosomes in TM. Centrifuge at 2500 × g for 10 min. Repeat three times. Store the suspension containing chromosomes in the TM solution at 0°C in ice buckets. Longer storage is recommended at -20°C, after adding glycerol, to obtain a final concentration of 20%.

For experimental purposes, the chromosomes can be maintained both in a contracted state and in monodisperse suspension in a buffer containing 0.005 M calcium chloride and 0.05% saponin in 0.02 M Tris at a pH of 7.0. Dispersion may be achieved by aspirating repeatedly through a syringe with a No 22 spinal tap needle.

From the HeLa metaphase cells, about 30-40% of the chromosomes are recovered. The average DNA content of isolated chromosomes has been estimated to be 0.5 μg per chromosome and therefore less than 10% of DNA is lost from the chromosomes during the isolation.

For permanent mounts

Add an equal volume of acetic-methanol (1:3) fixative to the suspension, mix thoroughly by aspirating through a pipette. Centrifuge at 2500 × g for 3 min. Resuspend the sediment in the fixative and centrifuge. Repeat several times. Disperse the final pellet in a small amount of fixative.

Prepare air-dried slides by spreading a drop of the concentrated solution over a grease-free slide and allowing it to dry. The preparations may be stained using the usual stains for chromosomes.

Isolation of Metaphase Chromosomes from a Large Population of Cells

Take Chinese hamster cells (2×10^{6}) in Roux culture bottles containing 90% Eagle's medium supplemented with 10^{-3}M sodium pyruvate, 2×10^{-4}M a-serine, $2\times^{-3}$M a-glutamine and 10% bovine serum.

After 48 h incubate the cells in the medium containing 0.025 μg/ml colcemid for 6 h at 37°C. Wash with Eagle's MEM to remove debris and collect the mitotic cells by pipetting. Chill at 4°C. To secure more mitotic plates, treat the rest of the cells with 0.025 μg/ml colcemid at 37°C for 4 h. Mix this population with the chilled ones after washing with Eagle's MEM. Repeat the process once more to secure large frequency of metaphases as compared to control harvesting, without colcemid.

After each step, take cell counts in haemocytometer, chromosomes being fixed in methanol and stained in Giemsa. Chilling decreases the plating efficiency from 4-8 h and there is a gradual increase in chromosome number. Addition of colcemid does not cause any damage or aberrations in chromosomes.

From the synchronized population, isolate metaphase chromosomes according to the method of Maio and Schildkraut (1969) described earlier with the following modifications:

Repeat centrifugation in the horizontal head of a centrifuge at 100-150 × g with gradual increase for 8 min while removing nuclei and unbroken cells. Pool chromosomes from the supernatant and sediment by centrifugation in horizontal head centrifuge at 550 × g for 10 min. Wash pellets by resuspending in 0.02 M Tris (pH 7.0) with 0.1% saponin. Centrifuge again at 550 × g for 20 min and collect purified chromosome populations.

B) Isolation of Chromosomes in Plants

Purification of chromosomes from plants may involve isolation of a single morphologically distinct chromosome through micromanipulation or the preparation of pure chromosome suspension. A modified method of the latter type (Schubert *et al* 1993) omits enzymatic digestion of cell walls and uses formaldehyde fixation before the rupture of cells to prevent stickiness of cytoplasm and aggregation of the chromosomes.

Protocol 1

Isolation of chromosomes by micromanipulation: Immerse freshly excised root-tips for 3 h in 0.05% colchicine solution, fix for 15 min in 45% acetic acid and immediately squash on coverslips by the dry ice technique. Dehydrate the preparations in 70% and 96% ethanol and air dry. It can be stored in glycerol for several weeks.

For micromanipulation, fill a chamber of 3 mm depth cut into a thick glass slide with liquid paraffin. Place the coverslip on the chamber with the squashed cells touching the paraffin oil, side by side with an empty siliconized coverslip which does not touch the surface of the paraffin oil and is positioned on the slide holder of the microscope (Edstrom *et al* 1987). Take up the chromosomes from the coverslip using tiny glass needles (with a 2 μm tip bent at an angle of 40°) of the micromanipulator, guided by a mechanical system laterally into the oil chamber and transfer onto the empty coverslip. Attach the coverslip onto a glass slide with Entellan. Wash the slides in chloroform and then 96% ethanol, to remove remnents of paraffin oil and air dry. Store in glycerol.

Protocol 2

Synchronization of meristems and preparation of chromosome suspensions

Incubate seedlings of *Vicia faba*, with about 2 cm long main roots, in aerated Hoagland solution containing 1.25 mM hydroxyurea, which blocks the cell cycle in S-phase, for 18 h at 24°C. Rinse in distilled water. Incubate root tips for 6 h in fresh Hoagland solution and for 3 h in 0.05% colchicine to arrest cells at metaphase. Rinse in distilled water. Fix the root tips in 6% (v/v) formaldehyde in 15 mM Tris buffer, pH 7.5 for 30 min at 4°C. Wash twice for 20 min in Tris buffer, at 4°C. Chop up the meristems of about 30 root tips with a scalpel in a petri dish containing 1 ml LB 01 lysis buffer (Dolezel *et al* 1989). The buffer contains 15 mM Tris-HCl, 80 mM KCl, 20mM NaCl, 2 mM disodium EDTA, 0.5 mM spermine, 0.1% Triton X-100, 15 mM mercapto-ethanol, pH 7.5. Pass the resultant suspension of released chromosomes and nuclei through a 50μm nylon filter to remove tissue and cellular fragments. Syringe twice through a needle of 0.7 mm diameter. In a glass tube, layer 0.7ml of the suspension on the top of 0.7ml 40% sucrose and centrifuge at 200rpm for 15min to remove nuclei and chromosome clumps. Carefully transfer the supernatant to Eppendorf tubes. The suspension contains upto 500 chromosomes per ml. Drop about 5μl portions immediately on clean ice cold slides. Air dry.

For scanning electron microscopy: Place one drop of chromosome suspension on a glass slide and cover with a coverslip. Remove coverslip by freezing on solid CO_2 or liquid nitrogen. Immerse the slides in 2.5% glutaraldehyde fixative buffer (50mM cacodylate, 2 mM $MgCl_2$, pH 7.2) and wash three times in buffer. Follow the usual schedules of osmium -TCH impregnation, dehydration, critical point drying and sputter coating (Wanner *et al* 1991).

In situ hybridization, *Giemsa banding* and *restriction enzyme banding* can be carried out on the air-dried slides. The slides can also be subjected to immunostaining of chromosomal antigens following the usual procedures described elsewhere.

This method gives a very good basis for production of chromosome- specific DNA libraries *via* microcloning and PCR amplification.

Protocol 3

Expose root-tips or fresh microsporocytes at room temperature to a solution of 0.05% colchicine, 2% cellulysin (Calbiochem); 1% macerase (Calbiochem); 0.25% pectinase (Sigma); 0.25% rhozyme (Rohm & Haas) and 13% mannitol at pH 5.7. The period of incubation ranges from 30 min for the meiotic tissue to 18h the mitotic tissue. Tease apart the digested material, pass gently through a pasteur pipette and incubate for another hour. Filter through several layers of cheesecloth to remove large debris. Collect the protoplasts by centrifuging at 200xg for 15 min. Wash twice with 20 vol. each time of 5mM 2 (*n*-morpholino) ethane sulphonic acid (MES) and 13% mannitol at pH 6.0. Resuspend in chromosome lysis buffer consisting of 15mM HEPES, 1mM EDTA, 15mM dithiothreitol (DTT), 0.5mM spermine, 80mM KCl, 20mM NaCl, 300mM sucrose and 500mM hexylene glycol at pH 7.0. Pass the protoplasts gently through a 27-gauge hypodermic needle 3 or 4 times until the cell membrane ruptures. Centrifuge at 200xg for 15min to remove cellular debris. Then centrifuge at 2500 × g for 10 min and collect chromosomes in the pellet.

For isolating chromosomes from cells in suspension culture, expose actively growing cells to 2μg/ml fluorodeoxyuridine and 1μg/ml uridine for one cell generation. Wash the cells in tissue culture medium supplemented with 2μg/ml thymidine to terminate the reaction. Then expose the cells to the colchicine- enzyme mixture as described above.

To confirm the yield of chromosomes, stain with specific dyes like Schiff's reagent and DAPI. In the lysis buffer, EDTA removes the divalent Ca,Mn and Mg cations which act as nuclease cofactors and also prevents cation-induced chromosome condensation. The polyamine spermine is added to prevent condensation. KCl and NaCl maintain the correct ionic equilibrium. DTT preserves the protein structure. Hexylene glycol helps to maintain protein structure and to rupture the membrane and disperse the chromosomes. Sucrose keeps interphase nuclei intact, preventing contamination of chromosome preparations. HEPES maintains the pH near neutrality (after Griesbach *et al* 1982).

Mass Isolation of Chromosomes from Plant Protoplasts

Culture cell suspensions from cell lines maintained in suitable medium. Culture suspensions in continuous light (3000 lx) on a rotary shaker (120rpm) at 25°C. Subculture at 2-day intervals, using 2ml of settled cells in 100ml culture medium.

For cell synchronization, harvest 5ml of settled cells from 1-day old suspensions and resuspend in 100ml fresh medium containing 2.5mM hydroxyurea (Sigma). Incubate for 24h. Remove hydroxyurea by three successive washes with fresh media. Incubate cells in fresh medium supplemented with 0.05% colchicine (Serva, Heidelberg) for another 11h on a shaker (180 rpm) in the dark. Alternatively, the step with hydroxyurea can be omitted.

For isolating the protoplasts, harvest colchicine-treated cells by centrifuging at 100g for 5min. Resuspend the pelletted cells in the supernatant medium in a ratio of 1:1. Mix cell suspension with an equal volume of enzyme solution [(5 mM $Mgcl_2$, 2mM $CaCl_2$, 3mM 2-*N*-morpholinoethane sulphonic acid, 170mM mannitol, 250mM glucose, 6% cellulase, (Onozuka R-10); 2% Rhozyme (Rohm & Haas); 2% pectinase (Serva); 5% Driselase (Fluka) at pH 5.6], containing 0.05 to 0.1% of colchicine. Incubate for 2-3 h on a shaker at 50rpm in the dark. Collect protoplasts by centrifuging at 100g for 3min. Wash pelletted protoplasts with a solution containing 100mM glycine, 2.5 mM $CaCl_2$ and 5% glucose at pH 6.0. Pellet protoplasts again by centrifuging at 100g for 3min.

For rupturing protoplasts, carry out the next steps on ice or at 0-4°C, unless otherwise noted and use siliconized glass tubes. Resuspend 2ml of pelletted protoplasts in 100ml of hypotonic glycine-hexylene glycol buffer (GHB) containing

100 mM glycine, 1% hexylene glycol at pH 8.4 - 8.6 adjusted by a saturated solution of calcium hydroxide. In some cases, it can be supplemented by 2.5% glucose. Incubate the protoplasts in GHB for 10min at room temperature. Chill the suspension in ice water. Add Triton X-100(Sigma) detergent from a 10% stock solution to a final concentration of 0.1%. Repeatedly pipette suspension to new tubes by plastic pasteur pipettes to disrupt protoplasts and liberate chromosomes into the suspension (after Hadlaczky 1984).

Purification of chromosomes: centrifuge suspension of ruptured protoplasts at 1000g for 20min. Resuspend the pellet containing chromosomes, nuclei and cellular debris in GH buffer supplemented with 0.1% Triton X-100 (GHT) and containing 1mM phenyl methyl sulphonyl fluoride (Sigma) and 1% isopropyl alcohol. Centrifuge at 200g for 10min. If necessary, repeat this differential centrifugation until nuclei and cellular debris are totally eliminated from the chromosome suspension. Mix the supernatant, which contains only chromosomes and slowly sedimenting contaminants, with a 1M sucrose solution in a ratio of 1:1. Layer onto top of a 1M sucrose solution made up in GHT buffer and centrifuge at 1000g for 20min. Repeat this step 2 or 3 times to purify the chromosome suspension.

For observation under light microscope, place a drop of sample on a slide, immerse in 0.1N HCl for 30 *s*, rinse in distilled water and stain with 2% Giemsa stain in Sörensen's phosphate buffer (pH 6.8) for 2-5 min at room temperature. Alternatively, place a drop of suspension on a slide, fix with acetic acid-ethanol (1:1), hydrolyse in 1N HCl at 60°C for 10min and stain with Schiff's reagent at room temperature for 1h.

For electron microscopy of whole mount preparations, spread concentrated chromosome suspension (10-15µl in GHT buffer) on a hypophase made from double glass distilled water. Pick up the samples on Formvar-coated 150-mesh copper grids and dehydrate through 30,50,70,90 and 100% ethanol, a mixture of ethanol and amyl acetate (1:1), and 100% amyl acetate. Dry the grids by a critical point dryer, using carbon dioxide and then coat with carbon in a vacuum evaporator. Observe by transmission electron microscopy.

For scanning electron microscopy, fix preparations onto a coverslip with 2.5% glutaraldehyde, dehydrate with ethanol series and dry by critical point drying, as described above. Cut coverslips carrying dry samples into small pieces with a diamond, mount on the preparation holder with a conductive paint and coat with gold.

Isolated chromosomes can be stored in GHT buffer in cold for over 48h.

C) Isolation of DNA from Animals

The majority of eukaryotic cells can be disrupted and lysed after thawing by freezing or by homogenizing in the extraction medium itself which may consist of sodium dodecyl sulphate (2%), paraminosalicylate (6%) and caesium chloride. Methods have been devised for different tissues such as mouse embryos, liver, spleen and testis; insect eggs and larvae, starfish oocytes; *Xenopus, Ascaris* embryos; ascites tumour and tissue culture cells with slight modifications whenever necessary (*see* Travaglini, 1973). Detergents and/or phenols are often used,

Method 1: Use of detergents

Suspend 2-3 g of cells in 25 ml STE (0.15 M NaCl, 0.05 M EDTA,0.05 M Tris, pH 8) and 2 ml 25% SDS. Incubate the mixture at 60°C for 10 min and cool at 20-25°C. Add 5 M sodium perchlorate to a final concentration of 1 M to the suspension, the whole mixture being shaken for 30 min with equal volume of chloroform-isoamyl alcohol. Centrifuge for 5 min at 3000-10 000 × g and separate the emulsion into three layers. Pipette out the upper aqueous layer and precipitate the nucleic acid by gently adding two parts of 95% ethanol. Spool the nucleic acid fibres with the aid of a stirring rod. Dissolve the precipitate in 10-15 ml 1 XSSC. Repeat the same procedure for deproteinization with chloroform-isoamyl slcohol several times, till no protein is noted at the interface. Precipitate the supernatant dissolved in the 1 XSSC (nearly two-thirds of the supernatant in volume). Add 50 μg/ml RNAase and incubate at 37°C for 30 min. Deproteinise again with chloroform-isoamyl alcohol, and precipitate the supernatant with equal volume of ethanol. Dissolve the nucleic acids in 9 ml of 1 XSSC. Add 1.0 ml of 3.0 M sodium acetate, 0.001 M EDTA (pH 7.0). Stir the solution and add 0.54 part of isopropanol dropwise. Precipitate DNA fibre.

Normally this is a useful method for extraction of DNA from eukaryotic nuclei but for tissues with high protein or polysaccharide content, repeated deproteinization with ethanol is substituted by combination with phenol or caesium chloride extraction methods.

Method 2: Use of phenols

Nucleases are inactivated by phenols, and when the latter are saturated with a lipophilic salt such as *p*-aminosalicylate, separation from protein can be carried out much more effectively than with SDS-chloroform octanol. Several combinations of phenol and salt are in use.

Add 600 ml of sodium-*p*-amino salicylate to 75 g of tissue and homogenise and break in a high speed mixer for 45 *s*. Filter the mixture through a Buchner funnel, to remove the debris and quickly add 600 ml of 90% phenol. Continue stirring for 1 h and centrifuge at 0°C (300 × g) for 1 h and remove the aqueous layer containing nucleic acid by suction. Wash the phenol and insoluble part with 6% sodium-*p*- aminosalicylate solution and separate the aqueous layer through centrifugation. Combine the two aqueous layers (400 ml), stir and add an equal volume of 2-ethoxyethanol. Remove the fibrous precipitate with a glass rod and dissolve in 100 ml of water. Add 6 g of sodium-*p*-aminosalicylate to the DNA solution and precipitate DNA again with 100ml of ethoxyethanol. The water-ethoxy ethanol mixture contains a flocculent precipitate of RNA. Dissolve quickly the DNA precipitate in 100ml water, add 4 g sodium acetate and precipitate DNA again with 100 ml 2-ethoxyethanol. Dissolve DNA in 50 ml water, add 2 g sodium acetate and 1-5 mg ribonuclease in 1 ml water, and keep the mixture at 2°C for 16h. Precipitate DNA again with 50 ml 2-ethoxyethanol and remove the solvent after precipitation, and then dissolve the precipitate in 33 ml water for 15-30 min. Add 33 ml 2.5 M dipotassium hydrogen phosphate and 1.65 ml 33% phosphoric acid followed by further addition of 33 ml 2-ethoxyethanol. Shake and keep at rest for separation of layers. Pipette out the topmost layer and centrifuge at 10000 × g for 1 h.Decant the clear organic layer, add a few drops of toluene and subject the mixture to dialysis twice against water and twice against sodium acetate taking 2 litres in each case. Remove the contents, centrifuge, make up to 4% with sodium acetate (100 ml) and finally precipitate DNA with 100 ml 2-ethoxyethanol. Take out the fibrous precipitate, wash with ethanol : water (3:1) twice and finally with ethanol and allow it to dry on calcium chloride in a vacuum desiccator.

DNA purification through 2-ethoxyethanol may be substituted by RNAse and amylase.

Precaution is necessary to ensure that all cells are disrupted during homogenization and the reagents used for extraction should be adequate. Extraction temperature should in no case be above 25°C. Moreover DNA is denatured even in presence of traces of phenol which lowers the *Tm* value. The tissue injury arising out of shearing forces because of the use of blender and vigorous shaking is also another disadvantage. Some of these limitations can be overcome with the use of a method in which both phenol and detergents are applied as follows:

Method 3: The technique combining both phenol and detergents has been applied for nucleic acid extraction from tissues of different organisms namely insects, amphibia, mammals and even plants. Suspend in 10 vol of buffer (w/v) at pH 7.6, Tris (0.05 M), KCl (0.025 M), Mg acetate (0.005 M), sucrose (0.35 M) and homogenize at 0°C.

Filter the homogenate through eight layers of gauze. Centrifuge the filtrate at 700 × g for 10 min. Suspend the pellet in 0.15 M NaCl, 2% SDS and 0.1 M EDTA and adjust to pH 8. Gently agitate for 10 min at 60°C as shaking may cause shearing of DNA.

Follow subsequent procedures as mentioned above for *detergent method* using 1 vol of cold ethanol instead of 2 vol. Collect the fibres in a glass rod. Centrifuge the residual flocculent DNA precipitate. Resuspend in 1 X SSC with the addition of 1 vol of ethanol. Collect the fibrous precipitate. Treat the fibrous DNA dissolved in SSC with RNase (150 μg/ml) at 37°C for 4 h, to remove RNA. To remove polysaccharides, digest the preparation with α-amylase (250 mg/ml) for 45 min and with pronase (50μg/ml) for 30 min both at 37°C. Treatment with pronase also removes the interfacial protein as far as practicable.

Add 1% SDS to the digest and treat twice with phenol, at 25°C. Deproteinize twice for 10 min each with chloroform- isoamyl alcohol for 10 min. Remove chloroform and phenol by shaking with ether. Precipitate DNA in 2 vol of ethanol.Collect the fibrous DNA and dissolve in 0.01 XSSC.

Method 4: Density gradient centrifugation for extraction

The method is based on the principle of separation of cell fractions through density gradient centrifugation. Several media such as sucrose, caesium chloride, Ficoll (a sucrose polymer), dextran, potassium tartrate, sodium bromide as well as silica gel have so far been used. The change in osmotic pressure as well as the induced chemical toxicity are often common problems with these techniques. For tissues with heavy cytoplasmic content, caesium chloride method for DNA extraction is suitable for separating different nucleic acids, polysaccharides and proteins.

Homogenize embryos (2 ml/10 ml of CsCl) in 4 M CsCl (density 1.40 g/ml) at 4°C. Centrifuge the homogenate at 40 000 r/min for 24 h at 20°C to equilibrium; polysaccharides and nucleic acids form a pellet. Separate the pellet and dissolve in 1 XSSC (9 ml SSC, to pellet from 2 ml egg white). Digest successively in the following enzymes for 1 h each at 37°C: (a) 0.01 vol α- amylase (10 mg/ml), (b) 0.1 vol ribonuclease (0.01 mg/ml), (c) 0.01 vol pronase (50 mg/ml). Centrifuge initially at low speed to remove insoluble material and then at high speed (40000 r/min) for 1 h at 4°C to pellet DNA. Dissolve DNA in 1 XSSC and store at −30°C.

Due to the use of strong concentration of CsCl which inhibits the action of nuclease, the method does not allow enzymic degradation of DNA. It may sometimes be necessary as in sea urchin embryo to shake again with chloroform-isoamyl alcohol before the use of enzymes to remove all traces of nuclease.

Method 5: Column methods for extraction

The methods are simple as compared to the previous procedures and involve preparation of the cell lysate and passing the lysate through a column for selective absorption from which the sample can later be eluted.

This method utilises absorption on hydroxyapatite (HAP) for separation of DNA directly from lysates. The technique involves, in addition to buffer, the use of (a) urea, needed for disrupting the cell, denaturing chromosome proteins, and inactivating enzymes, the last two properties being shared also by (b) sodium lauryl sulphate (SLS) and (c) ethylene diamine tetraacetate (EDTA)—the chelating agent for binding bivalent and polyvalent metal ions. Of these, SLS is normally purified by recrystallization from hot ethanol, dried after ether washing and stored as 25% at 4°C in a solid form. The only limitation of the technique is the limited capacity of HAP which can recover only a small amount of DNA from a large mass of tissue. The technique is as follows:

Suspend the tissue in a mixture of 8 M urea, 0.24 M phosphate buffer, 1% SLA and 0.01 M EDTA and homogenise in a blender. Pass the homogenate on HAP, and stir to check channelling. Wash HAP with heavy amount of urea buffer mixture (8 M urea, 0.02 M phosphate buffer). Wash with 0.14 M phosphate buffer to remove urea. Elute DNA with 0.04 M phosphate buffer.

Hydroxyapatite can distinguish between single and double stranded DNA at a wide range of temperatures. In view of this property, both thermal denaturation and reassociation kinetics analysis (*Tm* and *Cot*) are performed through the use of this column. The binding is not affected by the inclusion of solubilizing agents such as urea and detergents, which is an added advantage.

Take slurry of hydroxyapatite (Clarkson Chemical Co) stored in refrigerator, warm to 22- 24°C and pour a column (2 × 2 cm). Wash with 40 ml 0.05 M phosphate buffer prepared through equimolar proportions of NaH_2PO_4 and Na_2HPO_4. Prepare a sample of 10 ml in phosphate buffer (0.05 M) of 50 000 acid precipitable counts/min of tritiated sample DNA (crude lysate or purified form) and 10000 counts/min of heat denatured phosphorus labelled *E. coli* DNA. For crude lysate, buffer should be prepared in 8 M in urea. Add sample to the column and run 25 ml of 0.05 M phosphate buffer through the column. For crude lysate the first 5 ml of buffer should be 8 M urea. Run a continuous gradient and elute DNA using 100 ml each of 0.05 M and 0.35 M buffer and collect in 5 ml samples. Store all fractions in chilled state.

Count the refractive index (n) of all fractions for determination of salt concentrations and plot the refractive index against buffer concentrations (0.05 M and 0.35 M) for the stock solutions. Add to each fraction 0.2 ml of 2.5 mg/ml herring sperm DNA and then 2.5 ml cold 40% PCA which would form an acid precipitate. Filter the samples individually in Whatman (F/C) glass fibre filters, wash thrice with a total of 15 ml 5% TCA and lastly with 5 ml 95% ethanol. Dry the filters and count in a scintillation counter. Plot the data on a graph with counts/min against fraction number and buffer concentration against fraction number. Elution profiles of single and double stranded DNA are quite distinct.

Isolation of DNA by SDS—Proteinase K Treatment

This technique is applicable to most material and relies on proteinase K and SDS to dissolve the sample and digest the protein component without affecting the DNA (after Brown 1991).

A. Lysed cells, nuclei, and monolayers

Adjust fluid samples to 1% (w/v) SDS and 0.5 mg ml^{-1} proteinase K. Drained flasks containing confluent monolayers of tissue culture cells are incubated with 10 ml of digestion buffer [1% (w/v) SDS, 0.5 mg/ml^{-1} proteinase K; 50 mM Tris— HCl pH 9.0, 0.1 M EDTA, 0.2 M NaCl]. Incubate the samples at 55°C for 3-16 h with very gentle shaking . Add 1.0 vol. of phenol very carefully for 3 h at room temperature. Transfer the mixture to a Falcon tube and centrifuge at 4000 xg for 10 min at 25°C. Discard the lower phenol layer using an aspirator connected to a long pasteur pipette which passes through the aqueous phase. Add 1.0 vol. of phenol-chloroform, gently invert to mix the liquid phases, then centrifuge at 3000 *g* for 10 min at room temperature. Transfer the upper aqueous layer to a fresh tube, place on ice for 5 min, then add 2.5 ml 7.5 M ammonium acetate and 10 ml ethanol at 20°C. Hook the clump of DNA from the tube using a pasteur pipette, rinse with 70% (v/v) ethanol and then dry under vacuum. Place the sample in a microfuge tube with 1 ml TE pH 7.5 and allow to dissolve overnight.

B. Pulverized tissue samples

Transfer the frozen powder from 5-10 g tissue in small portions on to the surface of 20 ml buffer (50 mM Tris—HCl pH 9.0, 0.1 M EDTA, 0.2 M NaCl, 0.1 mg/ml^{-1}RNase A) in a 0.5-litre beaker. Swirl for 10 min, then transfer the homogeneous liquid to a Falcon tube. Adjust to 1% (w/v) SDS and 0.5 ml^{-1} proteinase K. Incubate for 16 h at 55°C.

Transfer the sample to a flat-sided bottle or a 1-litre beaker and add 20 ml of phenol. Leave at room temperature for 3 h, occasionally agitating the mixture. Place the sample in a Falcon tube and centrifuge at 3000*g* for 10 min at 20°C to separate the phases. Transfer the aqueous layer to a 30-ml corex tube and centrufuge at 10000*g* for 20 min at 25°C to sediment any undissolved materials. Dialyse the supernatant against 2 litres of TE pH 7.5, exchanging the buffer twice. Transfer the dialysed DNA sample to a 0.1-litre beaker and add 0.1 vol of 3 M sodium acetate pH 6.5 and 0.8 vol. isopropanol. Mix gently with a pasteur pipette and hook the clumps of DNA together. Rinse with 70% (v/v) ethanol and transfer to a 5 ml bijou bottle. Dry the sample under vacuum for 10 min to remove ethanol. Place 2.5 ml TE (pH 7.5) into the bottle and leave for several hours to dissolve the DNA.

Isolation of DNA from Pupa of Insects

Grind 3-day old pupae of insects (42.5 g) with a mortar and pestle at room temperature in 500 ml of Buffer A (0.05 M Tris-HCl pH 8.0; 0.1M disodium EDTA, 2% SDS and containing 50 ml toluene). Pour the brei into a silica-clad flask and shake gently on a rotary shaker (100 rpm) for 48h. Add solid $NaClO_4$ to a cencentration of 0.5M. Remove protein by extracting twice with an equal volume of chloroform: isoamyl alcohol (24:1). Centrifuge at 8000g for 30min at 20°C.

Precipitate the DNA from the supernatant aqueous layer with ethanol. Centrifuge. Resuspend the DNA pellet in 50 ml Buffer B(0.05M Tris-HCl, pH 8.0; 0.1M disodium EDTA; 0.15M NaCl) at 4°C. Add heat-treated RNAase A at 50µg/ml. Digest at 37°C for 2h. Add proteinase K at 200µg/ml. Incubate at 37°C for another 2h. Extract with chloroform-isoamyl alcohol and precipitate with ethanol as before. Resuspend pellet in 10ml of TE (10mM Tris-HCl, pH 7.6; 1mM EDTA). Precipitate with ethanol, resuspend in 5 ml of TE and again precipitate with ethanol as described before. Store at – 20°C.

For obtaining highly repetitive DNA, grind 20g of 7 day old pupae at room temperature in 250 ml Buffer A (including 0.15M NaCl). Pour into 250 ml polypropylene bottles and sonify using a Branson S-75 sonicator for 10 *s*. Pour into a silica-clad flask and shake for 72h at room temperature. Add $NaClO_4$ as before. Decant supernatant and extract as for the 3 day old pupa. Digest in Proteinase K at 37°C overnight. Resuspend DNA pellet in 15 ml TE and run through CsCl-ethidium bromide gradient. Collect the DNA fractions and clean by passing through urea-phosphate hydroxyapatite column at room temperature (after Hershfield and Swift 1990).

D) Isolation of DNA from Plants: Minipreparation

Rapid microscale method for isolation of plant DNA without ultracentrifugation with CsCl, adapted in higher plants (Dellaporta *et al* 1983) from the procedure commonly used for yeast DNA preparation (Davis *et al* 1980).

Weigh 0.5 to 0.75 g of leaf tissue, quick freeze in liquid nitrogen and grind to a fine powder in a mortar and pestle. Transfer powder and liquid nitrogen to a 30ml Oak Ridge tube. Add 15ml extraction buffer (EB = 100mM Tris pH 8.0; 50mM EDTA pH 8.0; 500mM NaCl; 10mM mercaptoethanol). Add 1.0ml of 20% SDS, mix thoroughly and incubate at 65°C for 10min. Add 5.0ml potassium acetate. Shake tube vigorously and incubate at 0°C for 20 min. Centrifuge at 25000g for 20 min. Transfer supernatant through a Miracloth filter (Calbiochem) into a clean 30ml tube containing 10ml isopropanol. Mix and incubate at – 20°C for 30min. Centrifuge at 20,000g for 15min. Remove supernatant and dry pellets by inverting the tubes on paper towels for 10min. Dissolve pellets with 0.7ml 50mM Tris, 10mM EDTA, pH 8.0. Centrifuge in a microfuge for 10 min to remove insoluble debris. Transfer supernatant to new tube, add 75µl 3M sodium acetate and 500µl isopropanol. Mix. Centrifuge in a microfuge for 30 *s*. Wash pellet with 80% ethanol, dry and dissolve in 100µl 10mM Tris, 1 mM EDTA, pH 8.0.

For difficult materials, like soybean, instead of the last step with sodium acetate and isopropanol, add 50µl of 3M NaOAc and 100µl of 1% CTAB (Cetyl trimethylammonium bromide), which will precipitate the nucleic acids. Pellet the precipitate for 30 *s* in microfuge and wash with 70% ethanol. Redissolve pellet in

400μl TE. Precipitate the DNA with 50μl 3M NaOAc and 1ml ethanol. Repeat this step to remove residual CTAB, leaving DNA in sodium form. Redissolve dry DNA pellet in 10mM Tris, 1mM EDTA pH 8.0 per g starting material. Such miniprep DNA can be stored for months and cut with a variety of restriction enzymes. However heat-treated RNase is needed to digest contaminating RNA. A typical reaction would contain Miniprep DNA 10.0μl; 10X restriction buffer 3.0μl; 0.5mg/ml RNAase 2.0μl; EcoRI 8.0 units and distilled water to make upto 30μl. Digest for 3h at 37°C.

Isolation of DNA by SDS—Phenol Extraction

The technique uses SDS with phenol to denature and dissolve macerated hyphae leaving the DNA intact, which is precipitated from solution with isopropanol. It is mainly employed for plant materials and fungi. Collect hyphae by filtration of a fungal culture through a Buchner funnel. Rinse with 20 mM EDTA pH 8.0, remove liquid, freeze in liquid N_2 and lyophilize. Grind the dried material in a small mortar. Resuspend 50 mg in a microfuge tube by stirring in 0.5 ml extraction buffer [0.2 M Tris HCl pH 8.5, 0.25 M NaCl, 25 mM EDTA, 0.5% (w/v) SDS]. 0.35 ml phenol and mix by inverting the tube several times. Add 0.15 ml chloroform and again invert to mix. Centrifuge for 1 h at 15 000*g*. Immediately transfer the upper aqueous layer to a tube containing 25 μl RNase A (20 mg ml^{-1}) and incubate for 10 min at 37°C. Add an equal volume of chloroform-isoamyl alcohol, mix, and centrifuge for 10 min at 15 000 *g*. Transfer the aqueous supernatant to a sterile microfuge tube and record its volume. Add a 0.54 vol. of isopropanol to the sample and invert to mix. The DNA precipitates and forms a clump in the tube. Centrifuge by pulsing for 5 *s* and remove the liquid using a pasteur pipette. Rinse the tube contents with 70% (v/v) ethanol, and recentrifuge to settle the pellet of DNA. Remove the liquid and dry the sample in vacuum. Dissolve the DNA in 100μl TE pH 8.0 by incubating at 4°C for several hours (see Brown 1991).

Isolation of Total Cellular DNA—CTAB Procedure

Prepare CTAB isolation buffer, according to the type of sample used (from Mark Chase at Kew). Preheat 10 or 20ml of isolation buffer containing 40 or 80μl of betamercaptoethanol in 50ml Blue Cap tubes in a 65°C water bath. Grind 0.5 to 1.5g of fresh leaf tissue in a mortar and pestle, preheated to 65°C, using a portion of the isolation buffer. Add remaining buffer and shake to suspend the slurry. Pour into the 50ml Blue Cap tube, incubate at 60-65°C for 15-20min. Alternatively, grind frozen leaf tissue in precooled mortar and pestle in liquid nitrogen and transfer to a tube containing isolation buffer. Shake to suspend and incubate at 60-65°C for 15-20 min. Extract once with equal volume of SEVAG (24:1 chloroform : isoamylalcohol), mixing gently but thoroughly. Extract for upto 30min for mucilaginous samples. Centrifuge at setting 7 in IEC clinical centrifuge for 5-10 min.

Transfer aqueous top phase containing DNA with a Pasteur pipette to a Blue Cap tube. Add 2/3 volume isopropanol at – 20°C and mix gently to precipitate DNA. Keep at – 20°C for 30-60 min. Centrifuge at speed 7 for 10min. Discard supernatant. Add 3ml wash buffer (70% ethanol, 10mM ammonium acetate). Shake to suspend pellet and wash for 5-60 min. Centrifuge DNA at speed 5 for 5min, remove supernatant and allow pellet to dry by evaporation of ethanol. Resuspend DNA in 3ml TE buffer (10mM Tris-HCl pH 8) and 0.25mM EDTA). Heat, if needed, to 65°C for a few min to suspend DNA. In case of DNA not dissolving, add 3.4g CsCl and continue to incubate at 65°C for upto a few hours. On the dissolution of the pellet, proceed with gradient centrifugation. Alternatively, for immediate restriction enzyme digest, resuspend DNA in 0.5-1.5 ml TE (see Doyle and Doyle 1987).

** For most samples; 2X CTAB buffer = 100mM Tris-HCl pH 8.0; 1.4M NaCl; 20mM EDTA, 2% CTAB (hexadecyltrimethyl ammonium bromide). For samples with very high water content, use 3XCTAB buffer. For samples with abundant mucilaginous polysaccharides use Wendel's CTAB = 100mM Tris-HCl pH 8.0; 1.4M NaCl, 20mM EDTA, 2%CTAB, 2% PVP 40 (polyvinypyrrolidone); 50mM ascorbic acid; 40 mM DIECA.

CTAB Total DNA Isolation in Plants

Rapid inexpensive method for total genomic DNA (nuclear, chloroplast and mitochondrial) for a wide variety of plant groups and some animals (after Doyle 1991).

Preheat 5-7.5 ml of CTAB Isolation buffer (2%CTAB Sigma H-5882; 1.4M NaCl; 0.2% 2-mercaptoethanol; 20mM EDTA; 100mM Tris-HCl, pH8.0) in a 30ml glass centrifuge tube to 60°C. Grind 0.5-1.0g fresh leaf tissue in liquid nitrogen in a chilled mortar and pestle and transfer to preheated buffer. Alternatively grind fresh tissue in 60°C CTAB solution in preheated mortar. Incubate sample at 60°C for 30min with occasional gentle swirling. Extract once with chloroform - isoamyl slcohol (24:1), mixing gently but thoroughly. Centrifuge at room temperature at 6000g for 10min. Remove supernatant, transfer to clean glass centrifuge tube, add 2/3 vol cold isopropanol, mix gently to precipitate nucleic acid. If possible spool out nucleic acids with a glass hook and transfer to 10-20ml of wash buffer (76% ethanol, 10mM ammonium acetate). Alternatively, centrifuge at low speed for 1-2min, remove supernatant, add wash buffer directly to pellet and resuspend by gently swirling. After 20min of washing, spool out or centrifuge nucleic acids (6000 rpm, 10min). Remove supernatant and dry residue briefly in air. Resuspend nucleic acid pellet in 1ml/TE (10mM tris-HCl, 1mM EDTA, pH7.4). Add RNAse A to a final concentration of 10μg/ml and incubate at 37°C for 30min. Dilute sample with 2 vol of distilled water or TE, add ammonium acetate (7.5M stock, pH 7.7) for a final concentration of 2.5M, mix, add 2.5 vol. of cold ethanol and gently mix to precipitate DNA.

Localization of Specific Repetitive DNA Sequences in Plant

Isolate total DNA from *Oryza sativa* var. IR36 and construct a lambda EMBL4 rice genomic library as described by Xie and Wu (1989). Isolate genomic clones from the library by using the repetitive sequence pOs48 (Os stands for *0. sativa*) as the probe(Wu and Wu 1987). For gel blots, digest DNA with EcoRI and SalI, fractionate by agarose gel electrophoresis and blot onto a nitrocellulose filter. Hybridize the filter to a 32_P-nick- translated pOs48 DNA fragment. Perform DNA sequencing by using the di-deoxynucleotide chain termination procedure adapted to single-stranded M13 phage DNA. (after Wu *et al* 1991).

Determine the copy number of repetitive DNA sequences by using slotblot hybridization with defined amounts of total rice DNA and recombinant plasmid DNA (Zhao *et al* 1989).

For chromosome preparations, disinfect dehulled rice seeds with 0.75% sodium hypochlorite for 10min, wash in running water for 30min; soak in tap water for 26h at 30°C. Transfer to petridish lined with cheesecloth and incubate at 30°C for 6h. Immerse two-thirds of the height of the seeds in 15 or 20mM 2′-deoxyadenosine for 16h at 30°C and wash thoroughly with distilled water. Immerse the roots and wash in distilled water. Pretreat roottips in 1.5mM 8-oxyquinolne for 2h at 20°C. Wash in distilled water, keep in a solution of 6% pectinase and 6% cellulase in 75mM KCl, pH 4.0 for 60min at 37°C. Wash with distilled water, fix in glacial acetic acid-methanol (1%), smear on a glass slide and flame dry.

Label DNA probe with 3_H- deoxynucleotides using nick translation. Specific activity of the probe should be approximately 4×10^7 cpm/μg DNA.

Incubate slides with metaphase preparations with 200μl of 100μg/ml RNAse per slide, cover with a coverslip, at 37°C for 1h. Denature with 70% formamide, 2 X SSC (0.30M NaCl, 0.03M sodium citrate, pH 7.0) at 70°C for 2min. Dehydrate through ascending ethanol series and airdry.

For *in situ* hybridization, apply 40μl of labeled and denatured DNA probe on each slide. Cover with a coverslip and keep in a moist chamber for 8 to 16h at 37°C. Wash three times, 2min each, with 50% formamide, 2 X SSC, pH 7.2 at 39-40°C and then five times (2min each) with 2 X SSC at 39- 40°C. Dehydrate by passing the slides through ascending ethanol series, keeping 2min in each. Air dry. Dip the hybridized slides in Ilford K2 emulsion diluted with 0.5% gelatin (1:1). Expose the slides at 5°C for 7 days. Develop in Kodak D19 (diluted two-fold with water) solution, fix in a Kodak Fixer at 15°C, stain with 4% Giemsa in phosphate buffer and rinse with tap water.

CHAPTER 11

MICRURGY

The isolation procedure is often adopted, not only for the chemical analysis of chromosomes but also for culturing the nuclei and chromosomes in natural or synthetic media as well as for cloning. The latter procedure requires micrurgical operation of the cell without the use of any chemical. It allows a study of the initial steps in chromosome metabolism and of the mechanism of genetic regulation of differentiation. On the other hand, the chemical method of chromosome isolation, in which only non-injurious compounds are employed, is useful for the study of the chemical make-up of the chromosomes at different stages of development.

MICRURGICAL METHOD OF ISOLATION OF NUCLEI, CHROMOSOMES AND MICRODISSECTION

The technique for micrurgical isolation was initially developed with respect to the salivary gland chromosomes of Diptera, where puffing at different segments in different phases of development provided adequate proof of the genetic control of differentiation and the change of pattern following treatment with different agents. Short term *in vitro* culture can be carried out following the hanging drop method, involving culturing in a drop of medium on the cover slip inverted over a depression slide, and sealed with oil, or on a slide covered with oil. Blowing through a pipette is recommended for adequate oxygen supply. The best medium for hanging drop culture is no doubt haemolymph but several other media, including TC 199 and Jones and Cunningham's medium for sciarids and chironomids, respectively, can also be used. One of the serious limitations of polytene chromosome culture is the fact that salivary gland cells have a strong ionic barrier against haemolymph, whereas owing to the presence of a special membrane, there is no such barrier between adjacent cells. Consequently, in the case of any cell leakage in haemolymph or lumen, there is a complete loss of ionic equilibrium, because all the cells are simultaneously affected. Therefore, when excising the gland, extreme care must be taken not to rupture, or shear, the ligaments adjacent to it. Moreover, because of the low regeneration rate of glands after injury, this method allows a study of DNA synthesis and transcription only, not of the entire process of mitosis.

ISOLATION AND CULTURE OF NUCLEI IN MEDIUM UNDER OIL

The essential requisites are, (a) a suitable medium, (b) a suitable oil and (c) siliconized slides. A good dissecting microscope serves the purpose for observation. Of the media used, the egg contents of *Drosophila* have been found to be very satisfactory. They have to be diluted at the later stages and culture can be prolonged even up to 4 h. Sugar medium with synthetic compounds, having the following composition, given very satisfactory results: Saccharose, 64.1805 g; Glucose, 3.3741 g; $MgCl_2$, 1.7866 g; NaCl, 1.6659 g; Tris buffer 0.51 (3.025 g Tris + 20.7 ml 1/N HCl) supplemented with polyvinylpyrrolidone or Luviskol-K 90. The oil needed to cover the culture for checking against desiccation is hydrofluorocarbon oil (Kal-F No. 10), the viscosity of which can be adjusted by mixing with paraffin. Another oil, often employed, is mixture of heavy mineral oil and Oronite Polybutane 128 (2:1).

The slide can be siliconized by dipping in a mixture of a few drops of silicone oil in 250 ml of acetone and drying for 24 h at 20-25°C. If required, the time, temperature and concentration can be varied.

The schedule followed for isolation and culturing is given below:

Cover the donor tissue with a drop of oil on a siliconized slide. From a second oil drop, on the slide containing the culture medium, transfer two spheres of medium to the first oil drop containing the donor tissue. Of the two spheres, sphere A should be about one-fifth to one tenth the volume of the other sphere, and B should be about 20 times larger in volume than the donor tissue. The transfer is carried out under a dissecting microscope, after bringing the medium in sphere A in contact with the donor tissue so that the medium in sphere A forms just a rim surrounding the donor tissue. The method is as follows:

For semi-isolation, gradually cut off and separate the cytoplasm and other extranuclear components from the nucleus in A, with the aid of a bent tungsten needle, leaving the nucleus surrounded by a very thin layer of cytoplasm. The tungsten needle must be sharpened previously by immersion in a hot mixture of potassium nitrate and sodium nitrite. Glass needles, drawn out in a gas microburner, can also be used. Draw the large droplet B and join it with A. Push the nucleus into the larger sphere B and finally cut off the connection between the two spheres.

Alternatively, *for complete isolation*, puncture the cell with a glass needle and squeeze out the nucleus and follow the procedure given for semi-isolation schedule.

For staining, after the incubation period, place a large drop of acetic-orcein solution on the material, avoiding conglomeration of cells, and stain for 20 min or more. Pass through acetone to remove the oil, rinse in water, blot off excess fluid, add a drop of acetic-orcein solution, wait for a few min and mount under a cover slip. Lactic-orcein solution, used alternatively, may be prepared by boiling 2 g of orcein in a mixture of 50 ml each of acetic acid and lactic acid, shaking and filtering. If necessary, a contrasting stain may also be applied, prepared by mixing light green (FS) 0.1% in 96% ethanol, and orange G, 0.2% in 70% ethanol, in the proportion of 55:45, followed by adding 1-2 drops of acetic acid to bring to pH 5.0. The method has also been extended to ultrastructure analysis from suspension culture.

Isolation of Chromosomes

To secure fixed chromosome preparations, the method is quite a simple one. The chromosomes are fixed as usual in 45% acetic acid and then removed from the cell with a needle.

To obtain unfixed chromosome preparation, the most convenient method is to puncture the salivary gland with a needle and squeeze out the chromosomes in sugar medium. There are two other methods, one involving treatment of the gland with pronase followed by homogenization and differential centrifugation whereas in the other, the glands are dipped in 0.25% solution of dried eggwhite for 2-3 h and then the stiffened chromosomes are isolated with needles.

In *transplantation experiments*, foreign cytoplasm is inserted within the host cell by piercing the cell without touching the nucleus, pressing the cytoplasm through the oil and then the slit with the help of a needle. Similarly, chromosomes can be donated to the host cytoplasm by taking out a chromosome, rolling it into a bundle and pushing it through the slit of the host cell. It is even possible to make a slit in the nucleus and transplant the chromosome inside the nucleus. The degree of perfection achieved in these procedures depends on experience, skill and steadiness of operation. Microsurgical methods have also been employed to secure mammalian somatic cell hybrids and their analysis and cloning.

MICRODISSECTION OF CHROMOSOMES

Microdissection of chromosomes in recent years has become a very elegant technique for chromosome manipulation and fragment cloning. Undoubtedly mechanical microdissection for microcloning is possible, but it is time consuming and depends to a great extent on the skill of the operator. Moreover, the harvesting of specific fragments may present difficulty. Microdissection of polytene chromosomes of Drosophila (Scalenghe *et al.* 1981) with fine glass needle and of human chromosome with laser beam (Monajembashi, 1986) has facilitated cloning of specific gene sequences. Lately, it has been shown that microdissection can be done more precisely if a laser coupled with microscope is used. Utilizing laser microbeam generated by oxcimer pumped dye laser, the chromosomes could be cut into less than 0.5 μ of polytene chromosomes (Ponnelles *et al.*, 1989). This method, followed by microcloning, can be utilized for microcloning of specific segments as done for telomeric sequences in Drosophila. It is also possible to isolate a single band of chromosome.

In the plant system, micromanipulation and quantitative chromosome map through image analysis have enabled dissection of specific regions of chromosomes and cloning of site-specific DNA sequences (Fukui *et al.*, 1991). An elegant method has been devised to secure fragments of chromosomes from barley, rice and *Crepis*, utilizing computerized argon ion laser dissection (Fukui *et al.*, 1992). The technique developed involves chromosome preparation on a plastic coverslip covered with a polyester membrane and C banding treatment following Giemsa staining. The next step involves identification of positive bands, dissection by irradiation with a micro laser beam, and recovery of fragments in Eppendorf tubes (Fukui *et al.*, 1992; Kamisugi *et al.*, 1993). The dissected fragment can be recovered within the tube by using tinted polyester membrane carrying the chromosome sample. The laser beam can cut a circle around the chromosome, which can easily be removed.

Protocol for Microdissection

Pretreat and fix root tips in methanol : acetic acid (3:1) following the usual schedule (after Fukui *et al.*, 1992). Wash the tips thoroughly and subject them to enzymatic maceration (2% cellulase Onozuka RS, 0.3% Pectolyase Y-23, Sheishin Pharma, Tokyo and 1.5% macerozyme R 200, adjusted to pH 4.2) on a glass slide as well as on a polyester membrane fixed at the bottom of a 35 mm plastic petridish at 37°C for 30 to 60 min. Wash off the enzymic mixture and macerate root tips with fine forceps into almost invisible fragments in a drop of the fixative. Air dry. Stain with 2% Giemsa solution covering the surface of the membrane and dipping in case of glass slides. Wash and air dry. Carry out microdissection in ACAS470 (Meridien Instruments, Okemos, Mich, USA) which consists of an argon-ion laser tube, an acausto optic modulator (AOM) and a controlling microcomputer (CPU 80286). The single laser beam of 488 nm is used for microdissection of chromosomes. The intensity of the laser beam is controlled by power supply and modulator. The laser beam which is introduced into the center of the axis of the inverted microscope is focussed to 1 μm by a 40 × objective and the target region is irradiated. In order to have a beam, less than 1 μm focussing on the chromosome, a 100 × objective is used. In general, with the increase in intensity of beam, the band cut will be wider.

For very fine chromosome micromanipulation, a band width of 0.5 μm obtained through regulating power supply and AOM is optimal. A beam intensity of 1 μm can be focussed in unnecessary regions—intra or extrachromosomal. Such laser treatment of unnecessary parts or segments can be monitored by negative result with DNA fluorescence staining of the remaining segments. The targetted region is dissected out after removal. The entire operation can be completed in 10 min.

Transfer of chromosome fragments or chromosome to Eppendorf tube is carried out by the dissection of the tinted polyester membrane carrying the chromosomes. For this purpose, a strong laser beam of 10 µm to 2 mm is used to divide the membrane in octagonal pieces. These octagonal pieces are picked up with fine forceps under stereomicroscope and placed in Eppendorf tube for storage at -20°C before use.

Microdissection of Animal Chromosomes

In this method devised by Ponnelles *et al.* (1989) the UV microbeam apparatus consists of an excimer pumped dye laser (Lambda Physik Gottingen, FRG), generating pulse energies at tunable wavelength about 340 nm. This wavelength has a good cutting efficiency with low DNA damage. Pulses of the laser at a repetition rate of 10 Hz are directed into an inverted microscope via the fluorescence illumination through the objective into the object plane, yielding power density of upto 10" N/cm^2. This energy is sufficient to evaporate all biological material. The selected chromosome region can be brought in the focus of the laser beam by moving the stage. Following microdissection, the chromosomes can be handled by the micromanipulator (Fonbrune micromanipulator - Bachofer Reutlingen) to be used for microcloning as well. In polytene chromosome of ***Drosophila***, the width of one average band of 0.2 µm corresponds to 0.1 pg of DNA. Following this method,the average amount of DNA per fragment would be 0.25 to 0.5 pg. The precision of the dissection is principally limited by the wavelength of the laser beam. Theoretically, it is even possible to dissect a single band of human chromosome.

Microdissection of Plant Chromosomes

Incubate seeds or seedlings on filter paper subsequently soaked with tap water at i) 4°C for 3 days; ii) 22°C for 5 h; iii) 1.25 mM hydroxyurea at room temperature, for 18h (after Schondelmaier *et al* 1993), iv) distilled water at room temperature for 5 h. v) 4 µM APM (O-methyl-O-(2-nitro-p-tolyl) N-isopropyl-phosphoroamido-thioate; amiprophosmethyl (Bayer-Leverkusen) at room temperature for 3h.

Cut roottips, rinse in distilled water, incubate in ice water overnight.

Store in 70% ethanol for one day

Wash in distilled water and incubate in enzyme solution (2.5% pectolyase Y23, 2.5% cellulase R10, 75 mM KCl, 7.5 mM EDTA, pH 4) for 45 min.

Treat in 75mM KCl for 15 min.

Wash protoplast suspension three times with 70% ethanol and centrifuge for 5 min at 75g.

Resuspend cell sediment in fresh fixative (ethanol-acetic acid, 3:1) and centrifuge for 2 min.

Remove supernatant, drop suspension on ice cold slides and use for microdissection after drying.

For microdissection, use an inverted microscope (Zeiss 1M 35) with programmable stage, micro-manipulator and phase contrast optics for selecting and manipulating suitable metaphase spreads at a maximum magnification of 640X.

Deposit a collection drop (2 ml) of 10 mM Tris-HCl, pH 7.5; 10 mM NaCl, 0.1% SDS, 1% glycerol, 500 μg/ml proteinase K on a depression slide overlaid with liquid paraffin (Merck spectroscopic) 7161 grade.

Collect the chromosome or fragments in that drop. For cloning, after lysis in the collection drop (2 ml), purify the DNA, followed by restriction digestion, ligation with vector containing universal sequencing primer, amplification in PCR and cloning in a standard plasmid vector.

CHAPTER 12

IDENTIFICATION OF CHROMOSOME SEGMENTS AND DNA SEQUENCES BY *IN SITU* MOLECULAR HYBRIDIZATION

The study of chromosome structure and the localization of functionally differentiated segments of chromosomes and gene loci have been greatly facilitated through the application of molecular hybridization technique at the chromosome level. It is possible to prepare a cytological map of the DNA sequences and the precise locus of transcription of specific messengers through this method. This knowledge of the operational mechanism of chromosomes at the molecular level is due basically to important advances in methodology, one involving the reannealing technique for analyzing the sequence complexity of DNA and the other the hybridization of RNA and DNA molecules.

The principle of the *in situ* technique is to utilize single stranded probes of known sequences or segments and their hybridization with complementary sequences on chromosomes. The detection of hybridized regions through autoradiography or fluorescence permits identification of specific gene loci in the chromosome. The entire operation is based on the principle that a single strand of DNA can always pair with complementary sequences of another RNA or DNA strand.

Pure labelled nucleic acid fractions obtained from the *in vivo* state or prepared from complementary DNA sequences *in vitro* are hybridized with previously denatured DNA of the chromosomes *in situ* . The detection of a particular DNA sequence in a chromosome is dependent on (1) the sensitivity of detection, and (ii) the sequence complexity of DNA. The strength of detection depends on the amount of radioactivity and the sensitivity of autoradiographic procedure or the strength of fluorescence. Nucleic acid of high specific activity is always desirable for hybridization, 10^6 dpm/µg being the lower useful limit (Pardue and Gall, 1975). Highly repeated sequences aggregated at one locus are ideal for *in situ* localization but there are methods for detection of low copy sequences as well .

The separation of the double helix into single strands of DNA, termed as denaturation, can be secured most effectively by heating the nucleic acid solution. Alternative methods of treatment include strong acid, alkali or even some other compounds such as formamide. The separation can be detected by the rise in ultraviolet absorbance at 260 nm. Melting temperature or *Tm* implies the temperature corresponding to the midpoint of the absorbance. The value of Tm is directly correlated with GC ratio of DNA as it confers a higher stability against thermal denaturation than the AT component. Ionic concentrations have a significant influence on the Tm value.

Reassociation of DNA, implying restoration of the duplex, is the reverse process and requires a stronger salt concentration and lower temperature than that required for denaturation. In general, the temperature required for optimal effect is much below the Tm value, through for RNA/DNA hybrid duplex it is near the Tm value. Repeated sequences associate more readily than unique sequences. As a corollary, reassociation time is a direct index of its sequence complexity. For measuring the rate of reassociation, the most convenient method is detection of fall in absorbance at 260 nm. Other methods include measurement of the amount of labelled DNA fragments immobilized on nitrocellulose filters or passing after reassociation through a hydroxyapatite (calcium phosphate) column which holds back the duplex. Britten and Kohne devised a method of comparing the rate of reassociation of samples of different concentrations at different periods . The term C_ot (C_o denoting the initial concentration and t implying the period of treatment for reassociation) is expressed in terms of titre i.e. mol/s/1, and the midpoint i.e.C_ot 1/2 is termed as the value.

The most convenient method is to fix the tissue in acetic-ethanol (1:3) , followed by squashing in 45% acetic acid and removing the slide by the dry ice method. In general, formaldehyde fixation should be avoided as it may interfere with denaturation of DNA. Removal of pre-formed RNA, both chromosomal and extrachromosomal, is desirable, as otherwise it may compete with the labelled RNA for hybridizing with DNA. In *in situ* technique, the denaturation is carried out normally through solutions of low and high pH, high temperature, formamide, acids and alkalies. Of all these methods, heat treatment followed by quenching does not yield more than 1% reannealing. The swelling in NaOH is often counteracted to some extent by SSC treatment.The comparatively mild method using HCl with low pH value, maintains chromosome structure intact with good stainability, but some amount of depurination has been reported.

In order to detect the degree of denaturation, Steffensen and Wimber (1972) recommended the use of fluorochromes, based on the principle that double stranded structures fluoresce green as against red fluorescence of single strands after excitation. After denaturation through formamide treatment, reannealing and consequent reversal to green fluorescence can be obtained by treating denatured unstained preparations in 2 XSSC at 65 °C for 20 min. The next step is the application of tritiated complementary nucleic acid solution to the denatured preparation on a grease free, clean slide and incubation at requisite temperature in a moist chamber. The entire set is incubated at annealing temperature for the requisite period.

Several factors control annealing. It is always desirable to add excess of non-radioactive non-competing nucleic acid to avoid non-specific binding of radioactive nucleic acid. The temperature mostly required for incubation is usually below the Tm value, except for poly U, where the optimal temperature is 30 °C . In cases where this value is not known, trial should be given around 60-65 °C. With organic solvents like formamide , even 25-40 °C may serve the purpose.

For hybridization with complementary RNA, competition between complementary strands of DNA is also to be taken into account. However, complementary single strands take nearly 20 min at 65 °C to form substantial amount of DNA duplex. Therefore, it is desirable to secure reannealing within the first few minutes with optimal concentration and temperature. Varying concentrations of tritiated RNA with high specific activity, such as 3.0 µg/ml, 2.3 µg/ml, 0.2 µg/ml and 0.01 µg/ml have been used to secure such optimal effect. Complex unique sequences require considerably more time for reannealing than simple and homogeneous repeated DNA. For *in situ* technique, in general, the ionic strength, nucleic acid concentration, temperature and period of incubation, should be determined for each material. The conditions are more or less identical to those needed for filter hybridization. *To remove non-specific RNA* from nuclear and cytoplasmic structures, after washing thoroughly in 2 XSSC, treatment with RNAse is needed. After non- specific complexes are removed 2 to 3 brief changes in 2 XSSC followed by keeping at 4 °C with continuous stirring in excess of SSC is desirable prior to dehydration and air drying.

Autoradiography : Principles of autoradiography have been discussed in detail in another chapter. Liquid emulsion procedure is generally applied in order to secure staining. Kodak NTB 2 or Ilford K2 diluted with equal amount of distilled water is suitable.

General protocol

A. For animal cells (after Pardue and Gall, 1975)

Take small pieces of fresh tissue, not more that 5 mm in diameter, tease and fix for 5-10 min in acetic- ethanol (1:3). For cell suspensions, gently centrifuge cultured cells in medium, resuspend cell pellet in 40 times vol of hypotonic solution (3 : 1 distilled water and medium), keep for 5 min at 26-28°C, gently centrifuge again and replace hypotonic solution by equal vol of 50 % acetic acid. Fix in ice chamber for 20 min . After fixation, resuspend the cells in a fresh change of 50% acetic acid.

Put a drop of 45% acetic acid on a siliconized 18 mm cover slip.

Transfer a small piece (maximum 1 mm) from the fixed and teased material to acetic acid on the cover slip and mince thoroughly. Remove any large piece. For fixed cell suspensions, just transfer a drop to 45% acetic acid.Place a subbed slide on the cover slip and apply slight pressure so that the cover slip adheres firmly. Subbed slides can be prepared by dipping detergent-washed, grease-free slides, for a few hours in subbing solution (0.1% gelation, dissolved initially in hot water to which 0.01% chrome alum is added later and dried for a few hours before use. Place the slide with the cover slip surface down on a filter paper and apply uniform pressure with the thumb to secure a well scattered monolayer smear. Keep the slide against dry ice for a few min for freezing and remove slip. Transfer the slide to ethanol for a few min, air dry and store.

For Drosophila salivary gland chromosomes, dissect glands in Ringer's solution, transfer to 45% acetic acid on the siliconized cover slip. Before squashing on the subbed slide, apply mild pressure for chromosome spreading. After detaching the slide through dry ice technique, keep in acetic-ethanol (1:3) for 2-3 min before transferring to ethanol and air drying.

Transfer the slide (with squashes or cell suspensions) to a moist chamber, apply a drop of RNAse solution in 2 X SSC at a concentration of 100 μg/ml, cover with a cover slip and keep in moist chamber at 37°C for 1 h or 26-28°C for 2 h. This step is necessary to remove pre-formed endogenous RNA. (Moist chamber can be prepared with 4-inch plastic petri plates with the bottom lined with moist filter paper and containing 5-10 ml 2 X SSC. The slide can be suspended above the liquid by U-shaped glass rods or two rubber pieces. The medium inside the cover slip and in the petri plate should have the same ionic strength to avoid any shift in concentration). Remove the cover slip by dipping in 2 X SSC. Give three changes of the slides in 2 X SSC and transfer first to 10% and then to 95% ethanol and air dry.

For denaturing DNA, transfer the slides to 0.07 M NaOH at 26- 28°C and keep for 2-3 min. Give three changes in 70% and subsequently two in 95% ethanol, keeping 10 min in each and air dry. Take the slides for hybridization and choose for incubation in SSC or SNB buffer depending on the need for sodium ions. For lower temperature (37°C), 40% formamide in 4XSSC and for higher temperature (65°C) 2 XSSC may be used. Add excess of non-radioactive non-competitive nucleic acid to prevent non- specific binding. *For hybridization*, put a drop of radioactive nucleic acid to be hybridized on the preparation coverslip and place the slide in a moist chamber as described before. Seal the chamber and incubate in the required temperature, for example 10-15 h in 2 XSNB at 65°C. The moist chamber must contain the same concentration of buffer as that under the coverslip.

After hybridization, remove the slide from the oven. To remove non-specifically bound RNA-DNA hybrids, wash in 2 XSSC for 15 min at 26-28°C. Incubate in pancreatic RNAse (20 μg/ml in 2 XSSC) for 1 h at 37°C, and rinse twice in 2 XSSC for a total of 10 min. To remove non-specific bound DNA-DNA hybrids, dip in 2 XSSC at a temperature 5°C below the incubation temperature. Give three more changes in 2 XSSC at the same temperature for a total period of 10 min. Pass through 70 and 95% ethanol and air dry.

For autoradiography, carry out the entire operation in absolute darkness or safelight. Melt the emulsion (Kodak NTB 2 mixed in equal proportions with distilled water and kept in 10 ml vials at 45°C for 10 min and pour in a plastic dipping chamber held at 45°C, taking care to avoid air bubbles. Dip the slides in the chamber, gently withdraw, drain off excess fluid and place vertically in an air drier for about 1 h for drying. Store the slides in sealed boxes (with silica gel inside for maintaining dryness) in the dark for a few hours to several days depending on the requirements.

Stain in Giemsa (stock solution diluted with 0.1 M phosphate buffer, pH 6.8 just before use) for a few minutes. Check at intervals to find out the exact period necessary for staining. Rinse in distilled water, air dry and cover with Permount under a coverslip.

Several authors have developed methods using radioactive iodine ^{125} I for labelling. In this technique (Attenburg, Getz and Saunders, 1975), the reaction mixture contains purified nucleic acid, $TlCl_3$ and a mixture of $Na^{125}I$ and KI carrier. Constituents are mixed in acetate buffer (pH 5.0) at O°C and an ionic strength of 0.05-0.10 M (Na). The mixture is kept in sealed container at 60-80°C for 15-20 min followed by chilling to 0°C. A reducing agent is added at the termination point, pH is raised to 8.5-9.0 and again heated for 20 min at 60°C. The nucleic acid is thus iodinated and recovered through gel filtration chromatography and radioactivity counted through counter. The labelled nucleic acid may then be used for hybridization on chromosomes.

In situ techniques have been modified for identifying DNA sequences of a particular genome in the target individual when the latter contains sequences from different sources. Some of these sequences, being related, may hybridize with each other. In order to eliminate this difficulty, as a prehybridization step with the genome probe, blocking unlabelled DNAs from other sources in high concentration, are added so that they hybridize with complementary sequences in the recipient. Only those sequences in the latter which are complementary to probe sequences would remain unattached or unblocked and could be visualized following hybridization with the fluorescence probe. For chromosome specific repetitive sequences, short incubation at 37°C is often sufficient. But for unique sequences, it is desirable to use large insert probes and prehybridize in a mixture with probe and excess unlabelled genome DNA for a long period so that repetitive sequences are blocked. Continued hybridization overnight may lead to binding of target and probe.

The lengths of the labelled and unlabelled sequences should be less than that of the total length of the DNA sequences in the chromosome (Schwarzacher *et al* 1989). Reduction of the sequences in size in the order of 50-1000 bp can be achieved by sonication and shearing as well as during labelling, which leads to 80-120 bp fragments. For radiolabelling with cloned DNA sequences the favoured method is to use *E. coli* RNA polymerase to prepare tritium-labelled complementary RNA. However, radioactive DNA can also be produced by nick translation, as mentioned later.

The method utilizing genome as probe does not involve cloning and screening and has been utilized effectively in different plant species, such as barley, rye and *Nicotiana.*

The *in situ* technique permits the localization of the distribution of tandem repeats, dispersed repeats and complete genes in the chromosome. Cloned probes can be utilized for detection of repetitive or single copy sequences in chromosomes. Chromosome specific sequences can be used in YAC vector as well and can be identified later in the chromosomes by *in situ* technique as done for chromosome 21 of the human genome (Chumakov *et al*, 1992).

Use of non-radioactive probe: Non-radioactive methods for indirect labelling of probes involve the use of compounds such as biotin as the primary label and later streptavidin which is used as a conjugate to signal generation system. Otherwise, antibody of a hapten is incorporated into the probe whose recognition leads to detection. Similarly, photobiotin is detected either through luminescence or fluorescence or alkaline phosphatase conjugate — colorimetrically with enzyme conjugate. In direct labelling however, the signal generation system is directly attached to the probe which is detected at post hybridization or with enzyme and enzyme conjugate. The examples include horse radish peroxidase where the detection is through high chemoluminescence (Mundy *et al*, 1991).

The hybridization sites through fluorescent or biotin labelling are more precise than with radioactive labelling as the detection is on chromosomes rather than on silver grains of the photographic emulsion. Moreover, it requires only a few hours of exposure as compared to the other which may require weeks. The entire genome, specific chromosomes or even fragments or even low copy sequences can be used as probes for localization.

Majority of the reagents for probe labelling and securing fluorescence are available commercially. An alternative method is to label with an enzyme and detect the label by a reaction catalyzer of the enzyme. For preparation of the probes, labelling with reporter molecules and cutting into 100-400 bp fragments are necessary. Such reporter molecules, bound to fluorescent affinity reagents after hybridization, include biotin, digoxigenin, dinitrophenyl and others. These are incorporated in the probe by nick translation, random primer labelling, and *in vitro* transcription. The advantage of biotin is that it is a water soluble vitamin capable of incorporation with nucleic acid bases. After hybridization, the sites can be detected through fluorescent avidin and can be amplified with biotinylated anti-avidin. A separate counter stain for chromosomes such as with propidium iodide can be used—the former fluoresces greenish yellow and the latter red.

Initially, nucleotide analogues with biotin were prepared with dUTP and UTP by attachment of biotin with 5 carbon of the pyrimidine (Langer and Ward 1989). The allylamine linker arm length of the carbon atom is utilized in designating the compounds such bio^{-11} dUTP. Later another procedure involving connection of the linker arm with amino - nitrogen at 6 position of adenine or 4 position of cytosine was adopted (vide Mundy *et al*, 1991). Incorporation of biotin may also be made through photoactive substances such as aryl azide. It is dark- stable but on photoactivation forms aryl nitrene which reacts with nucleic acid bases. A biotin aryl azide compound can be used for labelling nucleic acid by mixing in dark and exposing to short wavelength visible light.

In addition to antibodies, biotin probes are detected through two allied proteins like avidin or streptavidin. The former is a 58 kd glycoprotein and the latter is isolated from *Streptomyces avidinie*. Signal generation is achieved through covalent attachment with fluorescent or labelled streptavidin or Colloidal Gold.

In situ technique has been so modified as to visualize two plant DNA sequences in the same chromosome. In such cases two different probes containing different sequences and labelled with different fluorescent reporters are applied in the mixture for hybridization. Similarly, for detection, different and specific antibody conjugates are necessary followed by amplification. For observation, a counterstain is also required for the chromosomes. It is observed finally under epifluorescent lens of a fluorescent microscope. For instance, in *Secale cereale* a probe from rye and another from wheat were used, the former labelled with biotin and the latter with digoxigenin. The detection was carried out with red labelled avidin for biotin and sheep antidigoxigen fluorescein for digoxigenin. Similarly, signals were amplified later in biotinylated antiavidin D and FITC-conjugated rabbit antisheep in detection buffer. Chromosomes were counterstained with DAPI in buffer. Biotinylated probe emitted orange/red and digoxigenin label yielded green fluorescence.

In nick translation, template-dependent DNA polymerase and 5′–3′ and 3′–5′ exonuclease activities are utilized to add labelled nucleotides to the 3′ end of a nick sequentially while removing nucleotides from the adjacent 5′ end terminus. Due to the randomness of the nick, the labelling is uniform in any double stranded DNA substrate. Nick translation is utilized for both radioactive and non-radioactive labels. The size of the probe is to a great extent dependent on the concentration of the DNAse and the incubation period. Probes of about 500 bp give optimum effect. With shorter probes, reduced sensitivity and with longer probes, such as 1 kb, background effect may be noticed. An incubation temperature of 15-20°C is adequate.

Random primer labelling takes advantage of Klenow fragment of DNA polymerase I produced by proteolytic cleavage. This enzyme does not have 5′–3′ exonuclease activity and can be used to copy single stranded DNA molecule from the 3′ end of primer DNA annealed to the template. The absence of 5′–3′ activity prevents degradation of the primer from 3′ end and so the incorporated nucleotides are not excised.

In *.in situ* hybridization, purity of the probe and quick annealing are essential. Purity is normally assured by the use of recombinant clone whereas dextran sulphate accelerates the annealing process. The use of Denhardt's solution and addition of blocking DNA are checks against nonspecific labelling. The size of the probe is determined by taking into account the ratio of the primer to substrate concentration of nucleotide. Of the different substrate concentrations, it is advisable to go by the lower range. For example, with 25 μg DNA, the probe needed should be 2×10^9 dpm μg^{-1} specific activity for radioactive probes. It can be increased by using 6000 $Cimmol^{-1}$ of label to around 5×10^9 dpm μg^{-1} specific activity (Brown, 1991). With multiple labels, the rate of incorporation would be more as also the reaction time needed. Non-radioactive labels are specially suited where resolution is of greater importance than high sensitivity, such as in *in situ* hybridization.

Use of centromere-specific antibody: The monoclonal antibodies are also now widely available and help in the localization and quantitation of chromosomal components. The use of antibodies permits the localization of non DNA components, specially protein in the chromosomes. The variability of protein components during different phases of development can be analysed. It is based on the principle that structural changes affect accessibility of general chromosomal antibodies and as such there are major changes in antibody binding.

The antibody technique has been applied for locating the components of the kinetochore, which is regarded as a centromere associated protein structure, important for interaction of centromere with the spindle. It is localized using kinetochore antibodies which may be obtained from serum of scleroderma patients. The method involves the localization of primary antibody by a peroxidase-labelled secondary antibody followed by a nickel chloride modification of diaminobenzidine reaction. The presence of the antigen 19.5 kDA (CENP-A) protein has been demonstrated in human centromere (vide Kingwell and Rattner, 1987). Similarly, histone H has been localized in centromere heterochromatin which normally remains inaccessible to antibody due to condensation.

The use of polyclonal and monoclonal antibodies has helped in the detection of BrdU. This method has facilitated the study of replication banding of mammalian chromosomes or sister chromatid exchange (Vogel *et al*, 1989). The same antibody technique can be adopted using non-radioactive and non-fluorescent stages. It involves in principle BrdU incorporation followed by coating with anti BrdU antibody, biotin-conjugated antimouse antibody, mixing with avidin DH and biotinylated horseradish peroxidase H and visualization through a solution of diaminobenzidine -4HCl (DAB Bottite), $NiCl_2$ and 0.02% H_2O_2 in 100 ml Tris-buffer (pH 7.2) (Taniguchi and Tanaka, 1989).

Anticentromere antibodies from CREST scleroderma sera have helped in the identification of many centromere specific proteins in mammalian cells. Through SDS-gel electrophoresis, centromere specific human autoantigens have been identified. An example is autoantigen CENPA, an acid soluble component of highly specific nucleosome core particle from HeLa cells (Palmer *et al*, 1990).

CENP-A present in sperm nucleus functions as a centromere specific or more particularly kinetochore specific core histone. Similarly, monoclonal antibody against specific protein such as gliadin protein in wheat has enabled the identification of the locus of this specific gene in short arm of 1B chromosome (Knox *et al*, 1992).

The same method with antitopoisomerase II antibody has led to the demonstration of topoisomerase in the mitotic scaffold as well as mitotic chromosome core (Moens and Earnshaw, 1989).

REPRESENTATIVE SCHEDULES

In Situ Hybridization in Mammals Using Autoradiography

Objective of *in situ* hybridization is the detection or localization of specific RNA or DNA sequences in cells, cell nuclei or chromosomes using a radioactive single stranded nucleic acid 'probe' of complementary base sequece, followed by autoradiography.

Sample Protocols

1. DNA to DNA - repeated sequences:

Applicable for the localization of nucleolar organisers on chromosomes, nick-translated ribosomal DNA (rDNA).

A) *Preparation of probe*: Dissolve enough DNA probe to provide a total of 3 X 10^6 counts per min as determined from counting by liquid scintillation counter in 30 µl of 0.1M NaOH. Dissolve thoroughly by rolling the liquid several times at room temperature for several min. Add 150 µl of deionised formamide (95%) and mix well to give final concentration of 50%. Add 60 µl 2 X SSC and mix thoroughly to give a final concentration of 4 x SSC (pH 7.0). Add 30 µl of distilled water, mix, cool on ice for several min. Add 30 µl of 0.1M HCl, mix and keep ice cold for 10-15 min, within which time *in situ* hybrids must be set up.

B) *Preparation of slides*: Use air dried slides fixed in 95% ethanol or acetic acid-ethanol or methanol mixture (1:3). Remove endogenous RNA by treating with ribonuclease (Sigma). Place the slides for 2 h at 37°C in a solution containing 50 µg/ml of RNAse A (Sigma cat. no. R4875) and 100 units/ml of RNAse T1 (Sigma cat. no. R8251) in 2XSSC (pH 7.0). Alternatively, 30 µl of the enzyme solution can be placed on the slide, covered with a coverslip and incubated in a moist chamber at 37°C. Wash three times in 2XSSC, keeping 10 min in each. Transfer slides to 0.07M NaOH at 20°C for 3 min. Wash three times in 70% ethanol. Wash twice in 95% ethanol and airdry.

Transfer the slides to moist chamber, with the preparation uppermost. Using a micropipette, place 30 µl of the reaction mixture as prepared in A) as a drop in the middle of the preparation area of each slide. Cover with a clean coverslip. Incubate in moist chamber at 37°C for 6 - 12 h.

C) *Washing after hybridization*: Transfer slide to beaker containing 2XSSC to remove the coverslip. Keep slides in 2XSSC at 65°C for 15 min. Wash twice in 2XSSC at room temperature, keeping 10 min for each wash. Wash twice in 70% ethanol and then in 95% ethanol, keeping 10 min in each wash. Airdry. The slide is ready for autoradiography.

2) DNA (probe) to RNA (chromosomal repeated sequences)

For localization of RNA transcripts that are still attached to their chromosomal DNA template, using a nick-translated DNA probe.

Steps A, C and D are same as in protocol 1) for DNA-DNA. Step B) is not needed since it involves removal of endogenous RNA.

3)RNA (probe) to DNA (chromosomal) repeated sequence

Using ^{3}H-labelled complementary RNA (cRNA) as a probe for localizing chromosomal DNA sequences.

A) Preparation of probe: Make up the RNA solution in 2XSSC or in 4XSSC - 50% formamide at a concentration to provide 50,000 to 100,000 counts/min per each 30 µl of final reaction mixture.

B) *Prepare slide as for DNA to DNA protocol.*

C) *Hybridization*: The procedure is the same as DNA to DNA protocol. Keep at 65°C for 6-12 h if the probe is in 2XSSC and at 37°C if it is in 4XSSC - 50% formamide.

D) *Washing after hybridization*: Transfer slide to beaker containing 2XSSC to remove coverslip. Keep for 15 min at room temperature in fresh 2XSSC. Transfer to ribonuclease (50 μg/ml RNAse A, 100 units/ml RNAse T1 in 2XSSC) at 37°C for 1 h. Wash twice in 2XSSC, keeping 15 min in each wash. Treat with 5% trichloroacetic acid (TCA) at 5°C for 5 min. Wash two times each in 2XSSC, 70% ethanol and 95% ethanol, keeping for 10 min in each wash. Air dry. The slides are now ready for autoradiography, as described elsewhere.

4) DNA to DNA - single copy sequences: H-labelled DNA to mitotic metaphase chromosomes

Pretreat mitotic airdried preparations for 5 h at 60°C in 10X Denhardt's solution. 1X Denhardt's solution contains 0.02% bovine serum albumin (fraction V Sigma cat. no. A-4503), 0.02% Ficoll (Sigma cat. no. F-4375), 0.02% polyvinylpyrrolidone (Sigma cat. no. PVP-360). Denature chromosomal DNA with 70% formamide in 2XSSC, pH 7.0, at 70°C for 2 min. Prepare hybridization mixture: labelled DNA at not more than 50 μg/ml, 1000-fold excess of unlabelled denatured salmon sperm DNA (Sigma cat. no. D-1626); 50% formamide; 2 X SSC; 10% dextran sulphate pH 7.0; 1X Denhardt's solution. Denature DNA probe by incubation of the hybridization mixture at 70°C for 15 min. Hybridize at 37°C for 12 h. Wash after hybridization in 50% formamide and 2 X SSC at 39°C, followed by washing at room temperature in 2 X SSC and 70% and 95% ethanol. Air dry. The general procedure is similar to protocol 1) *for DNA-DNA repeated sequences.*

Methods for labelling probes

(A) Complementary RNA (Pardue and Gall 1971)

Prepare DNA solution in 0.04 m Tris, pH 7.9, to be have an optical density at 260 nm of between 1.0 and 2.0 (50-100 μg/ml). Place DNA in a boiling water bath for 5 min and cool on ice if transcription from single-stranded DNA is required. Prepare dried "XTPs" containing 100 μCi each of GTP, CTP, ATP and UTP, all tritium labelled with specific activities of 10-30 Ci/mM. Keep in small glass tube and vacuum dry. Add to the dried "XTPs" : DNA 100μl ; β- mercaptoethanol 50μl (Sigma cat. no. M-6250, about 14.3 M, diluted 20 μl in 10ml water); salt solution 50 μl (Tris 1M, pH 7.9, 4.0 ml; 2 M KCl 9.4 ml; 1 M $MgCl_2$ 0.58 ml; 0.01M EDTA 0.88 ml, made upto a final volume of 25 ml with water, 80 μl of 0.125 M $MnCl_2$ added per ml of salt solution before use); water 40 μl; RNA polymerase (*E. coli* Sigma cat. no. R5376) 10 μl. The total volume should be 250μl.

Cover tube with Parafilm and incubate at 37°C for 60-90 min, add 0.75 ml of 0.04 M Tris, pH 7.9 and 20 μl of DNAse I (Sigma cat. no. D-1126, 1 mg/ml in water), equivalent to 20 μg. Keep at 20°C for 15 min.

Remove two 10 μl samples and spot on nitrocellulose filter. Dry. Place one filter in cold 5% TCA for 5 min, wash in 70% ethanol and dry. Count both filters in a liquid scintillation system to determine the percentage of counts which are TCA insoluble (usually between the 15 to 40%). Add 200 μl of 5% Sarkosyl or sodium dodecyl sulphate (SDS) 5 % in water. Add 50μg of E. coli RNA in 1 ml of 0.04M Tris, pH 7.9 to act as carrier in cases where cRNA is being transcribed from eukaryotic DNA. Add 2 ml of redistilled water-saturated phenol. Alternatively, use a mixture of phenol 100 g, 8- hydroxyquinoline 0.1g, chloroform 100 ml and isoamyl alcohol 4 ml. Centrifuge for 10 min at 10,000 r/min. Remove supernatant, extract phenol with 200 μl 0.04M Tris. Gently add supernatant to Sephadex column (G-50, 15X1 cm) and collect 25 consecutive 30 drop fractions. Spot 10μl from each fraction on nitrocelluse filter, dry and count. Pool peak fractions, keep in 85°C water bath for 3 min. Filter through 0.45 μm pore nitrocellulose filter. Store at –20°C.

Nick Translation of DNA:

It is used specifically for 1 μg of DNA in 100 μl of reaction mixture, expected to yield DNA with specific activities of $2 - 6 \times 10^6$ counts min^{-1} μg^{-1}.

Mix together in a small vial 2×10^{-3} μM each of ^3H-labelled dATP, dCTP, dGTP and TTP. Freeze dry 0.018 ml of the mixture in a small vial. Prepare 5 ml of salt solution containing Tris-HCl 1 M, 2.5 ml pH 7.9; 0.5 ml of 1 M mercaptoethanol; 0.25 ml of bovine serum albumin (10 mg/ml); 1.5 ml distilled water; 0.25 ml of 1 M $MgCl_2$ (Maniatis *et al* 1975).

Prepare the DNA solution at a concentration of 200μg/ml (OD_{260nm}4). Add to the vial containing XTPs consecutively 0.01ml of the salts mixture, 0.005ml DNA (1μg); distilled water to make upto 0.094ml. Add 0.005ml (12.5 units) of DNA polymerase (Grade I, Boehringer Mannheim, 550units/0.1mg/0.22ml).

Dilute the 1mg/ml stock of DNase I (Sigma cat.no. d-1126, electrophoretically purified) down to 10^{-3}mg/ml with water. Add 0.001ml of the diluted solution to the reaction mixture and incubate at 15°C for 1h. Add 0.005ml of reaction mixture to 0.895ml of a solution containing 50μl of bovine serum albumin stock (1mg/ml) and 845μl water. Add 0.1ml of 100% TCA. Keep on ice for 15min. Pass 5ml of ice-cold 5% TCA through a glass fibre filter (Whatman GF/C, 2.5 diameter) and then through the reaction mix-serum albumin-TCA sample. Wash the filter by passing through three 5ml lots of cold 5% TCA. Dry in a 65°C oven for 20min and count in a liquid scintillation system using a toluene-based scintillation fluid.

Compare radioactivity by counting a control sample of reaction mix. Stop the reaction by adding to the vial 0.1ml of water- saturated phenol and mix well with a Pasteur pipette. Centrifuge at 5000g for 5min. Load the aqueous supernatant directly onto a G50 Sephadex column (1 x 35 cm) that had been prewashed with distilled water. Elute with distilled water, collecting 30 drop fractions.

Take 0.005ml from each fraction and count in liquid scintillation system using a tergitol scintillator (PPO 5.0g; tergitol 15-s-9 Union Carbide 350 ml; POPOP 50mg; toluene 650ml). Combine the fractions comprising the first of the two peaks of radioactivity to come off the column, freeze dry and redissolve DNA in 0.05ml distilled water.

The DNA is now ready for use.

Random Primer Labelling

The Klenow fragment of DNA polymerase I is produced by proteolytic cleavage of the enzyme or, as is now more common, from an over-producing recombinant *E. Coli* strain expressing the appropriate portion of the gene (after Brown 1991).

The enzyme lacks the 5′ → 3′ exonuclease activity of the intact DNA pol I enzyme and can be used *in vitro* to copy a single- stranded DNA molecule starting from the 3′-end of a primer DNA annealed to the template. A mixture of hexanucleotide primers of random sequence, derived either from digestion of calf thymus DNA or by chemical synthesis allows effectively any DNA template to be copied. As with nick-translation, nucleotides labelled with ^{32}P, ^{35}S, ^{3}H, ^{125}I, or biotin can be incorporated. The absence of the 5′ → 3′ exonuclease activity prevents degradation of the primer from its 5′-end and also ensures that incorporated nucleotides are not subsequently excised. This increases the efficiency of the reaction and allows a range of reaction times and temperatures to be adopted.

The technique is relatively insensitive to the purity of the substrate DNA and both mini-preparations and fragments in low- gelling-temperature agarose can be labelled. Random-primer labelling is now probably the method most frequently used for labelling insert DNA as opposed to intact recombinant vector molecules.

For Nick-translation, the material needed are: 10 x buffer (600 mM Tris-HCl pH 7.8; 100 mM $MgCl_2$; 100mM 2-mercaptoethanol; TE buffer, pH 8.0; 5 x nucleotide mix (100 μM dATP, 100 μm d GTP 100 μM dTTP in TE); enzyme mix (0.006 units ml^{-1} DNase I and 500 units ml^{-1} DNA pol I); [α – ^{32}P] dCTP at 3000 Ci $mmol^{-1}$; 0.5 M EDTA, pH 8.0.

Dilute the DNA to be labelled to a concentration of 5-50 μg ml^{-1} in either distilled water or TE.

Add the following in the order given, to a microfuge tube on ice: DNA solution, 50-500 ng; 5 x nucleotide mix, 10 μl; water to bring the volume to 30 μl; 10

x buffer, 5 μl; [α – ^{32}P] dCTP (3000 Ci mmol^{-1}), 10 μl (100 μCi); enzyme mix 5 μl; to make a total volume of 50 μl.

Mix gently by pipetting and cap the tube. If necessary, spin briefly (2 *s*) in a microcentrifuge. Incubate the reaction mix at 15°C for 60 min. Any incubation time between 30 min and 3 h may be employed. Terminate the reaction by addition of 2 μl 0.5 M EDTA, pH 8.0.

Store at –20°C in the presence of 20 mM EDTA, pH 8.0 before use in hybridization analysis. Prolonged storage can lead to substantial degradation and probes of high specific activity should be stored for no longer than three days. Denature the probe by heating to 95-100°C for 5 min immediately before use.

Calculation of the specific activity of the probe:

(a) During nick-translation, nucleotides are effectively excised and replaced, and there is usually no net synthesis of DNA. Thus the probe specific activity is calculated simply as:

$$\text{specific activity (d.p.m.}\mu\text{g}^{-1}) = \frac{\text{total activity incorporated(d.p.m)}}{\text{amount of substrate added(}\mu\text{g)}}.$$

(b) 50% incorporation in the above reaction would given a specific activity of 2 x 10^8 d.p.m μg^1 with 500 ng input DNA and 2 x 10^9 d.p.m. μg^{-1} with 50 ng. (1 μCi = 2.2 x 10^6 d.p.m.).

For Random primer labelling, materials needed are:

10 x buffer (600 mM Tris-HCl pH 7.8; 100 mM $MgCl_2$, 100 mM 2- mercaptoethanol; TE buffer, pH 8.0; 5 x nucleotide mix (100 μM d ATP, 100 μM dGTP, 100 μM dTTP in TE); Klenow fragment of DNA polymerase I; [α – ^{32}P]dCTP at 3000 Ci mmol^{-1}; random hexanucleotide primers (20 OD units ml^{-1} in TE containing 4 mg/ml^{-1} bovine serum albumin); 0.5 M EDTA. pH 8.0. Dilute the DNA to be labelled to a concentration of 2-25 μg ml^{-1} in distilled water or TE. If double stranded DNA is being labelled, denature by heating to 95-100°C for 2-5 min in a boiling water bath, then chill on ice.

3. Add the following, in the order given, to a microfuge tube of ice: DNA solution, 25-250mg; 5 x nucleotide mix, 10 μl; water sufficient for a final vol. of 50 μl; 10 x buffer 5 μl; random primers 5 μl; [α – ^{32}P] dCTP (3000 Ci mmol^{-1}) 10 μl (100 μCi); Klenow polymerase 2 units to make a total volume of 50 μl.

Mix gently by pipetting up and down once or twice, and cap the tube. If necessary, spin briefly (2 *s*) in a microcentrifuge to collect the contents at the bottom of the tube. Incubate the reaction mix at room temp for 3 h to overnight, or a 37°C or

30 min to 3 h. Terminate the reaction by adding 2 µl of 0.5 M EDTA, pH 8.0. Store at -20°C in the presence of 20 mM EDTA, pH 8.0 before use in hybridization analysis , Prolonged storage can lead to substantial degradation and probes of high specific activity should be stored for no longer than three days. Denature the probe by heating to 95-100°C for 5 min immediately before use.

Calculation of the probe yield and specific activity:

During random-primer labelling there is net synthesis of DNA, while the initial substrate remains unlabelled. Both participate in the subsequent hybridization. so:

- Probe yield = ng initial substrate DNA + ng DNA synthesized. As the average mol.wt of a nucleoside monophosphate in DNA is 350, for labelled nucleotide at $X \times 10^3$ Ci $mmol^{-1}$:
- $$\text{ng DNA synthesized} = \frac{\mu\text{ Ci incorporated} \times 0.35 \times 4}{X}$$

  Note that a multiplication factor of 4 is included as there are four nucleotides, only one of which is labelled.

  Once the probe yield has been calculated, the specific activity can be determined:

  $$\text{specific activity (d.p.m } \mu g^{-1}) = \frac{\text{total activity incorporated (d.p.m.)}}{\text{probe yield } (\mu g)}$$

A worked example:

- assume that in the reaction mix set up in step 3, 70% incorporation was obtained with 25 ng DNA
- this means that 70% of 100 µCi = 70 µCi was incorporated at 3000 Ci $mmol^{-1}$
- so the amount of nucleotide incorporated = $70 \times 0.35/3 = 8$ ng
- the amount of DNA synthesized = $8 \times 4 = 32$ ng ng
- The total DNA = 32 + 25 = 57 ng = 0.057 µg
- as 70 µCi = 1.5×10^8 d.p.m:

$$\text{probe specific activity} = \frac{1.5 \times 10^8}{0.057} = 2.6 \times 10^9 \text{ d.p.m } \mu g^{-1}.$$

Similarly, 70% of incorporation with 250 ng DNA gives a total DNA concentration of 250 + 32 = 282 ng, and a probe specific activity of

$$\frac{1.5 \times 10^8}{0.282} = 5 \times 10^8 \text{ d.}p.m. \ \mu g^{-1}.$$

***Primed in situ labelling (PRINS)* in human chromosomes with non radio-active probe**

Obtain acetic acid: methanol fixed metaphase plates from normal human leucocytes by standard procedure.

Denature slides in 70% formamide, 2 x SSC at 70°C for 2 min, dehydrate in an ascending ethanol series (70,80,90,95,99% v/v) and air dry. Pre-incubate the slides at the annealing temperature (37-53°C) for 20-30 min. Preheat coverslip and reaction mixture as well. The reaction mixture contains 2-400 pmol oligonucleotide; 10nmol each of dATP, dCTP and d GTP; 5 nmol bio-11-dUTP in 25µl 1 x NT buffer (1 x NT buffer = 50mM Tris-HCl, 7.2; 10mM $MgSO_4$; 100µM dithiothreitol; 50µg/ml bovine serum albumin BSA). Add 1U Klenow polymerase after preheating the reaction mixture. Transfer mixture on slide, spread evenly and cover with a coverslip. Polymerize for 15-20min at the annealing temperature. Terminate reaction by washing in 100ml 50mM EDTA, 50mM NaCl at 65°C for 1min. Transfer slide to 100ml 1 x BN buffer (100mM $NaHCO_3$, pH 8.0, 0.01% Nonidet p-40) at 4°C.

Visualize the regions biotinylated by primer extension by the FITC-avidin biotinylated anti-avidin sandwich technique as described in earlier schedule. The FITC-conjugated avidin binds specifically to incorporated biotin. Counterstain with fluorescent propidium iodide binding with DNA (Koch *et al* 1989).

** Oligonucleotides were synthesized on an Applied Biosystem DNA synthesis machine. They were used immediately after elution.

This method with modification has been used for mouse meiotic chromosomes (Moens and Pearlman 1990). The synthesized hybridization primer was the 17 nucleotide sequence (5'A TTCGT TGG AAA CGGGA 3').

DNA In Situ Hybridization (Human)

Skin

Use sections of skin from paraffin blocks after removing paraffin though xylene-ethanol grades and air drying at 42°C.

In situ hybridization using photobiotin as labelling agent for cDNA probes:

Label cDNA probes (20µg lots) by combining equal volumes of probe and photobiotin acetate dissolved in distilled water (20µg at 1µg/µl). Keep in icewater bath and irradiate for 20min at a distance of 20cm from a 250Watt lamp (Mercury vapour, Philips BHRF). Add 60µl of 100mM Tris-HCl, 1.0mM EDTA, pH 9.0 and mix well. Add 100µl butan-2-ol and mix thoroughly. Centrifuge at high speed for 1 min. Remove and discard supernatant. Repeat extraction with butan-2-ol. The aqueous phase will now be concentrated to about 30-40 µl and unreacted photobiotin removed by extraction. Add 5µl of 3M sodium acetate and mix well. Add 100µl 100% ethanol. Keep at - 20°C overnight. Centrifuge at -20°C for 10min. Decant supernatant and rinse pellet of DNA with cold 70% ethanol. Dry in vacuum. Dissolve in 400µl sterile water and store at - 20°C (after EM Nicholls, personal communication). c-DNA probes were obtained from Research Laboratories and manufacturers. Place 100µl of photobiotin-labelled probe (5 µg) and 20 µl of 10 x SSC in a sterile tube and heat at 90°C for 10min to dissociate DNA. Mix and chill in ice. Add 900µl hybridization buffer (1.0g dextran sulphate Sigma; 5.0ml formamide BDH; 2.0ml 10 x SSC; 1.0ml 10mM EDTA; ;1.0ml Herring sperm DNA, 2.5mg/ml, Boehringer Mannheim) and mix. Store at -20°C.

Add 40-50µl of the appropriate DNA probe in hybridization buffer to the air-dried sections, cover with a coverslip, place in a moist chamber and incubate for 22-28h at 42°C for hybridization of the single stranded DNA probe with mRNA sequences of the tissue. Remove coverslip by rinsing in 2 x SSC and 0.1% SDS at 42°C. Incubate for 15min in 2 x SSC-0. 1% SDS and then in 0.4 x SSC- 0.1% SDS at 42°C. Repeat twice. Rinse successively in two changes of 2 x SSC and 2 x PBS at room temperature. Keep slides in two changes of Buffer 1 (0.1M Tris, 2mM $MgCl_2$; 1.0 M NaCl; 0.05% Triton x-10, pH 7.5), for 5min in each. Drain and add 50µl of ***Blocking solution*** (3.0g bovine serum albumin dissolved in 70ml sterile water and adjusted to pH 3.0 with conc HCl. Heat in boiling water bath for 20min, cool to room temperature and adjust to pH 7.5 with 10M NaOH. Add 10ml of a mixture of 1.0M Tris-HCl, 20mM $MgCl_2$, 0.5% Triton X100, pH 7.5 and 5.8 g NaCl and make up to 100ml with sterile water). Cover section with Blocking solution, incubate for 15min at room temperature. Rinse slides in Buffer 1 and drain. Add 50µl of avidin-alkaline phosphatase (5µg/ml in Buffer 1, Bresatec, Adelaide) and incubate for 10min. Rinse in Buffer 1. Incubate twice in Buffer 2(0.1M Tris, 0.1M NaCl, 10mM $MgCl_2$, pH 9.5) for 10 min each. Drain slides and add 40-50µl of Developing reagent (5ml, Buffer 2, 50µl NBT and 50µl BCIP, Bresatec, Adelaide). Cover with a coverslip. Store in dark for 1h. Wash in water several times. Counterstain with Nuclear Fast Red (0.1g nuclear fast red Chroma in 100ml of 5% aluminium sulphate, a few crystals of thymol added) for 3min and Metanil Yellow (0.25g metanil yellow Chroma in 0.25% acetic acid) for 2min. Wash twice in water. Airdry overnight at 42°C. After thorough drying add immersion oil for observation.

In Situ Hybridisation and SCE Combined Method in Human Leucocyte

Two procedures; (i) hybridization with a cosmid probe labelled by nick translation in the presence of digoxigenin dUTP and (ii) FPG banding. It allows simultaneous observation of the hybridization signal visualized by an alkaline phosphatase- conjugated antibody and the banding pattern. A single photographic step is needed.

Chromosome preparations: Prepare metaphase spreads from phytohaemaglutinin stimulated lymphocytes, after treatment with 30 µg/ml 5-bromo-2′-deoxyuridine (BrdU) for the last 7h of culture.

Probe: CD3H12 isolated from a cosmid library of total human DNA after screening with the human DNA insert isolated from plasmid pD3 H12 (D20S6). The purified cosmid DNA was labelled by nick translation with digoxigenin-dUTP, as given by the supplier (Boehringer-Mennheim).

In situ hybridization: Treat slides with 100 µg/ml RNAse in 2 x SSC, at 37°C for 1h. Wash thrice in 2 x SSC pH 7.0; dehydrate in an ethanol series (50, 75 and 100%, 1min in each). Proteinase K treatment in *not* needed. *For DNA denaturation*, immerse the slides in 70% formamide 2 x SSC pH 7.0 for 2min at 70°C. Dehydrate in the same ethanol series as above. *For probe hybridization* incubate together 50% formamide, 0.3M NaCl, 30mM sodium citrate, 10% dextran sulphate, 100µg/ml sonicated human DNA; 1% Denhardt's solution, 0.1% SDS, 20mM NaH_2PO_4, 20mM Na_2HPO_4, pH 6.8 and 1 to 2 µg/ml of digoxigenin-dUTP labelled cosmid DNA overnight in a moist chamber at 40-42°C (20-25µl of the mixture under a coverslip). Rinse the slides twice for 20min at 42°C in a mixture of equal volumes of 4 x SSC, pH 7.0 and formamide and wash twice for 10min each in 2 x SSC at room temperature (Zhang *et al* 1990).

Immunological detection is carried out according to the procedure given by the manufacturer. Dilute the anti-digoxigenin antibody phosphatase conjugate 1: 7500 in buffer 1(100mM Tris- HCl, 150mM NaCl, pH 7.5. Incubate slides for 90-120 min with the diluted antibody-conjugate. Wash twice (15 min each) in buffer 1. Rinse in buffer 3(100 mM Tris-HCl, 100 mM NaCl, 50 mM $MgCl_2$, pH 9.5) for 2 min. Incubate for 1-2 h in colour solution supplied by manufacturer. Terminate treatment by washing twice (15 min each) in buffer 4 (10 mM Tris-HCl, 1 mM EDTA, pH 8.0).

Chromosome staining and banding: Stain the slides after immunological detection directly for 2min in a 2% Giemsa solution. Photograph the metaphases. Band using FPG staining technique, described under Chapter on Banding Patterns of this book. Denature the slides for 10min in Earle's solution, pH 6.5 at 97°C to enhance banding. Photograph again.

Alternatively, carry out the FPG treatment after hybridization but before immunological detection.

In Situ Hybridization in Insects

Using conventional *in situ* protocols, it is possible to map both moderately and highly repeated sequences of low reiteration on dipteran polytene chromosome Modifications of this technique permit the mapping of single copy sequences on mitotic chromosomes. Protocols have been developed for carrying out the procedure on specimens prepared for transmission electron microscopy.

The 5S system in *Xenopus laevis* represents a much lower repetition frequency then mouse satellite. In addition, probes form intergenic space regions are available which discriminate between oocyte clusters (20,000 per haploid genome) and a somatic cluster (400 genes per haploid genome). Oocyte clusters occur at or near the tips of most Xenopus chromosomes, whereas the somatic clusters appear at the same location as an oocyte cluster on a single chromosome which also carries and oocyte cluster.

In Situ Hybridization for Insects

Dissect out salivary glands from third instar larvae of *Drosophila* and squash in 45% acetic acid. Remove coverslip by freezing in liquid nitrogen. Dehydrate in ethanol and air dry.

Incubate slides in 2 x SSC at 65°C for 30 min, dehydrate in ethanol and airdry. immerse in 0.07N NaOH for 2 to 3 min to denature chromosomal DNA and then wash thrice in 2 x SSC, keeping 5min in each. Dehydrate and airdry. Label DNA probes by nick- translation, as described for mammals, using Bio-11-dUTP or Bio- 14-dATP (Enzo and BRL). For 10 slides, dissolve 1 μg of biotinylated probe in 200 μl solution containing 50% formamide, 10% dextran sulphate, 0.4μg salmon sperm DNA in 2 x SSC. Incubate this mixture at 90°C for 10min and apply to the squashes. Cover with a coverslip, seal and incubate in a moist chamber at 80°C for 10min. Then hybridize at 37°C for 16-20h. Wash the slides in 2 x SSC for five times, keeping 10min in each, first at room temperature, twice each at 42°C and lastly twice at room temperature. Incubate slides for 5min in PBS containing 0.1% Triton X-100 and wash twice in PBS,keeping 5min in each. To detect biotin, incubate the slides with a streptavidinperoxidase complex (Detek 1-hrp Signal Generating System, Enzo) for 90min. Wash in 2 x SSC twice for 5min each; in PBS containing 0.1% Triton X-100 and finally PBS, keeping 5min in each. Stain with Giemsa, air dry and mount in DPX. Peroxidase activity can be detected using diaminobenzidine as a chromogen (after Visa *et al* 1991).

Two-Colour Hybridization in Human Leucocyte Culture

Selected whole chromosomes are stained in one color by hybridization with composite probes whose elements have DNA sequence homology along the length of the target chromosomes. All chromosomes are counterstained with a DNA-specific dye so that structural aberrations between target and non-target chromosomes are clearly visible (Weier *et al* 1991).

Centromeric probe production and biotin labelling: Generate a collection of nucleic acid probe sequences that has elements homologous to alpha satellite DNA sequences at or near the centromeres of all human chromosomes, by the amplification of selected elements of genomic DNA using the polymerase chain reaction (Saiki *et al* 1988). Select degenerate PCR primers to be homologous to two regions of the 171bp alpha satellite repeat that are highly conserved among the various human chromosomes (Murray and Martin 1987; Willard and Waye 1987).

Produce specifically the primers WA1 and WA2, that bind to positions 37 to 52 and 11 to 26 respectively, of the published census sequence.

Synthesize such regenerate oligonucleotide primers using phosphoramidite chemistry on a DNA-synthesizer (Applied Biophysics, Foster City Calif model 380B). Synthesize and further purify the oligonucleotides by C_{18} reverse phase chromatography and HPLC according to the specifications of the manufacturer (Waters Chromatography, Milford, Mass, USA). Both primers WA1 and WA2 are 23-mers and carry 6bp restriction enzyme recognition sites to facilitate molecular cloning of amplification products. WA1 carries a 5′ HindIII site and WA2 a Pst1 site. Add an extra base to the 5′ end of each primer to ensure reproduction of the restriction sites by *Thermus aquaticus* (Taq) DNA polymerase. Perform *in vitro* DNA amplification using approximately 200 ng of human genomic DNA (100ng/μl) as amplification template per 200μl reaction min. Perform PCR for 30 cycles using an automated thermal cycler (Perkin Elmer Cetus, Norwald, Conn, USA). Denature the DNA template at 94°C for 1 min (1 min 30 *s* during the first cycle). Anneal the primer at 45°C and perform extension at 72°C using 5 units of Taq DNA polymerase (Bethesda Research Laboratories, BRL Gaithersburg, Md), and 1.2μm primer.

For biotinylation and further amplification of the probe, add a 5μl aliquot of the product to 200μl reaction mix containing 0.25mM biotin-11-dUTP (Sigma, St. Louis, Mo) in the absence of dTTP and 10units of Taq polymerase. This reaction was carried out during an additional 20 PCR cycles. Amplification of *das* DNA (degenerate alpha satellite probe) can be confirmed visually by gel electrophoresis of 10μl aliquots of the PCR reaction in either 1.8% or 4% agarose (BRL) in 40mM Tris-acetate, 1mM EDTA buffer, pH 8.0 containing 0.5μg ethidium bromide (EB). Store labelled probe and amplified DNA at -18°C.

Probe labelling with acetyl aminofluorene (AAF): Precipitate approximately 20μg unlabelled PCR products in 50% isopropyl alcohol, 0.1M sodium acetate at 20°C for 30min. Centrifuge to pellet DNA and resuspend in 80μl of TE (10mM Tris-HCl, pH 7.6, 1mM EDTA). Add 20μl ice cold ethanol and raise sample temperature to 37°C. Add 5 μl of N-acetoxy-2- acetylaminofluorene (10mg/ml in DMSO, Chemsyn Science Lab, Lenexa, Kan) and incubate reaction mixture at 37°C for 1 h. Extract DNA three times with phenol/chloroform/isoamyl alcohol (Maniatis *et al* 1986) and then extract with ether twice at room temperature. Precipitate out DNA in 2.5 vol ethanol, 0.1 M sodium acetate, dry and resuspend in 500 μl 10 mM Tris-HCl, pH 8.0, 1 mM EDTA.

Composite Probe Production: Generate biotinylated composite probe by nick translation of the human chromosome 4-specific HindIII Library pBS-4. Isolate plasmid DNA from the DH5α MAX host cells (BRL) by a cesium chloride gradient. Cut 100 μg aliquots repeatedly with 0.1 unit of DNAse I(BRL) until the double stranded fragments range in size from 100 bp to 3kb, with an average size of 700 bp. After each digestion, stop the reaction with a 10 min incubation at 70°C. Assess size distribution by electrophoresis of small aliquots on agarose gels. Nick translate the cut DNA with 5 units of *E.coli* DNA polymerase I(BRL), incorporating biotin-11- dUTP (Enzo diagnostics, New York, NY) as reported elsewhere.

Cell culture and metaphase spread preparation: Irradiate heparinized whole blood cultures at ambient temperature with 0.66 MeV gamma rays from a Cs-137 source to a total dose of 4.0Gy. Prepare metaphase plates 48 h after irradiation. Store slides at -20°C.

Fluorescence in situ hybridization (FISH) with centromeric satellite DNA: Denature metaphase chromosome preparations for 4 min at 70°C in 70% formamide (IBI, New Haven, Conn); 2 x SSC pH 7.0. Prepare hybridization mixture (55% formamide, 10% dextran sulphate, 1μg/μl sonicated herring sperm DNA, 2 x SSC, pH 7.0). Add biotinylated *das* probe (1 μl, 20ng/μl) without purification to 9 μl of the hybridization mixture. Denature DNA in hybridization mixture at 72°C for 5 min. Chill on ice. Add chilled mix to metaphase spread, cover with coverslip and hybridize overnight at 37°C in a moist chamber. Wash slide in 50% formamide, 2 x SSC pH 7.6 at 42°C for 10 min each, followed by two washes of 20 min each in PN buffer (0.1M sodium phosphate, pH 8, 0, 0.1% NP40) at 37°C. Detect biotinylated probe by incubating for 20 min in 5μg/ml avidin - FITC (fluorescein isothiocyanate, Vector Lab, Burlingame, Calif) in PN buffer plus 5% non-fat dry milk and 0.01% sodium azide at room temperature. Wash twice in PN buffer at room temperature. Counterstain DNA with 0.5 μg/ml DAPI (Calbiochem, La Jolla, Calif) in antifade solution. Observe under Zeiss Universal fluorescence microscope with a Plan-Neofluar 63 x 1.20 oil objective using appropriate filters to simultaneously observe FITC and PI fluorescence and separate filters for DAPI. Wash the slide briefly in PN buffer and add PI (2 μg/ml in antifade solution, Sigma).

Dual staining of chromosome 4 and centromeric satellite DNA: For each slide, mix 0.5 μl of chromosome 4-specific biotinylated composite probe (pBS-4, 20ng/μl) with 3.5 μl Mastermix 1.0 (71.4% formamide, 14.3% dextran sulphate, 2.9 x SSC, pH 7.0); 0.2 μl herring sperm DNA (1μg/μl Sigma), 0.3 μl of sonicated human placental DNA (1 mg/ml Sigma) and 0.5 μl distilled water. Heat the mix to 72°C for 5 min to denature the DNA. Incubate at 37°C for 1 h. Mix centromeric probe (0.5μl of AAF-labelled *das* probe DNA, approx. 40 ng/μl) with 3.5 μl Mastermix 1.0; 0.5μl herring sperm DNA and 0.5μl water and denature at 70°C for 5 min. Immediately add the satellite probe to the 5μl composite probe mix. Add the hybridization cocktail to the slide with metaphase spreads from irradiated lymphocytes that had been denatured as described above. Hybridize in a moist chamber at 37°C for overnight. Wash twice in 50% formamide, 2 x SSC, pH 7.0 for 15 min each at 42°C and then two changes of PN buffer for 5 min each. Incubate in PNM (PN buffer containing 5% nonfat dry milk) for 5 min at room temperature. Incubate the slides in avidin - FITC (5 µg/ml) in PN buffer at room temperature. Wash twice in PN buffer. Amplify signals by incubation with biotinylated goat anti-avidin antibodies (Vector Lab.) at 1:250 dilution in 1X Dulbecco's PBS (Ca, Mg-free) + 0.05% Tween 20 + 2% normal goat serum. Wash with PN and incubate with avidin-FITC. Wash in PN for 15 min at room temperature and cover with 50 µl of undiluted supernatant from murine hybridoma cells producing antibody against AAF. Incubate for 1 h at room temperature. Wash the slides twice in PN buffer, block for 5 min in PNM and incubate with 25 µl 7-amino-4-methylcoumarin-3-acetic acid (AMCA) - conjugated rat-anti-mouse-antibody (Jackson Immuno Research, West Grove, Pa., 1:50 dilution in 1 x Dulbecco's PBS (Ca, Mg free) + 0.05% Tween 20 + 2% normal goat serum). Remove unbound antibody by washing twice in PN buffer for 2 min each at room temperature. Treat slides with PNM for 5 min and incubate with mouse-anti-rat antibody conjugated AMCA (Jackson Immuno Research 1:50 dilution in 1 x Dulbecco's PBS Ca, Mg free + 0.05% Tween 20 + 2% normal goat serum) for 1 h at room temperature. Remove unbound antibody by washing in PN. Counter-stain DNA with PI(0.5µg/ml in antifade). Use fluorescence microscope using an FITC filter set for simultaneous observation of FITC and PI to determine the binding distribution of the whole chromosome 4 probe. It allows rapid detection and characterization of structural chromosome aberrations.

Multiple Banding and In situ Hybridization in Mammal

Chromosome Preparations: Expose cultured cells to 0.1µg/ml colcemid for 30 min, treat with hypotonic 1% sodium citrate for 15 min and fix with three changes of acetic acid- methanol (1:3). Prepare air dried slides.

Triple Fluorescent Staining: Stain the chromosome preparations first with 0.5 mg/ml chromomycin A_3 (Sigma) in the presence of 2.5 mM $MgCl_2$ for 30 min and rinse with McIlvaine's buffer (pH 6.8, 0.15 M phosphate - citrate). Stain the preparation again with 1.0 mg/ml distamycin A (DA, Sigma) for 20 min, wash with the same buffer and stain with 1.0 µg/ml DAPI (4'-6-diamidino-2-phenylindole, Sigma). Mount in a 1:1 mixture of glycerol and McIlvaine's buffer. Observe under a Zeiss fluorescence microscope equipped with vertical illuminator 100 and HBO 50 mercury lamp. Observe with the help of BP365 11 exciter filters for DA - DAPI bands and BP436 8 for chromomycin A_3 R bands.

Restriction Enzyme Induced Banding: Apply about 30 units of the restriction enzyme needed (EcoRII, HpaI from Bethesda Research Lab; MspI, HpaII, from Bochringer Mannheim; AluI, BamHI, EcoRI, HaeIII, HinfI, RsaI, Sau3A, ThaI from Toyobo Co) dissolved in 30 µl of the appropriate buffer to the air dried chromosomes. Mount with coverslip and incubate in a moist chamber at 37°C or the temperature suggested by the manufacturer for 4 to 20 h. Remove coverslip and rinse slide in distilled water, treat with 100 µg/ ml ribonuclease A for 1 h and stain with 10µg/ml propidium iodide for 15 min.

Chromosome Hybridization: Treat air dried chromosome spreads with 100 µg/ml of RNAse A in 2 x SSC for 1 h at 37°C, denature with 70% formamide in 2 x SSC at 70°C for 3 min, dehydrate in a graded ethanol series, starting from 70% upwards and air dry. Dissolve DNA probes, labeled with biotin - dUTP (Bio-11-dUTP, Enzo-Biochem) by nick translation in the hybridization mixture (10% dextran sulphate, 4 x SSC, 40% formamide, 0.3 µg/ml sonicated salmon sperm DNA, 1 x Denhardt's solution, 10 mM sodium phosphate, pH 6.8) at 1.0 ng/ml. Denature at 70°C for 5 min and store at 4°C. Apply approximately 30µl of the probe solution to each slide. Hybridize by incubating the slides with a coverslip at 37°C in a moist chamber overnight. Wash the slides successively in 2 x SSC, in 0.5 x SSC (twice for 5 min each) and in phosphate - buffered saline (PBS). Treat with 10% normal goat serum for 30 min at 37°C for blocking. Treat sequentially with the goat antibiotin-antibody (Enzo Biochem Inc, 1:50 dilution) biotinylated rabbit anti-goat IgG (Vector Lab, 1:50 dilution and finally with avidin peroxidase conjugates (ABC reagent, Vector Lab), keeping in each reagent for 1-2 h at 37°C. Wash thoroughly with PBS between each treatment.

For colour development: Immerse the slides in a freshly prepared solution of 0.5 mg/ml 3, 3' - diaminobenzene (DAB, Sigma) in 10 mM Tris-HCl, pH 7.6, containing 0.01% H_2O_2.

MICRO-FISH — Rapid Regeneration of Probes by Chromosome Microdissection

Chromosome Microdissection: Harvest metaphase cells from tissue culture or biopsy samples following conventional cytogenetic techniques. Prepare metaphase spreads on 24 x 60 mm No 1.5 coverslips and stain by trypsin-Giemsa banding. For microdissection, use glass microneedles controlled by a Narashige micromanipulator (Model MO 302) attached to an inverted microscope. Before use treat microneedles with UV light for 5 min (Stratalinker, Stratagene). Dissect out chromosome fragment and transfer to a 20 µl collecting drop (containing proteinase K 50 $\mu g/ml^{-1}$) in a 0.5 ml microcentrifuge tube. Use fresh microneedle for dissecting out each fragment (after Mettzer).

PCR Amplification: Incubate collecting drop with chromosome fragment at 37°C for 1 h and then at 90°C for 10 min. Add the components of the PCR reaction to a final volume of 50 μl in the same tube (1.5 μM universal primer containing sequence, CCGACTCGAGNNNNNNATGTGG, as suggested by Telenins HGM 11 abstract No 27106, p. 316; 200 μM each dNTP, 2mM $MgCl_2$, 50 mM KCl, 10 mM Tris-HCl pH 8.4; 0.1 mg/ml^{-1} gelatin and 2.5 U Taq DNA polymerase (Perkin Elmer Cetus). Overlay the reaction with oil and heat to 93°C for 4 min and cycle for 8 cycles at 94°C for 1 min; at 30°C for 1 min, at 72°C for 3 min, followed by 28 cycles at 94°C for 1 min, at 56°C for 1 min and at 72°C for 3 min with a final extension at 72°C for 10 min.

To prepare probes: Label 2 μl of the microdissection PCR in a secondary PCR reaction identical to the first except for reduction of the TTP concentration to 100 μM and the inclusion of 100 μM biotin-11-dUTP cycled for 8 cycles at 94°C for 1 min, at 56°C for 1 min and at 72°C for 3 min with a final extension at 72°C for 10 min. Purify the products of this reaction with a Centricon 30 filter and use for *in situ* hybridization. Alternatively, probe label through biotinylation with biotin-11-dUTP in a nick translation reaction.

In situ Hybridization of Micro-FISH Probes: based on standard Imagenetics (Naperville, Illinois protocol after Pinkel *et al* 1986): Denature slides with metaphase plates at 72°C for 2 min in a bath of 70% formamide, 2 x SSC, dehydrated through ethanol series and airdry. Place in a slide warmer at 37°C, with 20 μl of hybridization mixture (10 ng μl^{-1} probe, 50% formamide, 10% dextran sulphate. 1 x SSC, 3 μg Cot1 DNA BRL). Place a coverslip, seal with rubber cement and incubate overnight at 37°C in a moist chamber. Wash slides three times in 2 x SSC/50% formamide at 42°C for 3 min, then rinse for 3 min in 2 x SSC at ambient temperature.

To detect probe, wash slide in 0.1M sodium phosphate, 0.1% NP-40, pH 8.0 (PN) at 45°C for 15 min, at ambient temperature for 2 min and then incubate in a bath containing 5 μg/ml^{-1} fluorescein-conjugated avidin (Vector Laboratories) in PN; with 5% nonfat dry milk and 0.02% sodium azide for 20 min at ambient temperature. Wash slide twice in PN for 2 min each. Incubate in a bath containing 5 μg/ml^{-1} biotinylated anti-avidin (Vector Laboratories) in PN with 5% nonfat dry milk and 0.02% sodium azide for 20 min at ambient temperature. Wash twice in PN for 2 min each and repeat incubation in fluorescein-conjugated avidin. Finally apply 10 μl of fluorescence antifade solution (10 μg/ml^{-1} p-phenylamine dihydrochloride in 90% glycerol pH 8, 0) containing 0.2 μg/ml^{-1} propidium iodide and cover with a coverslip.

In an alternative schedule to localise micro FISH signal on chromosome bands, first process the slides by trypsin-Giemsa banding, photograph and destain by sequential washing in 70% ethanol (1min), 85% ethanol (1 min), Carnoy's fixative (10 min), 3.7% formaldehyde in PBS (10 min) and PBS (5 min twice) before hybridization. For hybridization including fluorochrome-labelled *whole chromosome paints* (WCP, Imagenetics, Naperville IL), include 10 ng μl^{-1} WCP in the hybridization mixture. Examine slide with a Zeiss Axiophot microscope equipped with a dual bandpass fluorescein-rhodamine filter.

Sensitive High Resolution Chromatin and Chromosome Mapping *in situ*

The method is derived from *in situ* hybridization and non- isotopic detection techniques described in this chapter with the following modifications: frozen storage and hardening of nuclear preparations before processing; treatment with RNAse H after hybridization instead of with RNAse A before hybridization; use of probe concentrations 100-1000x greater than that used in standard protocols; monitoring of probe fragment size after nick translation to use probe molecules approximately 300-1000 nucleotides long and staining with fluorescein-avidin in 4 x SSC rather than PBS (Lawrence *et al* 1988).

In Situ Hybridization Using Radioactive Probe in Plant Pachytene

Freeze slides containing chromosome spreads in liquid nitrogen and remove coverslip. Rinse with a freshly prepared mixture of acetic acid: 95% ethanol (1:3). Treat with 100 µg/ml RNAse in 2 x SSCP (1 x SSCP = 0.15M NaCl, 0.015M sodium citrate, 0.02M sodium phosphate pH 7.0) at 37°C for 1 h under a siliconized coverslip in a moist chamber. Wash slide thoroughly in 2 x SSCP, dehydrate sequentially in 70, 80 and 95% ethanol and air dry.

For *in situ* hybridization, denature chromosome preparations in 70% formamide in 2 x SSCP at 70°C for 4 min. Transfer quickly to chilled 70% ethanol and dehydrate in 80, 95 and 100% ethanol. Denature probe mixture, containing 50% formamide, 10% dextran sulphate, 100ng/100 µl probe DNA, 10µg/100µl carrier DNA (sheared *E. coli* DNA) at 70°C for 6 min and quickly cool in an ice bath. Add 20 µl probe mixture to each slide and cover with a coverslip. Incubate in a moist chamber at 37°C for 20 h. Rinse in two changes of 50% formamide in 2 x SSCP at 39°C, keeping 15 min in each. Wash in 2 x SSCP at 39°C for at least 7 changes, keeping 10 min in each.

For Autoradiography, dehydrate slides and coat with photographic emulsion (diluted L4 photographic emulsion, Polyscience, L4:water = 1:1) at 45°C. Airdry slides in dark for 1 h, and keep in light-tight box at -20°C for 2-10 weeks. Develop in Kodak D-19 for 4 min, treat with 1% acetic acid, fix with Kodak fixer, rinse thoroughly in distilled water and stain with Giemsa for light microscopy (after Shen *et al* 1987).

Single Copy Gene from Cell Culture in Plants

Maintain cell suspension cultures of parsley in dark and subculture every 7 days.

For chromosome preparation, treat three day-old cultures with aphidocoline (7.5 μg/ml) and 5-bromodeoxyuridine (BrdU, 5 x 10^{-4}M) for 24 h, wash and reincubate for 6 h with fresh medium containing deoxythymidine (3 x 10^{-5}M).

Treat for 2 h with colchicine (0.00625%). Isolate protoplasts by the method of Murata (1983) as follows: Collect colchicine- treated cells by centrifuging at 100xg for 5 min. Reduce supernatant to one-fourth of the original volume. Add an equal volume of enzyme solution (2% cellulysin, 1% macerase in 0.7M mannitol, 10 mM MES, pH 5.6), Continue protoplast isolation for 1 h at 25°C in the dark on a rotary shaker (50rpm). Filter the protoplasts through 80 μm nylon mesh and wash twice with a solution containing 100 mM glycine, 2.5 mM $CaCl_2$ and 5 % glucose at pH 6.0 (Hadlaczky *et al* 1983). Hypotonize in 75 mM KCl solution for 10 min and fix overnight in Farmer's solution (glacial acetic acid - ethanol 1:3). Change fixatives till a clear supernatant is obtained. Spread the suspension on ice-cold wet slides and air dry at room temperature.

The steps needed for *in situ* hybridization are similar to those described earlier, using the relevant probes (after Huang *et al* 1988).

Genomic In Situ Hybridization with Non-radioactive Probe (GISH) for Plants

I. Isolation of Genomic DNA (after Appels and Dvorak 1982, Maniatis *et al* 1982, Bennett *et al* 1992)

1. Surface sterilise 0.5 to 1.0 g fresh leaf material in 2% bleach in deionized water for 10 min. Briefly rinse in water.

2. Crush to fine powder in mortar and pestle, with a little sterilized water and liquid nitrogen.

3. Add 1 ml isolation buffer to form ice slurry (buffer: 2 x CTAB, 100 mM Tris-HCl pH 8.0, 1.4M NaCl, 20 mM EDTA, 2 % CTAB (hexadodecyl (cetyl) trimethyl ammonium chloride). Continue brisk homogenization.

4. Add 100 μl of 5% SDS (sodium dodecyl sulphate), followed by 100 μl proteinase K (0.2 mg/ml stock solution) and stir vigorously,.

5. Transfer homogenate to a 15 ml centrifuge tube.

6. Rinse pestle and mortar with 1 ml isolation buffer and add to centrifuge tube.

7. Incubate at 37°C for 3 h.

8. Extract DNA by adding an equal volume of phenol/chloroform to mixture in centrifuge tube. Gently invert several times.

9. Centrifuge at 13000 rpm for 10 min.

10. Transfer supernatant fluid to a clean centrifuge tube. Discard bottom layer.

11. Repeat steps 8, 9 and 10.

12. To centrifuge tube add an equal volume of chloroform to the contents. Gently invert several times and repeat steps 9 and 10.

13. precipitate the DNA by adding 3 M sodium acetate equal to 1/20th of total volume of mixture in tube and then absolute ethanol about two and a half times the present volume of mixture in tube. Leave at - 20°C for overnight.

14. Centrifuge at 13000 rpm for 10 min. Discard supernatant, wash pellet with 70% ethanol 1 ml and drain. Resuspend pellets in TE (pH8)- 500 µl aliquots until dissolved.

15. Incubate in 10 µg/ml RNAse at 37°C for 3 h.

16. Add SDS (5%) equal to 1/20th of total mixture in tube.

17. Repeat step 13, leaving to reprecipitate for at least 4 h.

18. Centrifuge at 13000 rpm for 10 min, discard supernatant, wash in 70% ethanol and resuspend pellet in TE pH 8.0, as in step 14.

19. Store in 50 µl aliquots at -20°C for short term or - 70°C for longterm.

II. Assessment of Isolated DNA and Shearing

Prepare 2 µl DNA, 8 µl TE (pH 8.0) and 5 µl loading buffer (40% sucrose, 0.25% bromophenol blue; 1kb ladder with 30% loading buffer (100 ng/µl). Run on a 0.5% (or 1%) agarose gel in 1 x TBE, 60 v, 0.5 mA for 2h.

For shearing DNA from *Milium montanium*, a shearing strategy included use of hypodermic syringe coupled with brief pulsing (1-2 *s*) in a microcentrifuge and longer periods (up to a minute) on a vortex mixer. The method is as follows:

Vortex 50 µl of genomic DNA for 1 min; pulse briefly; suck up-and-down 25 times using hypodermic apparatus; vortex; pulse; suck up-and-down 25 times; vortex; pulse; repeat several times. For optimal labelling, genomic DNA should be sheared into fragments 10 to 12 kb in length.

III. Biotinylation of genomic DNA
(adapted from manufacturers of the BRL labelling system)

Shear genomic DNA to sizes between 10-12 Kb. Add 2-3 µg DNA to 70 µl water, mix well, add 10 µl 10x dNTP mixture, mix and then and add quickly 10 µl 10x enzyme mixture from the freezer. Mix and pulse in benchtop centrifuge. Incubate at 16°C for 1 h. Add 10 µl stop buffer, making a total of 100 µl. Add 60-90µg of sonicated salmon sperm DNA which reduces non-specific binding and helps to bring down biotinylated DNA. Add 3M sodium acetate equal to 1/10th volume in tube (11µl). Mix well. Add 300 µl cold absolute ethanol. Close tube tightly and invert gently several times.

Precipitate the DNA at -20°C for about 1 h or at -70°C for 20-30 min. Centrifuge at 15100 rpm for 10 min. Discard supernatant. Wash pellet with 70% ethanol (500µl). Invert tube several times. Dry pellet gently under vacuum for 15-20 min. Dissolve pellet in 21 µl of the TE (pH 8.0), keeping overnight at room temperature.

Test Biotinylation on Nitrocellulose Strips

Prepare buffers 1, 2 and 3 as instructed by the manufacturers BRL. Pipette 1 µl of the 21 µl "Labelled" DNA in TE onto nitrocellulose strip and dry at 80°C under vacuum for 1- 2 h. Make up 100 µl of buffer 2 and heat 50 µl to *c.* 40°C. Wash strip in buffer 1 for 1 min for rehydration; then incubate in buffer 2(3% BSA in buffer 1) for 20 min at 40°C. Gently blot strip between filter paper; dry at 80°C for 15-20 min; remove and store at room temperature overnight.

Rehydrate in buffer 2 for 10 min. Incubate in 2 µg/ml BRL streptavidin (2 µl in 998 µl) in buffer 1 for 10 min with gentle shaking. Make it up to 1 ml and wash strip in Petri dish. Wash strip three times in 40 ml buffer 1, for 3 min each time with gentle agitation. Incubate in 1 µg/ml BRL biotin AP (1 µl in 999 µl) in buffer 1 for 10 min, agitating gently. Wash twice in 40 ml buffer 1 for 3 min each. Wash in buffer 3 for 3 min twice.

For Detection: to 7.5 ml buffer 3, add 33 µl NBT and 25 µl BCIP solution. Incubate in sealed plastic bag for upto 4 h in dark. Well-labelled DNA turns dark. Wash strips in TE, blot between filter paper and dry in oven for 10 min.

IV. RNAse Treatment

Wash slides in 2 x SSC for 10 min. Pipette 100 µl of RNAse solution onto slides (100 µg/ml boiled RNAse in 2 x SSC), cover with coverslip and incubate in moist chamber (with 2 x SSC) in incubator at 37°C for 1 h. Carefully remove coverslip and place slide in metal rack. Wash slides thrice in 2 x SSC, keeping for 5 min each. Dehydrate slides through ethanol series of 70% and absolute on ice, keeping twice for 5 min in each. Dry slides at approx. 45°C, usually for a few hours.

V. Probe Mixture Preparation

Prepare probe mixture containing per slide (20 µl) the following: DNA (in TE pH 8.0), 100-200 ng; deionized water (20% vol); formamide (Fisons electrophoresis grade), 50% vol; 50% dextran sulphate (20% vol); 3M NaCl 10% vol. Mix and leave on ice. Add dextran last. Denature in water bath at 85°C for 15 min.

After 13 min, add salt, vortex briefly to mix and return tube to water bath. Transfer tubes immediately to ice to quench for 2 min. Pipette 20 µl of probe mixture onto denatured slide. Cover with plastic coverslip. Place slide in moist chamber, seal with masking tape and incubate overnight at 37°C.

VI. Chromosome Denaturation and Hybridization

Transfer slides after RNAse treatment (step IV) to 70% formamide in 2 x SSC and keep in water bath (68-72°C) for 2 min. Dehydrate slides through ethanol series kept at -20°C (70%, 90%, absolute, absolute), keeping 2 min in each. Air dry slides, add probe mixture (step V above) and leave to hybridize.

VII. Post-hybridization Washes

Remove coverslip and keep slides in 2 x SSC for 10 min at 42°C. Wash slides in 50% formamide in 2 x SSC for 10 min at 42 °C. Keep in 2 x SSC for 10 min at 42°C, followed by 5 min in 2 x SSC at room temperature on an orbital shaker.

VIII. Detection of Hybridization and Microscopic Examination

Prepare: BN buffer (0.1M sodium bicarbonate, 0.05% v/v; Nonidet P-40 Sigma, pH 8.0) or BT buffer replacing Nonidet with Tween 20; 5% (w/v) BSA (bovine serum albumin) in BN buffer; 5 µg/ml F- avidin (Vector Labs) in 5 % BSA-BN; 10 µg/ml F-avidin (Vector Labs) in 5% BSA-BN; 5% (v/v) Normal goat serum (NGS) in BN buffer; 25 µg/ml propodium iodide in phosphate-buffered saline (PBS); 1-2 µg/ml DAPI in Mclvaine's buffer (pH 7).

Wash slides in BN for 10 min. Block with 5% BSA-BN (10 µl/slide) for 5 min *without coverslip*. Incubate slides in 5 µg/ml F-avidin in 5% BSA-BN (50-100 µl) *under coverslips* for 1 h at 37°C in moist chamber. Remove coverslips. Wash slides twice in BN at 40°C, for 5 min each. Wash in BN at room temperature for 5 min. Block with 5% NGS (100 µl/slide) for 5 min *without coverslip*. Incubate slides *under coverslip* in 25 µg/ml biotinylated anti- avidin in 5% NGS-BN (50-100 µl/slide) for 1 h at 37°C in moist chamber. Remove coverslip and wash in BN twice at 40°C and once at room temperature, keeping 5 min in each.

Incubate slides in 10 µg/ml F-avidin in 5% BSA-BN under coverslips for 1 h at 37°C. Remove coverslip and wash slides twice in BN at 37°C. Incubate slides in 1-2 µg/ml propidium iodide in PBS for 10 min in dark. Rinse in BN buffer (30-60 *s*). Incubate slides in 2 µg/ml DAPI (100 µl under coverslip) for 30 min in dark. Remove coverslip, rinse in BN (30 *s*) and then distilled water. Dry slightly. Mount under 22 x 22 mm coverslip with BN: Citifluor (Agar Scientific) mountant (1:1, v/v).

Prepare the hybridization mixture according to table below		
Solution	**Amount per slide (total 40 μ l)**	**Final concentration**
100% formamide high grade	20μl	50%
50% (w/v) dextran sulphate in water	8μl	10%
20 x SSC	4μl	2x
Probe	4μl	25 to 100 ng per slide
Blocking DNA (autoclave from salmon sperm)		2 to 100 x probe concentration
10% (w/v) sodium dedecyl sulphate in water	0.2 to 4 μl	0.05 to 1%
water	in requisite amount to make final volume per slide	40 μl

Scan under uv light (365 nm), locate cells and photograph FITC fluorescence under blue light (450-490 nm). Usually half of automatic exposure time gives good results.

The method, with slight alterations, has been used for localization of parental genomes in a wide hybrid between *Secale africanum* and *Hordeum chilense* (Schwarzacher *et al* 1989) and genetic divergence in *Gibasis* (Parokomy *et al* 19992).

In Situ Hybridization with Automated Chromosome Denaturation in Plants

A programmable temperature controller was modified for denaturation of the chromosome spread preparations on slides prior to DNA-DNA hybridization (Make: Cambio from Genesys Instruments, Cambridge after Heslop Harrison *et al* 1991).

Protocol 1. Slide Preparation and Pretreatment for DNA:DNA In Situ Hybridization:

Dry chromosome preparations in an oven at 35 to 60°C overnight. Add 200 μl of x 100 μg/ml (w/v) RNAse A in 2 x SSC (0.3M NaCl, 0.03MNa citrate pH 7.0), cover and incubate for 1 h at 37°C in a humid chamber. Wash slides thrice for 5 min each in 2 x SSC at room temperature. Transfer slides to 4% (w/v) freshly depolymerized paraformaldehyde in water and incubate for 10 min at room temperature. Wash thrice for 5 min each in 2 x SSC at room temperature. Dehydrate through successive grades of 70, 90 and 100% ethanol, keeping 3 min in each. Air dry at room temperature.

* Stock RNAse A: dissolve 10 mg/ml of DNAse-free RNAse in 10 mM Tris-HCl, pH 7.5, 15 mM NaCl. Boil for 15 min, cool, store frozen. For use, dilute 1:100 in 2 x SSC.

** Depolymerised paraformaldehyde: Add 2g of paraformaldehyde to 40 ml water, heat to 60 to 80°C, clear with 10 ml 0.1M NaOH, adjust to 50 ml.

Protocol 2: Denaturation, Hybridization and Post-hybridization Washes

Saturate pad at bottom of temperature controller with 2 x SSC. Set at 80 to 85°C.

Denature the hybridization mix at 70°C for 10 min. Transfer to ice for 5 min. Add 30 to 40 µl of denatured hybridization mix to each slide and cover with plastic coverslip. Transfer to preheated humid chamber. Denature at 80°C for 10 min and slowly cool to 37°C. It can be done automatically using the programmable temperature controller.

Keep the chamber in an oven at 37°C overnight for hybridization. Float coverslips off in 2 x SSC at 42°C. Wash slides twice for 5 min each with 20% (v/v) formamide in 0.1 x SSC at 42°C. Wash four to six times more as follows: thrice in 2 x SSC for 3 min each at 42°C; cool for 5 min; thrice in 2 x SSC for 3 min each at room temperature. These schedules can be varied.

** *Suggested material; Secale cereale*, root-tips pretreated in colchicine, fixed in acetic acid-ethanol (1:3); enzyme softened and squashed in 45% acetic acid.

*** *DNA probe used* : Clone pSc 119.2, contains a 120 base pair tandemly repeated DNA sequence isolated from rye. It was labelled with digoxigenin-11-dUTP (Boehringer-Mannheim) by nick translation (Leitch *et al* 1991). Biotin-labelled probes can also be used (Schwarzacher *et al* 1989).

Protocol 4: Visualization of Chromosomes and Labelling Sites

Add 100 µl of 2 µg/ml (w/v) DAPI (4', 6-diamidino-2- phenylindole) in McIlvaine's buffer (9mM citric acid, 80mM disodiumhydrogen phosphate) to each slide, apply a coverslip and stain for 5 to 10 min at room temperature.

Remove coverslip, drain, add 100 µl of propidium iodide (PI, 5 to 10 µg/ml) in 2 x SSC per slide, replace coverslip and stain for 5 to 10 min at room temperature. Wash in 4 x SSC/Tween and use 10 µ antifade mountant (AF-2, Citifluor, London, UK). Place thin glass coverslip over specimen on slide, squeeze out excess antifade by blotting under filter paper. DAPI stains DNA and fluoresces blue under ultraviolet excitation. Propidium iodide stains DNA and is used as counterstain for FITC. Both are excited by blue light but FITC emits green and PI orange-red light. Since FITC label overlays PI counterstain, the sites of label appear yellow and the unlabelled chromatin red-orange.

DAPI stock solution; 100 µg/ml in water. propidium iodide stock solution: 100 µg/ml in water; both stored at -20°C.

For studying fluorescence, epifluorescence microscope is needed (suggested filters Zeiss 09. Leitz 12/3 for FITC and PI; Zeiss 02, Leitz A for DAPI).

For photography, suggested colour print films: Fujicolor Super HG 400 and Kodak Ektar 1000.

DNA-DNA hybridization methods are generally based on those of Pinkel *et al* (1986) for mammalian chromosomes but have been modified for plants almost at each step.

Mapping of Low Copy DNA Sequence in Plants

Prepare chromosome spreads from fixed root-tips after digesting with cellulase and pectolyase. Nick translate the cDNA clone pSc 503 corresponding to a rye γ-secaline gene, containing a Hind III insert of 900 bp in a pUC8 vector (supplied by PR Shewry, Rothamsted Experiment Station, England) with biotin-11-dUTP as follows:

Add one and one half of plasmid DNA to a 50 µl mixture of 50 mM Tris-HCl, pH 7.5/5 mM mgCl_2/30 µM each of dATP, dGTP, dCTP (Pharmacia) and biotin-11-dUTP (Enzo Diagnostics), also containing 20 ng of DNAse I (Sigma) and 12 units of DNA polymerase (BRL) and incubate at 15°C for 2 h. Terminate labelling by adding 5 µl of 0.2 M EDTA (pH 8.0). Purify labelled probe from unincorporated nucleotides by passing through a Sephadex G-50 spin column. Evaluate incorporation of biotin-11- dUTP by means of dot blot using a BRL streptavidin-alkaline phosphatase-detection system. Hybridize biotinylated probes to chromosome preparations following instructions given by Enzo Diagnostics. Incubate preparations in an RNAse A (1 µg/ml) in 2 x SSC solution at 37°C for 45 min. Keep slides in 70% formamide/2 x SSC at 70°C for 3.5 min. Immediately dehydrate in a series of ethanol washes (70, 90, 100%), at -20°C for 5 min each. Air dry and place in plastic high humidity chambers. Prepare hybridization mixture as 50% (v/v) formamide/10% (w/v) dextran sulphate/2 x SSC with salmon sperm DNA at 0.1 mg/ml. Denature in a boiling water bath for 10 min and store on ice. Add a measured amount of the mixture in a droplet to the chromosome preparation on a slide, cover with a coverslip. Keep the chamber at 80°C for 10 min to denature the preparations and the probe. Transfer to 37°C and incubate overnight.

After hybridization overnight, rinse off extra probe with 2 x SSC, and keep successively in fresh changes of 2 x SSC for 5 min at room temperature, for 10 min at 37°C and again 5 min at room temperature. Wash in 0.1% Triton/PBS for 4 min and then in PBS alone for 5 min at room temperature. Use primary and secondary antibodies to amplify the signals from the hybridized biotinylated probes. Apply goat antibiotin IgG (Sigma) at a concentration of 30 µg/ml and incubate slides at 37°C for 30 min. Remove unbound antibody by washing thrice in PBS, for 1 min each at room temperature. Next apply biotin-conjugated rabbit anti-goat IgG (Sigma) in a 1:400 dilution of the supplier's stock. Incubate and wash as for goat anti-biotin treatment. Incubate the slides with 0.7% secondary antibody horseradish peroxidase (BRL) for 30 min at 37°C. Streptavidin horse radish peroxidase conjugate was used as the reporter molecule. Wash slides in 2 x SSC for 5 min at room temperature and then in 0.1% Triton/PBS for 4 min. To visualize the hybridization complex, add 0.05% diaminobenzidine hydrochloride (BRL) and 0.03% hydrogen peroxide (Sigma) in PBS. Keep for 5 min at room temperature. Rinse with PBS and counterstain in 2% Giemsa (EM Science) for 1 min. Measure arm lengths of each labelled chromosome from a high resolution television screen with calipers (after Gustafson *et al* 1990).

In Situ Hybridization (Plants) with Biotin-labelled Probe

Pretreat fresh root-tips in ice water for 24 h and fix in glacial acetic acid: 95% ethanol (1:3) for 2 to 5 days. Squash in 1% aceto carmine and store at -70°C (Rayburn and Gill 1985).

Nick translate probe pSc 199 (Bedbrook *et al* 1980) with kit form obtained from Enzo Biochem Inc (325 Hudson Sreet, New York, NY 10013) and label with biotinylated UTP (Detek I-hrp kit from Enzo Biochem Inc).

For nick translation, mix 400 pg of DNAse 1; 12 units of DNA polymerase 1; 1μg of clone DNA and the nucleotides as given in the kit in a 50 μl reaction mixture of 0.5M Tris (pH 7.5) in 5 mM $MgCl_2$. Keep 2 and half hours at 18°C and pass through Sephadex G-50 spin column to separate the nucleotides from the labeled DNA.

For in situ hybridization, place 0.4 μg of labeled pSc 119 in a mixture of 50% formamide, 10% dextran, and 30 μg of sheared carrier DNA in 2 x SSC (0.6M NaCl, 0.06M sodium citrate). Denature the DNA at 85°C for 10 min. Place the slides in 70% formamide in 2 x SSC at 70°C for 2.5 min. Rapidly dehydrate in an ethanol series (70, 95, 100%) at - 20°C. Apply 20 μl of probe mixture to each slide. Cover with a coverslip. Incubate slides in moist chamber at 38°C for 6 h. Remove coverslip. Rinse slides in 2 X SSC at room temperature for 5 min; at 37°C for 10 min and again at room temperature for 5 min. Rinse in 0.1% Triton X-100 in PBS (0.13 M NaCl; 0.007M dibasic sodium phosphate and 0.003M monobasic sodium phosphate) and then briefly in PBS alone. Drain the slide, *without* drying. Place 120 μl of a complex of strepavidin-biotinylated horseradish peroxidase on each slide and coverslip. Incubate at 37°C for 30 min. Remove coverslip, rinse slide in 2 x SSC for 5 min and then in 0.1% Triton X-100 in PBS for 2 min both at room temperature. Rinse briefly in PBS, place 500 ml of a solution of 0.05% diaminobenzidine tetrahydrochloride (DAB) and 1% hydrogen peroxide on the slide for 5 min. Rinse slide with PBS and stain immediately with 2% Giemsa for 1 min. Air dry overnight and mount in Permount. Observe hybridization sites under bright field optics to distinguish hybridization sites from Giemsa bands. Photograph under both bright field and phase contrast optics in a Zeiss Photosystem III technical pan 2415.

Minor modifications have been used for different plants. For ISH of *Triticum aestivum* cv Chinese, denature slides in 70% formamide/2 x SSC at 70°C for 2 min, dehydrate in an ethanol series at -20°C. Denature labeled pScT7 DNA at 100°C for 10 min. Hybridise in a mixture containing 50% formamide, 2 x SSC, 10% dextran sulphate, salmon sperm carrier DNA at 500 μg/ml and biotin-labeled pScT7 at 4 μg/ml for 6 h in a moist chamber at 37°C (Mukai *et al* 1990).

Two Colour Mapping of Plant DNA Sequences

Prepare chromosome squashes of plant root-tip *Secale cereale* in this case, (after Leitch *et al* 1992) in 45% acetic acid.

DNA Probes:

pSc119.2 is a 120bp tandem repeat unit of DNA isolated from *Secale cereale* and labelled with digoxigenin dUTP (Boehringer-Mannheim). pTa71 is a ribosomal DNA sequence isolated from wheat *Triticum aestivum* labelled with biotin-11-dUTP.

Mix the probes immediately before use (each to a final concentration of 5 μg/ml) in a solution of 50% (v/v) formamide, 10% (w/v) dextran sulphate, 0.1% (w/v) SDS (sodium dodecyl sulphate) and 2 x SSC. Incubate slides in 100 μg/ml DNAse-free RNAse in 2 x SSC for 1 h at 37°C, wash twice in 2 x SSC for 10 min at room temperature, dehydrate in a graded ethanol series and air dry. Denature the probe mixture at 70°C for 10 min, load onto the slide preparation and cover with a plastic coverslip. Place the slides in a humid chamber and denature chromosomes and probe together at 90°C for 10 min.

DNA-DNA hybridization: Incubate slides overnight at 37°C. Then wash in 2 x SSC twice for 5 min each at 40°C. Wash thoroughly in 50% (v/v) formamide in 2 x SSC for 10 min at 40°C to allow DNA sequences with more than 85% homology to the probe to remain hybridized. Wash slides in 2 x SSC twice for 5 min at 40°C and then twice for 5 min each at room temperature.

Detection of hybridization: Biotin is detected with Texas Red-labelled avidin and digoxigenin with sheep antidigoxigenin- fluorescein (Boehringer Mannheim) simultaneously. Transfer slides to detection buffer (4 x SSC, 0.2% (v/v) Tween 20) for 5 min, treat with 5% (w/v) BSA (bovine serum albumin) in detection buffer for 5 min and incubate in 5 μg/ml Texas Red-labelled avidin and 20 μg/ml antidigoxigenin-fluorescein in detection buffer containing 5% (w/v) BSA for 1 h at 37°C. Wash the slides afterwards thrice in detection buffer at 37°C, keeping in each for 8 min. The signals from biotin and digoxigenin are amplified. Keep in 5% (v/v) normal goat serum in detection buffer for 5 min for blocking. Then incubate with 25 µg/ml biotinylated anti-avidin D and 10 µg/ml FITC-conjugated rabbit anti-sheep in detection buffer containing 5% (w/v) normal goat serum for 1 h at 37°C. Wash in detection buffer three times at 37°C, keeping 8 min each time. Reincubate the slides with Texas-Red labelled avidin as before. Counterstain the slides with 2 µg/ml DAPI in McIlvaine's citrate buffer. Mount in antifade solution (AFI, Citifluor) to reduce fading of fluorescence. Examine with a Zeiss epifluorescence microscope with filter sets 02, 09 and 12.

The sites of pSc119.2 hybridization show green fluorescence near the ends of the chromosomes. pTa71 hybridization is shown by orange/red fluorescence.

This method with simultaneous use of multiple probes helps in physical mapping of genes with relation to each other on the same chromosome and may also determine the direction of chromosome walking in long range gene cloning experiments. Available from Boehringer-Mannheim are: DIG DNA labelling kit (cat. no. 1175033); Anti-DIG fluorescein, Fab (cat. no. 1207741); Anti-DIG Rhodamine, Fab (cat. no. 1207750); Biotin-16-dUTP (cat. no. 1093070); Streptavidin-Alk Phos (cat. no. 10932660) Fluorescein- 12-dUTP (cat. no. 1373242); RNAse inhibitor for removing RNAse contamination (cat. no. 799017 and 799025).

Genome In situ Hybridization in Sectioned Plant Nuclei

Fix root-tips for 1 h at about 20°C in freshly prepared 1% (v/v) glutaraldehyde (diluted from Taab vacuum distilled 10 ml sealed ampoules) and 0.25% (v/v) saturated aqueous picric acid solution in phosphate buffer (0.05M Na_2HPO_4, 0.05M KH_2PO_4). Dehydrate through a graded ethanol series, embed in LR white (medium) resin and polymerize at 65°C for 16 h. Cut ultrathin (0.1 µm, gold) and thin (0.25 µm, blue) sections with a Reichert Ultracut for electron microscopy and light microsopy respectively. Expand all sections in knife trough containing 1% (v/v) benzyl alcohol in water. Pick up ultrathin sections on 200 gold mesh grids and transfer thin sections to glass slides coated with poly-L-lysine solution (Sigma). Incubate slides overnight at 60°C prior to hybridization. The subsequent steps are similar to those for *in situ* hybridization of squashes (after Leitch *et al* 1990).

In Situ Hybridization with Immunofluorescent Antibodies

CREST Antiserum in Indirect Immunofluorescence Staining

Human serum (LU851) containing anticentromere antibodies (ACA) was obtained from a patient suffering from a form of progressive systemic sclerosis (PSS or scleroderma) known as CREST (Calcinosis, Raynard's phenomenon, Esophageal dysfunction, Sclerodactylia and Telangiectasia). Using human, rat-kangaroo or Indian muntjac cells, this antiserum gives a centromere-specific immunostaining, which can be detected by the indirect immunofluorescence technique at dilutions of 1:15000.

Rinse cells grown on coverslips briefly in PBS and flatten by cytocentrifugation. Rinse twice for one min each in PBS, expose to 0.1% Triton X-100 detergent in PBS for 5 min, wash in PBS for 10 min and then treat with the anticentromere serum diluted 1:1000 in PBS. Wash three times in PBS, for 10 min each time. Expose slides to rhodamine-conjugated (TRITC) rabbit immunoglobulins to human IgG (Dakopatts, Denmark) diluted 1:10 in PBS. Wash extensively in PBS. Mount coverslips on slides in a PBS-glycerol (1:9) mixture and observe under a Zeiss fluorescence microscope (after Hadlaczky *et al* 1986).

Immunofluorecent Staining of Kinetochores in Human Micronuclei

Trypsinize cells from human fibroblast cultures. Seed onto sterile glass slides in growth medium. Incubate at 37°C for 72 h; replace the medium with prewarmed (37°C) 0.075M KCl and incubate for 30 min. Remove hypotonic solution and fix the cells *in situ* with prechilled (-20°C) 95% ethanol for 20 min at -20°C. Dry the slides.

Dilute scleroderma CREST antiserum (DK serum), which is the primary antibody in the immunofluorescence procedure, with phosphate-buffered saline (PBS) to different concentrations needed. Rehydrate the fixed cells in PBS for 5 min, incubate with an appropriate dilution of the primary antibody (200 µl/slide) for 30 min at room temperature in a humidified chamber. Wash twice in PBS for 10 min each. Dilute the secondary antibody (biotinylated goat antihuman IgG, Bethesda Research Lab) to 6 µg/ml of PBS. Apply 200 µl of this diluted solution to each slide.

Incubate for 30 min at room temperature in moist chamber, wash as for primary antibody. Incubate each slide with 200 µl of a solution (4 µg/ml of PBS) of streptavidin-Texas Red (Bethesda Res Lab) for 30 min at room temperature in a moist chamber. Wash as before. Counterstain with 0.00012% w/v acridine orange (Sigma) in phosphate buffer pH 6.86 (Fisher Gram-Pac) for 3 min. Mount coverslips with PBS: glycerol mixture (3:1). Keep slides at 4°C for 1 to 2 h. This method is very useful for detecting aneuploidy and chromosome breakage in cultured cells. Various modifications are used (Sunmer 1987, Degrassin and Tanzarella 1988, Hennig *et al* 1988, French and Morley 1989).

Monoclonal Antibody against Kinetochore Component - Schedule for Preparation in CHO Cells

Several polypeptide components of the kinetochore and centromere region have been identified using specific antibodies in the serum of patients suffering from the CREST syndrome. Monoclonal antibodies (mAbs) can be prepared directly against components of the kinetochore. A new component, for example, was characterised using a mAb in Chinese hamster ovary cells (Pankov *et al* 1990).

Separation of cells, nuclei and chromosomes: Grow Chinese hamster ovary (CHO) cells in monolayer cultures in McCoy's 5A medium with 10% fetal calf serum. Separate mitotic cells by shaking growing cultures. Replace medium with fresh medium containing 0.06 µg/ml colcemid 2 h after the preliminary shaking.

Prepare nuclei from the cells detached by trypsin, wash in PBS (0.15M NaCl, 0.1M sodium phosphate, pH 7.5) and suspend in nuclear isolation buffer (NB: 50 mM Tris-HCl, pH 7.4; 0.25 M sucrose; 5 mM $MgSO_4$; 1 mM phenylmethyl sulphonyl fluoride, PMSF; 100 units/ml Trasylol Miles). Add to Triton X-100/(0.5%) and rupture in a Dounce homogenizer. Centrifuge the nuclei (3000g, 20 min). Precipitate cytoplasmic fractions in 20% TCA.

Preparation of monoclonal antibodies: Use a nuclear fraction, enriched in several kinetochore polypeptides as immunogen to generate mAbs. Confirm the presence of these polypeptides in different fractions extracted from the nuclei, using CREST sera to probe aliquots dotted on nitrocellulose membranes.

To prepare the fraction used as immunogen, digest nuclei of bovine thymocytes with DNAse I and RNAse A (both at 250 µg/ml, 2 h at 4°C) in NB and centrifuge (3000g, 20 min). Extract the pellet in 10 mM Tris-HCl, pH 8.5; 0.8M NaCl; 0.2 mM $MgSO_4$; 1mM PMSF; 100 units/ml Trasylol, for 30 min at 4°C and centrifuge at 10,000g for 20 min. Dialyse the supernatant overnight against 20mM Tris-HCl, pH 8.0; 0.15 M NaCl and 1mM PMSF. Remove precipitated histones by centrifugation at 10,000g for 30 min. Detect polypeptides in the supernatant by CREST sera, after subjecting it to SDS-PAGE and transferring to a membrane.

Inject aliquots of the material with about 100 µg protein in complete Freund's adjuvent into Balb/c mice. Boost after 21 days with 50 µg of antigen in PBS and sacrifice after 3 days. Fuse splenic lymphocytes with an equal amount of mouse Sp2/0 myeloma cells. Screen supernatants from the hybridomas by immunofluorescence for antibodies recognising the kinetochore regions of CHO cells, as described below. Purify the mAb 37A5, an IgM, from ascitic fluid by gel exclusion chromatography on Sephacryl S-300.

Immunofluorescence microscopy: For localizing antigens on cells grown on microscope slides and on mitotic cells and metaphase chromosomes centrifuged onto slides with a cytocentrifuge, fix the samples with 3% formaldehyde in PBS for 30 *s*, wash with PBS, permeabilize with PBS, 0.5% Triton X-100 (PBST) for 5 min and rinse with PBS. Incubate with antibodies for 1 h in a humid chamber. Overlay slides at room temperature with hybridoma supernatants, or with 15 µg/ml purified mAb 37A5 in PBS containing 1% bovine serum albumin (PBSA).

Wash in PBS thrice for 10 min each, overlay slides at 4°C with fluorescein-conjugated rabbit antimouse Ig (M+G+A) from Zymed Lab, diluted 1:40. For double labelling, incubate first with a 1:1 mixture of 15 μg/ml mAb 37A5 and an antikinetochore CREST serum (diluted 1:1000) and then a 1:1 mixture of fluorescein-conjugated anti-mouse antibodies and Texas Red- conjugated rabbit anti-human Ig (M+G+A), diluted 1:40. Wash in PBS thrice for 10 min each time. Mount in 50% glycerol, 0.15M NaCl, 0.1M glycine, pH 8.6, containing the fluorescent DNA stain Hoechst 33258 (10μg/ml). Examine under a Zeiss fluorescence microscope.

Indirect Immunofluorescence in Mammalian Sperm

Incubate purified bull sperm nuclei for 60 min at 22°C and centrifuge onto glass coverslips in 10 mM dithiothreitol. Wash coverslips with PBS, fix in PBS containing 2% paraformaldehyde for 20 min at 22°C, wash with PBS. Probe with whole anticentromere serum GD diluted 500-fold in PBS containing 0.05% Tween 20 and 0.05% sodium azide (PBS/Tween/Azide). Use the latter mixture as wash buffer. Use as secondary probe affinity purified fluorescein isothiocyanate (FITC) - conjugated goat anti-human immunoglobulin (Tago), Stain DNA with 0.25 μg/ml propidium iodide in PBS. Mount with a coverslip using a medium containing 1, 4-diazabicyclo (2.2.2)-octane. Observe under an MRC-500 Laser Scanning Confocal microscope (Biorad Microscience, Cambridge, Mass) with a Nikon Optiphot (Nikon, Torrance, Calif, after Palmer *et al* 1990).

Anti-Z-DNA Antibody in Mammals

Indirect immunofluorescence procedure using an antibody specific to the Z-DNA conformation gives reproducible results only when the chromosomes are fixed in 45% acetic acid but not in acetic- methanol (1:3) or 75% ethanol. Intense fluorescence may be observed both at C-band heterochromatin and NORs. This phenomenon may be attributed to the removal of chromosomal proteins by acidic fixation (after Ueda *et al* 1990).

Grow cells from male Indian muntjac cell line CCL157 in Ham's F12 medium supplemented with 10% fetal calf serum. To arrest cells in metaphase, treat culture cells with 5mM thymidine for 40 h. Wash with PBS and reculture with fresh medium. Add colcemid at a concentration of 0.5 μg/ml after 8-10 h. Harvest cells at 16-20 h after the release of the thymidine block.

Isolate metaphase chromosomes by washing mitotic cells first in PBS and then in ice-cold isolation buffer (25mM Tris- HCl, pH 7.0; 0.5mM $CaCl_2$; 1mM $MgCl_2$ and 1M hexylene glycol). Resuspend cells in 10 ml of isolation buffer, incubate at 37°C for 10 min and disrupt by passing through a 22 gauge needle. Centrifuge the lysate at 100g for 5 min. Recentrifuge the supernatant at 1500g for 10 min to pellet chromosomes. Resuspend chromosomes in a small volume of the isolation buffer.

For immunofluorescence detection of Z-DNA, place a drop of isolated chromosome suspension on the poly-L-lysine-coated glass slide and keep for 5 min at 4°C for the chromosomes to be attached. Rinse the slides in PBS and fix with 45% (v/v) acetic acid solution for 5 min. Give two changes in the fixative, keeping 5 min each time. Rinse fixed chromosomes. Treat isolated chromosomes for 30 min with 1:50 diluted rabbit normal serum in PBS for blocking. Then treat successively with goat anti-Z-DNA antibody (polyclonal, Gle, 1:50 dilution); biotinylated rabbit anti-goat IgG (Vector Lab, 1:50 dilution) and finally with fluorescein - conjugated streptavidin (Amersham, 1:50 dilution) at 37°C, keeping in each for 1 to 2 h. Photograph through a Zeiss Fluorescence microscope on Kodak Tri-X film.

Modified Banding of Human Chromosomes for Electron Microscopy

Leucocyte culture: Grow human peripheral leucocytes for 48 to 54 h at 37°C in RPMI 1640 supplemented with glutamine, antibiotics, fetal calf serum and phytohaemagglutinin. Block the cells at S-phase by adding thymidine (final concentration 300 μg/ml) or BrdUrd (200 μg/ml) for 17 h. Wash twice with Hank's BSS to remove the synchronizing agent. Resuspend in complete medium 199. Grow the cells for a further period (4.5h after thymidine, synchronization in the presence of thymidine, 3μg/ml or 5 h after BrdUrd exposure in the presence of BrdUrd 30 μg/ml). In this experiment BrdUrd is incorporated in late replicating DNA following the combination of thymidine as synchronizing agent and BrdUrd as releasing agent. BrdUrd is incorporated in the early replicating DNA when it is used as synchronizing and thymidine as releasing agents.

Harvest culture without using colchicine. Treat with hypotonic (0.075M) KCl for 15 min at 37°C and fix in three changes of acetic acid-methanol (1:3), keeping for 15 min in each at 20°C. Prepare metaphase spreads by dropping two drops on an ice-cold slide and air-drying.

Immunochemical banding for electron microscopy: Cover well spread metaphase plates without cytoplasm with a 100 mesh copper grid. Draw a circle under each plate on the underside of the slide with a diamond pencil. Destain the slides in an ethanol series (50, 70 and 90%). Denature double-stranded DNA with deionized formamide (70%) diluted in 2 x SSC at pH 7.0 for one min at 70°C. Pass the slides through a graded series of ethanol (70, 80, 90 and 100%) at -20°C. Air dry. Dilute the mouse monoclonal anti-BrdUrd antibody from Becton Dickinson (USA) 1:10 in phosphate-buffered saline (PBS, pH 7.6) containing bovine serum albumin (BSA 0.5%) and Tween 20(0.5%).

Incubate the slides with the diluted antibody for 45 min at 37°C under a coverslip in a moist chamber. Wash three times with PBS-BSA, keeping for 5 min each time.

For gold-labelling two methods can be used:

i) Incubate the slides for 45 min at 37°C with an undiluted secondary antibody, a goat antimouse immunoglobulin (IgG) labelled with gold (EM grade, 10nm, Janssen, Pharmaceutica).

ii) Alternatively, first incubate the slides for 60 min at room temperature, after adding 100µl of a rabbit anti-mouse antiserum (Dako, Stockholm, Sweden) diluted 1:20 in PBS containing 0.5% BSA to each slide and covering with a coverslip. Wash thrice in PBS-BSA. Add a protein A-gold complex (15nm) diluted with PBS and incubate for 30 min at room temperature. Wash the material three times in PBS-BSA and fix in 3.0% glutaraldehyde for 30 min at room temperature. Transfer the previously selected and marked metaphase plates from the glass slide to an EM grid (Messier *et al* 1986). Observe the unstained mitosis in a Siemens Elmiskop 1A EM operated at 60 kV (after Drouin *et al* 1989a, b).

Antibody Labelling and Flow Cytometry in CHO Cell Line

Block cells in metaphase from monolayer culture of Chinese Hamster Ovary grown in RPMI 1640 medium, supplemented by fetal calf serum (2-5%), 2mM l-glutamine and 100U/ml each of penicillin and streptomycin (Gibco). Block cells in metaphase by adding 0.05 µg/ml Colcemid (Gibco) 2 to 3 h before harvesting. Collect cells by centrifugation for 8 min at 4°C (800 rpm).

For chromosome isolation, suspend cells in hypotonic isolation buffer (50 mM KCl, 10 mM $MgSO_4$, 5mM Hepes, 0.1mg/ml RNAse A, pH 8.0. Pellet by centrifugation and resuspend in the same buffer prewarmed to 37°C for 10 min. Adjust the cell concentration to 100,000/ml. Cool suspension on ice for 10 min and add 1/10 vol of Triton X-100. Keep on ice for 10 min. Release chromosomes by passage through a 22g needle using a 5ml syringe. Repeat passage 4 or 5 times for adequate cell lysis (after Turner and Keohane 1989).

Antibody labelling of chromosomes in suspension: The mouse monoclonal IgM antibody HBC-7 to histone 2B is derived from a Balb/c mouse immunized with human chromatin and used in the form of tissue culture supernatant from a subclone of hybridoma HBC-7. Since the supernatant contains 15% fetal calf serum, no extra protein is needed to suppress non-specific binding. Mouse monoclonal antibody 33 to double-stranded DNA is of the IgG2a subcloss and is used in the form of partially purified immunoglobin appropriately diluted in tissue culture medium plus calf serum.

Mix 100 μl of antibody with 500 μl of chromosome suspension. Leave for 30 min on ice. Add 200 μl of a 15% sucrose solution made up in the isolation buffer. Spin at 1200 rpm for 12 min at 4°C.

The sucrose solution minimises chromosome damage and provides more complete separation of the chromosomes from first and second antibodies. Remove supernatant after centrifugation, leaving about 50 μl of fluid. Resuspend chromosomes in 200 μl of FITC-conjugated sheep antibody to mouse IgM (Sigma). Keep 30 min on ice. Then gently mix with 200 μl of the isolation buffer, underlayed with sucrose and centrifuge as before. Resuspend in 400 μl of the isolation buffer and keep on ice. Immediately before flow cytometric analysis, counterstain with propidium iodide (PI) or ethidium bromide (EB) at a final concentration of 50 μg/ml.

Flow cytometric analysis: Analyse the labelled chromosomes on a Becton Dickinson FACS 440 flow cytometer equipped with a Spectra-Physics argon ion laser and interfaced with a consort 40 computer system. Tune the laser to 600 mW and 488 nm. Collect fluorescein (green) emission on fluorescence channel 1 via a 530 nm band pass filter and PI emission on channel 2 via a coambination of 520 and 590 nm long pass filters. Analyse using a 50 μm nozzle and a flow rate of 200-400 events. Record at least 100,000 events for each sample.

Fluorescence microscopy: Mix labelled chromosomes with an equal volume of isolation buffer containing 10% formaldehyde. Keep for 10 min at 4 °C or at room temperature. Spin the chromosomes onto glass slides in a cytocentrifuge (Shandon Eliot Cytospin). Increase speed gradually to 1200 rpm over the first minute and maintain for another 3 min. Mount in 50% glycerol in Dulbecco's phosphate-buffered saline and seal with nail varnish. Observe under a Zeiss epifluorescence microscope equipped with an HPO 50 mercury vapour lamp.

Antigen Reaction with Human Autoantibody in Diptera

Sera from patients with rheumatic diseases, containing anti-Ku antibodies, recognize antigens in HeLa and Trichosia pubescens cells, indicating that the antigen is highly conserved and may have a similar role in the cellular metabolism of both insect and mammalian cells.

Preparation of polytene chromosome spreads: For ordinary light microscopy, fix the salivary glands for 2-5 min in acetic acid: ethanol (1:3) and squash in 50% acetic acid.

For immunofluorescence with human sera: fix the salivary glands for 30 *s* in buffered saline (PBS) containing 1.75% freshly prepared formaldehyde and immediately transfer to another saline solution containing 0.5% NP-40. After 10 min disrupt the glands by passing several times through a glass pipette with a small diameter. Transfer the released chromosomes with a siliconized glass needle to a clean slide. Remove excess of isolation medium by aspiration. Wash the chromosomes stuck on the slides with several changes of buffered saline (PBS). Incubate for one h at 36°C with 40 µl of 1/20 dilution of an affinity purified goat anti-human IgG conjugated with fluorescein isothiocyanate (FITC). Wash in three baths of PBS and mount in glycerine: PBS (9:1) (after Amabis *et al* 1990).

Immunofluorescence in Nematodes

To observe gonial meiotic divesion, dissect out small pieces of testes in a drop of 0.7% NaCl at 37°C. Cover with a coverslip and squash. Freeze slides in liquid nitrogen to remove coverslips and fix in 3.7% formaldehyde in PBS at room temperature for 10 min. Wash slides thrice in PBS, keeping 5 min each time. Permeabilize in 1% Triton X100 in PBS for 7 min. Wash again in PBS three times and incubate for 30 min at 37°C with monoclonal anti-β-tubulin antibody (Amersham) at a 1:1000 dilution in PBS. Wash the slide three times for 5 min each in PBS and incubate for 30 min at 37°C with a fluorescein-conjugated goat anti-mouse IgG (Cappel) at a dilution of 1:50 in PBS. Wash the slide twice in PBS and mount in a glycerol: PBS mixture (9:1) containing a drop of Hoechst 33258 staining solution for comparative chromosome visualization.

Anti-topoisomerase II in recognising Meiotic Chromosome Core in Bird

Two polyclonal antibodies against topoisomerase II, which recognise the mitotic metaphase chromosome scaffold, give at pachytene a positive immunocytological reaction with the chromatin and predominantly with the cores and centromeric regions of the paired chromosomes. Apparently during meiotic prophase, this DNA-binding enzyme, implicated in transient double strand breaks, chromosome condensation and anaphase separation, is associated with the chromatin and synaptonemal complex of the pachytene and diplotene chromosomes.

Antibodies required are: i) Anti-topo II 2b2 raised in a guinea pig against an 170,000 gel band from chicken mitotic chromosome scaffold (Earnshaw *et al* 1985). It is highly species specific. ii) Anti-topo II RG prepared in a rabbit against a fusion protein containing 42500 M_r of the chicken topo II carboxy-terminus fused to 32000 M_r of the bacterial TrpE protein (Hoffman *et al* 1989).

For preparation of Rooster (Gallus domesticus) spermatocyte spreads, macerate a testes in minimal essential medium (MEM) pH 7.3. Gravity-sediment the tissue in 15 ml MEM in a plastic centrifuge tube. Collect supernatant and spin at low speed. Resuspend the pellet and spin down the cells again. Shake loose the pellet. Touch 7 µl samples of the dense cell suspension to the surface of a 0.5% NaCl solution. Pick up the cells on glass slides for light microscopy and on plastic-covered slides for electron microscopy (0.5% plastic in chloroform). Fix cells in 1% paraformaldehyde, pH 8.0, with 0.06% SDS at 4°C for 3 min and in the same without SDS for another 3 min. Wash three times for one min each in 0.4% Photoflo (Kodak) and air dry.

In preparation for primary antibody, wash the slides for 10 min each in PBS twice, and once each in mix (0.05% Triton, 0.025% gelatin, 0.5mM EDTA in PBS) and in PBS. Keep for 10 min in holding buffer (1:10 diluted antibody dilution buffer; 3% bovine serum albumin with 0.05% Triton) and air dry. Incubate slides with primary antibody at 1:100, 1:500 and 1:1000 dilution in antibody dilution buffer (*adb*) for 1 h at room temperature and overnight at 4°C. Wash slides in PBS, mix and holding buffer again. Incubate with secondary antibody diluted in .*adb* for 45 min at 37°C. Use goat anti-rabbit IgG (GAR)-conjugated to peroxidase for light microscopy and GAR conjugated to 5nm gold for electron microscopy. For guinea pig antibodies, use an intermediate rabbit anti-guinea pig IgG. Wash off the peroxidase-GAR twice in PBS, once in mix and again once in PBS. Incubate the slides for 5 min in degassed Tris-imidazole buffer (50mM Tris, 10mM imidazole, pH 7.5) with diaminobenzidine (0.06% DAB) and for 30 min in the same solution after adding hydrogen peroxide (0.03%). Wash slides in water, rinse in Photoflo and dry. Wash the gold-GAR-treated slides successively in PBS, mix, PBS, water, Photoflo and air dry, Float off the plastic coating with cells and place the grids on the plastic. Pick up the plastic film with grids on a sheet of Parawax. Postfix immunogold preparations after 24 h in 0.1% OsO_4 to improve contrast.

BrdU Antibody Technique (BAT)

A method was developed to study chromosome replication without synchronization in human lymphocyte and amniotic cell cultures visualizing very short BrdU pulses by an immunologic technique (BAT) (after Vogel *et al* 1986, 1989).

To a standard lymphocyte culture add BrdU to a final concentration of 10 µg/ml. Remove the medium after 5 min. Rinse culture once with Hank's solution and add fresh medium containing 0.2 mM thymidine for pulse chase experiments. Prepare metaphase spreads after treatment with hypotonic 0.056M KCl and fixation in acetic acid - methanol (1:3) and then air dry.

Denature DNA in chromosomes by immersing the slides in 0.07N NaOH and ethanol (5:2, v/v) for 30 *s*. Rinse with PBS and process *without drying*. Dilute the BrdU antibody (Partec, Switzerland) 1:500.

Use an anti-mouse IgG coupled with horse radish peroxidase as second antibody. Follow the schedule given under *in situ* hybridization earlier. Visualize the antibody by the diaminobenzidine/H_2O_2 reaction and use methylene blue as counterstain for the chromosomes.

In a modified vesion to study separation in man-mouse: Treat cultured cells with colcemid for 30 min and then hypotonic solution (0.075M KCl) for 10 min.

Deposit cells onto a clean slide with the help of a Shandon cytocentrifuge. Fix in 80% ethanol at -70°C and keep at -20°C for 30 min. Wash in Dulbecco's phosphate buffer (DPB). Place one drop of antinuclear (anti-kinetochore) antibody (Davis antibody) on the slide. Cover with a coverslip. Incubate the slide at 37°C for 30 min in a moist chamber. Wash thoroughly in DPB. Incubate the cells again in fluorescein-(FITC) isothiocyanate-conjugated goat anti-human IgG for another 30 min. Wash again. Stain with 0.01% ethidium bromide and mount in a 3% solution of *n*-propylgallate in glycerine. Examine under a Zeiss microscope fitted with excitation filter for epifluorescence.

For the detection of DNA replication, treat the exponentially growing cells with 10^{-6} M BrdUrd (5-bromodeoxyuridine) for 4-6 h. After treatment with colcemid and hypotonic solution, fix the cells in acetic acid: methanol mixture, prepare flame-dried slides. Incubate for 30 min with IU4 - an anti-BrdUrd antibody from the Lawrence Livermore Laboratory. Wash the slides in DPB and incubate again in FITC - conjugated goat - antimouse for 30 min. Wash in DPB, stain with 0.01% ethidium bromide and mount in a 3% solution of n-propylgallate in glycerine (Vig and Athwal 1989).

Products available for In situ Hybridization

1. *Spectrum CEP direct-labelled chromosome enumeration system* - from Imagenetics - used in molecular cytogenetic research that uses fluorescence in situ hybridization procedures. Chromosomes can be counted in both metaphase and interphase cells. Spectrum Orange and Spectrum Green fluorophores are used, which are covalently linked to chromosome-specific satellite DNA probes, eliminating the difficulties of multistep detection in conventional indirectly labelled systems based on biotin/avidin or digoxigenin. In a single step, Spectrum CEP satellite DNA probes simultaneously hybridize and fluorescently tag each target sequence. A complete test kit and manual are supplied, including standard four hour hybridization protocol, the use of "touch prep" procedures for solid tissue samples, modified ASG banding procedures and methods for utilizing multiple probe hybridization.

2. *Chromosome Detection Kit fluorescence In-Situ hybridization*: The Chromoprobe DNA probe kit for the detection of chromosomes in interphase uses fluorescence *in-situ* hybridization, or FISH, which means that trisomies, such as Down's Syndrome, can be detected in hours in amniocytes or other cells, without the need to culture. The FISH technique has been considerably simplified by the use of a patented method to reversibly attach directly labelled DNA probes to special glass coverslips. This means that denaturation and hybridization take place under the coverslip in only one hour. The kit contains ten special coverslips coated with a fluorescein labelled alphoid centromeric DNA probe, hybridization fluid, counterstain/antifade and rubber solution. The only equipment required is a standard fluo-

rescent microscope and a heating block capable of maintaining a temperature of 75°C. Kits are available for detection of chromosomes, 13/12, 18, X and Y.

Cytocell Ltd. Home Farm, Hill Rd, Lewknor, Oxon OX9 5TS, UK.

3. *A full range of probes are* available including all three major types of human satellite DNA: alpha, beta and the classical satellites. The range includes an expanding number of whole chromosome paint probes. The probes are available separately or in kit form for fluorescence or light microscopy for fixed tissue, whole cells or metaphase spreads. The probes are used in nonisotopic *in situ* hybridization techniques. Alpha Laboratories Ltd, 40 Parham Drive, Eastleigh. Hants S05 4NU, UK.

Biotinylated genomic DNA is hybridized to cDNA probes in solution and the hybridized complexes are captured using *streptavidin- coated magnetic beads*. It can be used to isolate cDNAs rapidly from large genomic intervals. The method is PCR-based (Tagle *et al* 1993).

4. Fluorescence activated cell sorting using monoclonal antibody. Cells containing human chromosome 5 in a human-mouse monochromosomal hybrid could be sorted very effectively using monoclonal antibody 6G12 using a fluorescence activated cell sorter (FACS) 440 (Becton Dickinson Immunocytometry Systems), connected to a VAX 11-730 computer (Yoneda *et al* 1991).

Kinetochore Visualization with Crest Serum in Plants

Germinate pollen grains. After 7 h, harvest the pollen tubes, fix and process. Expose overnight at 4°C to a x15 dilution of CREST serum E.K., in phosphate-buffered saline containing 3% bovine serum albumin and 0.02% sodium azide, followed by a x50 dilution of a fluorescein-conjugated goat antihuman secondary antibody (Sigma) for 1.5 h.

For double localization, apply the CREST serum and fluorescein- linked secondary probe first and follow with an antitubulin protocol, employing a rhodamine-conjugated antimouse secondary antibody (Sigma). Mount the pollen tubes in a solution containing 1mg/ml p-phenylenediamine (Sigma), 10 μg/ml Hoechst 33258 (Sigma), 50% glycerol, and 0.1M Tris, pH 9.3 and view on a Universal micros-cope (Carl Zeiss), equipped with Olympus 60 and 100x DApo UV objectives for epifluorescence.

Stain pollen tubes with 100 μg/ml mithramycin A(Sigma) in deionized water and view on a laser scanning confocal microscope (MRC 500, BioRad Instruments, Cambridge, MA) using fluorescein excitation and a 60x planapochromat objective.

Kinetochores are shown by CREST antibodies to be paired and single fluorescent dots, located at the ends of microtubule bundles previously identified as kinetochore fibres (Palevitz 1990).

Antibody Method for Replicating Bands in Plants

Treat seedlings (two-day old) with 1 µg/ml 5- bromodeoxyuridine (BrdU) containing 1 µg/ml fluorodeoxyuridine (FdU) for 30 min in dark at 22°C. Rinse with 10 mM thymidine containing 1µg/ml FdU in dark. Incubate the seedlings in dark on wet filter paper after rinsing thrice with deionized water. The period of incubation depends on the species, varying from 4 h for *Crepis capillaris* to 10 h for *Allium fistulosum*. Add 0.02% colchicine to the seedling for the final hour of incubation.

Fix seedlings in acetic acid-methanol (1:3) for 1-7 days at - 20°C. Excise the root-rips, hydrate, macerate with enzymes on slides and flame dry (Hizume *et al* 1980).

For denaturation, incubate the slides in 1N.HCl at 60°C for 10 min. Wash gently with deionized water. Incubate in a phosphate-buffered saline (PBS, pH 7.2), containing 0.1% Tween 20 at 0°C for 5 min. For antibody treatment, coat the slides with 5000-fold diluted anti-BrdU antibody (mouse IgG, IMT Co.). Incubate at room temperature for 30 min in a moist chamber. Rinse in cold PBST twice for 5 min each. Coat the slides with a biotin- conjugated anti-mouse antibody (Vector Laboratories Inc) at a dilution of 1:200. Incubate a room temperature for 30 min in a moist chamber. Coat the slides with a complex (Vector Lab.), obtained by mixing avidin-DH and biotinylated horseradish peroxidase H. Incubate at room temperature for 30 min in a moist chamber. Rinse the slides twice in cold PBS, keeping in each for 5 min.

Visualise the peroxidase with a solution of 1 mg/ml 3. 3′ diaminobenzidine-4HCl (DAB; Dottite); 0.4 mg/ml $NiCl_2$ and 0.02% H_2O_2 in 100mM Tris-buffer (pH 7.2) at room temperature for 7 min in a moist chamber. Rinse the slides in distilled water twice for 5 min each. Counterstain the chromosomes with 0.1% neutral Red (Chroma-Gesellschaft Schmid GmbH & Co.) for 15-30 min. Rinse gently with distilled water, air dry, dip in xylene for 10 min and mount in Eukitt (Kinder Co. Germany).

The positively stained bands indicate the replication sites labelled during the S phase and SCE patterns.

Methylation of DNA

Methylation of DNA in the chromosome is supposed to play an important role in gene expression. It is necessary therefore to work out the distribution of methylated cytosine sites (5m C) at the chromosome level. Restriction endonuclease activity is undoubtedly dependent on specific sequences but in some cases it is also dependent on methylation of restriction sites. As such, restriction enzymes differentially sensitive to methylation at cleavage sites may serve as identification of such loci. A method is described here demonstrating the precise location of methylatable CCGG sites in the chromosome complement (Santoz and Fernandez-Piqueras 1989).

In situ methylation of chromosomes

1) Make squash preparation of mitotic and meiotic cells from gonadal tissues in 45% acetic acid, preinjected with 0.03% colchicine in insect saline solution, 6 h prior to fixation in ethanol-acetic acid (3:1).

2) Remove the coverslip by freezing in liquid nitrogen and air dry.

3) Incubate fresh preparation for less than 24 h at 4°C, 20μl of prokaryotic methylase M. HpaII (Bochringer) soln at a concentration of 0.5 units μl in 50mM Tris-HCl, pH 7.9; 10mM β-mercaptoethanol; 5 M S-adenosylmethionine in a moist chamber at 37°C for 16 h.

4) Rinse twice in incubation buffer and twice in the assay buffer, as prescribed by the supplier of HpaII endonuclease.

5) After methylation, incubate the preparations at with HpaII (0.5 units/μl) in the assay buffer for 16 h at 37°C in a moist chamber.

6) Use 10mM EDTA, pH 7.0 as a stop bath.

7) Stain preparations with 4% Giemsa in phosphate buffer, pH 6.8. Stain some slides with propidium iodide (5 μg/ml in PBS, pH 7.4) after incubation with RNAse A (100 μg/ml in PBS, pH 7.4) for 30 min at 37°C.

Treat control preparations with buffer without the enzymes along with the treated slides. Add 25% glycerol to methylase buffer since stock solution of the enzyme had been 10 units/ml in 50% glycerol. The comparison of the banded portions gives an index of methylatable sites.

CHAPTER 13

SPECIAL MOLECULAR TECHNIQUES NECESSARY FOR CHROMOSOME ANALYSIS

This chapter deals with certain molecular biology techniques, which are being utilized in chromosome research. The protocols of section A, Restriction Fragment Length Polymorphism have been outlined in detail. In sections B and C, PCR and Chromosome Walking, the principles of the technology have been described.

SECTION A : RESTRICTION FRAGMENT LENGTH POLYMORPHISM

The discovery of enzymes which can cut discrete sequences of DNA and the use of DNA fragments as marker probes of known sequences have made restriction fragment length analysis a powerful tool for mapping gene loci in chromosomes. The commonly used restriction enzymes break phosphodiester bonds. Endonucleases mostly have 4-6 base pair recognition sequences. Over 900 restriction enzymes have been isolated from prokaryotes (Micklos and Bloom, 1989). The endonucleases cut DNA at internal positions whereas exonucleases digest DNA progressively from the two ends. On the basis of random arrangement of nucleotides, a 6bp sequence has the chance of cutting once in 46 or 4096 base pairs. In general, the fragment size of chromosomal DNA varies greatly, upto several thousand base pairs. Even if the recognition sequences of different enzymes are identical in the number of base pairs involved, the frequency and sites of cleavage vary to a great extent. However, the cleavage sites may be reduced due to methylation. The polymorphism in size of restriction fragments is due to variation in restriction site distribution. Such alterations in sites can arise out of substitution, deletion or insertion of sequences or even genomic rearrangements. As sites of cleavage represent nucleotide loci, the variations can be utilized as parameters of genetic divergence (see table 1).

Following cleavage of DNA by different restriction enzymes, which may often be facilitated by addition of spermidine, separation is carried out by gel electrophoresis so that the fragments are separated according to their size. Ultimately, the loci and sequences in fragments are identified by Southern transfer and hybridization with known probes.

Table 1 :Some Restriction Enzymes and Their Cleaving Sequences

Microorganism	Enzyme Abbreviation	Sequence 5′ → 3′ 3′ → 5′
Bacillus amyloliquefaciens H	*Bam*HI	G↓GATCC CCTAG↑G
Escherichia coli RY 13	*Eco*RI	G↓AATTC CTTAA↑G
Haemophilus aegyptius	*Hae*II	PuGCGC↓Py Py↑CGCGPu
Haemophilus aegyptius	*Hae*III	GG↓CC CC↑GG
Haemophilus haemolyticus	*Hha*I	GCG↓C C↑GCG
Haemophilus influenzae Rd	*Hind*II	GTPy↓PuAC CAPu↑PyTG
Haemophilus influenzae Rd	*Hind*III	A↓AGCTT TTCGA↑A
Haemophilus parainfluenzae	*Hpa*I	GTT↓AAC CAA↑TTG
Haemophilus parainfluenzae	*Hpa*II	C↓CGG GGC↑C
Providencia stuartii 164	*Pst*I	CTGCA↓G G↑ACGTC
Streptomyces albus G	*Sal*I	G↓TCGAC CAGCT↑G
Xanthomonas oryzae	*Xor*II	CGATC↓G G↑CTAGC

Agarose Gel Electrophoresis and Hybridization

The sieving of DNA fragments according to size is performed by electrophoresis through a medium such as agarose gel. The molecules of DNA have a uniform charge/mass ratio and in a resistance-free medium have the same migration pattern on electrophoresis. The viscosity of both polyacrylamide and agarose allows DNA molecules to separate according to size and pattern during electrophoresis. The basic principle is the inverse relationship between the log of the molecular weight of the fragment with the distance that the fragment travels in the gel. The matrix of the highly purified agar can separate fragments ranging in size from 100 to 50,000 nucleotides. Low concentrations produce loose gels which can separate longer fragments whereas higher concentrations produce stiff gels permitting movement of small fragments. The consistency of the gel can thus be adjusted depending on the size of fragments analysed. Following staining with ethidium bromide—a fluorescent dye—the gel can be viewed under UV light. After electrophoresis, the gel is treated with dilute acid for depurination and alkali for denaturation. Smaller fragments are generated at single strand break sites facilitating speedy movement of the fragments. The next step is the transfer of the gel to the filter or membrane. The DNA is bound to the filter at the single stranded stage. This single stranded state permits the binding of complementary labelled probe strands by annealing. In the transfer of DNA from gel to membrane, partial depurination through weak acid and alkaline hydrolysis of phosphodiester bond at that site is essential. The fragments (-1 kb or so) can be conveniently transferred from gel to support.

For hybridization, the probe may originate from different sources. It may be homologous i.e. originating from individuals of the species to be analyzed as well as from related sources or heterologous, that is, from less related sources.

The strength of the signal in hybridization is proportional to the specific activity of the probe and inversely proportional to its length. The optimum effect can be secured with radiolabelling of very high specific activity, long exposure and adequate target DNA on the filter.

Blocking agents are necessary to check nonspecific attachment of all probes to filter. Normally this is achieved by Denhardt's reagent and heparin which are used in combination with denatured fragmented salmon sperm or yeast DNA or SDS detergent. If the hybridization signal to noise ratio is rather high, indicating abundance of DNA sequence, the use of 0.25% non-fat dried milk CO.05 x BLOTTO is recommended (vide Sambrook *et al*, 1989). Blocking agents are often omitted from the hybridization solution as high concentrations of protein may hamper probe DNA annealing. The addition of dextran sulfate or polyethylene glycol (PEG) increases the rate of reassociation of nucleic acids.

Blocking Agents

(i) *Denhardt's Reagent* - made up as 50 × stock solution - filtered and stored at —20°C. The reagent contains 5g—Ficoll (Type 400 Pharmacia); 5g polyvinylpyrrolidone; 5g bovine serum albumin (Pentextraction 5) and H_2O, 500 ml.

(ii) *BLOTTO* - 5% non-fat dry milk dissolved in .02% aqueous sodium azide, stored at 4°C, diluted 25-fold in hybridization buffer.

(iii) *Denatured Fragmented Salmon Sperm* (Sigma Type III sodium salt), dissolved in water at 10mg/ml concentration. The solution is stirred with magnetic stirrer for 2-4 h at room temp to help DNA dissolve. The solution is adjusted to 0.1 M NaCl, extracted once with phenol and once with phenol + chloroform. The aqueous phase is recovered, DNA sheared by passing rapidly 12 times through 17 gauge hypodermic needle, precipitated by adding 2 vol of ice cold ethanol; recovered by centrifugation and redissolved at a concentration of 10 mg/ml in water. The OD260 and the exact conc. of DNA are calculated. The solution is boiled for 10 min, stored at -20°C in small aliquots. Before use, it is heated for 5 min, in boiling water bath and chilled quickly in ice water. Denatured, fragmented DNA is used at 100 μg/ml in hybridization solutions.

iv) *Heparin* - (Sigma H-7005 porcine grade II or equivalent) - 50 mg/ml in 4 x SSPE or 4 x SSC and stored at 4°C; used with dextransulfate, 500 mg/ml; used without dextransulfate, 50 μg/ml.

The probe is normally prepared by initially nicking the DNA by DNAse I followed by resynthesis for the nick by labelled nucleotides with the aid of DNA polymerase. The sites are identified by autoradiography following Southern blotting and hybridization. Prior to hybridization, it is necessary to treat the filter in prehybridizing solution before transfer. Filters are then washed before exposure to X-ray film with intensifier screens. The sequences which hybridize with the given probe belong to discrete chromosome loci. The variations in restriction sites often characterise distinct alleles. The use of restriction fragments as genetic markers has made this study an invaluable tool in plant breeding.

In order to identify polymorphism within smaller taxonomic units, it is always desirable to use a large number of enzymes, all of which have specific sites of recognition. However, for working out the copy number and clone polymorphism, it is sometimes desirable after developing, to wash again in higher concentrations of x SSC at 65°C before further exposure.

Significance

Restriction markers have the added advantage over conventional genetic markers that their identification is of nucleotide sequences. As such they are not dependent on gene expression and remain unaffected by environment. They represent clearly the presence or absence of certain base sequences. Moreover, unlike conventional gene markers whose expression in the phenotype depends on the dominance or recessive nature, the restriction markers can be identified in both homo and heterozygous states, irrespective of the genetic background. The RELP markers can be directly hybridized *in situ* on chromosomes to locate the position of sequences in the chromosome structure.

Construction of maps

RFLP Maps: In order to construct a map of genetic loci, genetically divergent individuals are selected, which are otherwise compatible. The intraspecific variants are specially suitable. In order to estimate variation, it is always desirable to survey with a number of cloned probes. An idea of the extent of variation can be obtained by consulting data from systematic and biochemical studies as well.

After selecting the plants, DNA from each individual is extracted, digested with restriction enzymes and screened for polymorphism with the use of cloned probes. For hybridization, it is desirable to choose two such accessions showing sufficient polymorphism. After hybridization and segregation in selfed F_2 progeny, the individuals can be utilized for mapping. However, back crossing with one of the parents of F_2 individuals, also yields the desired result.

The DNA for each segregant individual needs to be isolated and the plants be maintained for repeated DNA extraction, if necessary. The choice of enzymes for the cut is dependent on the degree o polymorphism existing between the strains. The number of enzymes needed is more or less inversely proportional to the extent of visible polymorphism. In maize, only one or two enzymes can reveal polymorphism with most of the probes (Kodert, 1989).

The next step is to prepare survey or mapping filters from a series of agarose gels. Each filter set must contain a single restriction digest of all individuals to be scored. For example, each filter must contain either Pst I or EcoR for all individuals. This enables detection of polymorphism in all individuals with an enyme/ probe combination. With regard to single RFLP, the plant, when tested with the probe following a single RFLP gel, can reveal its homo or heterozygous state.

The sequential scoring of the mapping population with different probes from the library ultimately helps in analysis of linkage and mapping the distance. The linkage analysis is based on the extent of cosegregation. In sequential analysis with different probes, linked loci show cosegregation. The homozygosity of an allele for one probe, reveals homozygosity and cosegregation for the second probe as well, in case of linked loci. Thus each probe can be used with all other probes. Recombination data can be utilized to map the distances. The objection of RFLP study is ultimately to prepare a saturated map with RFLP markers spaced about 16-20 centimorgan part. This calculation is based on the premise that conventional genetic maps are about 1500 cm apart and as such about 150 well spaced markers are adequate for a saturated map.

Preparation and Examination of Agarose Gels (after Sambrook *et al* 1989)

Seal the edges of a clean, dry, glass plate (or the open ends of the plastic tray supplied with the electrophoresis apparatus) with autoclave tape so as to form a mold. Set the mold on a horizontal section of the bench.

Prepare sufficient electrophoresis buffer (usually 1 x TAE or 0.5 x TBE; see table 2 to fill the electrophoresis tank and to prepare the gel. Add the correct amount of powdered agarose (see table) to a measured quantity of electrophoresis buffer in an Erlenmeyer flask or a glass bottle with a loose-fitting cap. The buffer should not occupy more than 50% of the volume of the flask or bottle. Loosely plug the neck of the Erlenmeyer flask. When using a glass bottle, make sure the cap is loose. Heat the slurry in a boiling-water bath or a microwave oven until the agarose dissolves. Cool the solution to 60°C, and, if desired, add ethidium bromide (from a stock solution of 10 mg/ml in water) to a final concentration of 0.5 μg/ml and mix thoroughly.

Using a Pasteur pipette, seal the edges of the mold with a small quantity of the agarose solution. Allow the seal to set. Position the comb 0.5—1.0 mm above the plate so that a complete well is formed when the agarose is added. If the comb is closer to the glass plate, there is a risk that the base of the well may tear when the comb is withdrawn, allowing the sample to leak between the gel and the glass plate. Pour the remainder of the warm agarose solution into the mold. The gel should be between 3 mm and 5 mm thick. Check to see that there are no air bubbles under or between the teeth of the comb. After the gel is completely set (30—45 min at room temperature), carefully remove the comb and autoclave tape and mount the gel in the electrophoresis tank. Gels cast with low melting temperature agarose and gels that contain less than 0.5% agarose should be chilled to 4°C and run in the cold room.

Add just enough electrophoresis buffer to cover the gel to a depth of about 1 mm. Mix the samples of DNA with the desired gel- loading buffer (table 2). Slowly load the mixture into the slots of the submerged gel using a disposable micropipette, an automatic micropipettor, or a Pasteur pipette.

Close the lid of the gel tank and attach the electrical leads so that the DNA will migrate toward the anode (red lead). Apply a voltage of 1—5 V/cm (measured as the distance between the electrodes). If the leads have been attached correctly, bubbles should be generated at the anode and cathode (due to electrolysis) and, within a few min, the bromophenol blue should migrate from the wells into the

Table 2 : Gel-loading Buffers

Buffer type	6 x Buffer	Storage temperature
I	0.25% bromophenol blue 0.25% xylene cyanol FF 40% (w/v) sucrose in water	4°C
II	0.25% bromophenol blue 0.25% xylene cyanol FF 15% Ficoll (Type 400; Pharmacia) in water	room
III	0.25% bromophenol blue 0.25% xylene cyanol FF 30% glycerol in water	4°C
IV	0.25% bromophenol blue 40% (w/v) sucrose in water	4°C
V	Alkaline loading buffer 300 mN NaOH 6nM EDTA 18% Ficoll (Type 400; Pharmacia) in water 0.15% bromocresol green 0.25% xylene cyanol FF	4°C

body of the gel. Run the gel until the bromophenol blue and xylene cyanol FF have migrated the appropriate distance through the gel. Turn off the electric current and remove the leads and lid from the gel tank. If ethidium bromide was present in the gel and electrophoresis buffer, examine the gel by ultraviolet light and photograph the gel. Otherwise, stain the gel with ethidium bromide and then photograph.

Separation of Restriction Fragments of Mammalian Genomic DNA by Agarose Gel Electrophoresis

Digest an appropriate amount of DNA with one or more restriction enzymes. For Southern analysis of mammalian genomic DNA, approximately 10 μg of DNA should be loaded into each slot of the gel, when probes of standard length (> 500 bp) and high specific activity (10^9 > cpm/μg) are used to detect single-copy sequences; 30—50 μg of DNA are needed when oligonucleotides are used as probes. Proportionately less DNA may be used when the sample contains higher concentrations of the sequences of interest. The concentrations of DNA in preparations of high-molecular-weight mammalian genomic DNA are often so low that it is necessary to carry out restriction digests in relatively large volumes. The chief problem encountered during digestion of high-molecular- weight DNA is unevenness of digestion caused by variations in the local concentrations of DNA. Clumps of DNA are relatively inaccessible to restriction enzymes and can be digested only from the outside. To ensure homogeneous dispersion of the DNA.

a. Allow the DNA to stand at 4°C for several h after dilution and addition of 10 x restriction enzyme buffer.
b. *Gently* stir the DNA solution from time to time using a sealed glass capillary.
c. After addition of the restriction enzyme, *gently* stir the solution for 2—3 min at 4°C before warming the reaction to the appropriate temperature.
d. After digestion for 15—30 min, add a second aliquot of restriction enzyme and stir. At the end of the digestion, add the appropriate amount of gel-loading buffer and separate the fragments of DNA by electrophoresis through an agarose gel. For genomic DNA, a 0.7% gel cast in 0.5 x TBE containing ethidium bromide [0.5 μg/ml] may be used.

It is important to include controls containing DNAs of known size that can serve as molecular-weight standards (e.g., bacteriophage λ DNA cleaved with *Hind* III). These marker DNAs are usually run in the two outside wells of the gel. Sufficient marker DNA to be detectable by staining with ethidium bromide (~200 ng) should be applied to the gel. If the DNA has been stored at 4°C, it should be heated to 56°C for 2—3 min before application to the gel, to disrupt any base pairing that may have occurred between protruding cohesive termini.

After electrophoresis is completed, photograph the gel. Place a transparent ruler alongside the gel so that the distance that any band of DNA has migrated can be read directly from the photographic image. If desired, the gel may be stored at this stage before the DNA is denatured and transferred to the filter. Wrap the gel in Saran Wrap and store it on a flat surface at 4°C. Because the bands of DNA diffuse during storage, the gel should not be put aside for more than 1 day.

Transfer of DNA to Nitrocellulose Filters

For Capillary Transfer of DNA to Nitrocellulose Filters: After electrophoresis, transfer the gel to a glass baking dish and trim away any unused areas of the gel with a razor blade. Cut off the bottom left-hand corner of the gel to orient the gel during the succeeding operations.

Denature the DNA by soaking the gel for 45 min in several vol of 1.5 M NaCl, 0.5 N NaOH with constant, .*gentle* agitation. Rinse the gel briefly in deionized water, and then neutralize it by soaking for 30 min in several vol of a solution of 1 M Tris (pH 7.4), 1.5 M NaCl at room temperature with constant, gentle agitation. Change the neutralization solution and continue soaking the gel for a further 15 min.

While the gel is in the neutralization solution, wrap a piece of Whatman 3MM paper around a piece of Plexiglas or a stack of glass plates to form a support that is longer and wider than the gel. Place the wrapped support inside a large baking dish. Fill the dish with transfer buffer (10 x SSC or 10 x SSPE) until the level of the liquid reaches almost to the top of the support. When the 3MM paper on the top of the support is thoroughly wet, smooth out all air bubbles with a glass rod. Cut a piece of nitrocellulose filter (Schleicher and Schuell BA85 or equivalent) about 1 mm larger than the gel in both dimensions. Use gloves and blunt- ended forceps to handle the filter.

Float the nitrocellulose filter on the surface of a dish of deionized water until it wets completely from beneath, and then immerse the filter in transfer buffer for at least 5 min. Using a clean scalpel blade, cut a corner from the nitrocellulose filter to match the corner cut from the gel.

Remove the gel from the neutralization solution and invert it so that its underside is now uppermost. Place the inverted gel on the support so that it is centered on the wet 3MM papers. Make sure that there are no air bubbles between the 3MM paper and the gel. Surround, but do not cover, the gel with Saran Wrap or Parafilm to prevent liquid from flowing directly from the reservoir to paper towels placed on top of the gel.

Place the wet nitrocellulose filter on top of the gel so that the cut corners are aligned. One edge of the filter should extend just over the edge of the line of slots at the top of the gel. Make sure that there are no air bubbles between the filter and the gel. Wet two pieces of 3MM paper (cut to exactly the same size as the gel) in

2 x SSC and place them on top of the wet nitrocellulose filter. Smooth out any air bubbles with a glass rod.

Cut a stack of paper towels (5-8 cm high) just smaller than the 3MM papers. Place the towels on the 3MM papers. Put a glass plate on top of the stack and weigh it down with a 500-g weight. The objective is to set up a flow of liquid from the reservoir through the gel and the nitrocellulose filter, so that fragments of denatured DNA are eluted from the gel and are deposited on the nitrocellulose filter.

Allow the transfer of DNA to proceed for 8—24 h. As the paper towels become wet, they should be replaced. Remove the paper towels and the 3MM papers above the gel. Turn over the gel and the nitrocellulose filter and lay them, gel side up, on a dry sheet of 3MM paper. Mark the positions of the gel slots on the filter.

Peel the gel from the filter and discard it. Soak the filter in 6 x SSC for 5 min at room temperature to remove any pieces of agarose sticking to the filter. To assess the efficiency of transfer of DNA, the gel may be stained for 45 min in a solution of ethidium bromide (0.5 μg/ml in water) and examined by ultraviolet illumination. The intensity of fluorescence will be quite low because the DNA remaining in the gel has been denatured.

Remove the filter from the 6 x SSC and allow excess fluid to drain away. Place the filter flat on a paper towel to dry for at least 30 min at room temperature. Sandwich the filter between two sheets of dry 3MM paper. Fix the DNA to the filter by baking for 30 min to 2 h at 80°C in a vacuum oven. Overbaking can cause the filter to become brittle. If the gel was not completely neutralized before the DNA was transferred, the filter will turn yellow or brown during baking and chip very easily.

Hybridize the DNA immobilized on the filter to a ^{32}P-labeled probe.

For Capillary Transfer of DNA to Nylon Membranes under Neutral Conditions, fractionate the DNA by gel electrophoresis, and process the gel as described earlier. While the gel is in the neutralization solution, prepare the nylon membrane as follows:

Using a fresh scalpel or a paper cutter, cut a piece of membrane about 1 mm larger than the gel in both dimensions. Use gloves and blunted forceps to handle the membrane. Float the membrane on the surface of a dish of deionized water until it wets completely from beneath, and then immerse the membrane in transfer buffer (10 x SSC or 10 x SSPE) for at least 5 min. Using a clean scalpel blade, cut a corner from the membrane to match the corner cut from the gel. Transfer the denatured DNA from the gel to the membrane by capillary action.

To fix the DNA to the membrane either,

- place the dried membrane between two pieces of 3MM paper, and bake the membrane for 30 min to 2 h at 80°C in a vacuum or conventional oven, or
- expose the side of the membrane carrying the DNA to a source of ultraviolet irradiation (254 nm).

Since: Ultraviolet radiation is dangerous, particularly to the eyes. To minimize exposure, make sure that the ultraviolet light source is adequately shielded and wear protective goggles or a full safety mask.

Hybridization of Radiolabeled Probes to Nucleic Acids Immobilized on Nitrocellulose Filters or Nylon Membranes

Prepare the prehybridization solution appropriate for the task at hand. Approximately 0.2 ml will be required for each square centimeter of nitrocellulose filter or nylon membrane. The prehybridization solution should be filtered through a 0.45- micron disposable cellulose acetate filter (Schleicher and Schuell Uniflow syringe filter No. 57240 or equivalent).

Prehybridization solutions used for detection of low- abundance sequences contains:

6 x SSC (or 6 x SSPE); 5 x Denhardt's reagent; 0.5% SDS and 100 µg/ml denatured, fragmented salmon sperm DNA. 50% formamide may be added if needed.

Formamide: Many batches of reagent-grade formamide are sufficiently pure to be used without further treatment. However, if any yellow color is present, the formamide should be deionized by stirring on a magnetic stirrer with Dowex XG8 mixed-bed resin for 1 h and filtering twice through Whatman No. 1 paper. Deionized formamide should be stored in small aliquots under nitrogen at -70°C.

For detection of moderate or high-abundance sequences; use 6 x SSC (or 6 x SSPE) and 0.05 x BLOTTO, 50% formamide may be added if needed.

Float the nitrocellulose filter or nylon membrane containing the target DNA on the surface of a tray of 6 x SSC (or 6 x SSPE) until it becomes thoroughly wetted from beneath. Submerge the filter for 2 min. Slip the wet filter into a heat-sealable bag (e.g., Sears Seal-A- Meal or equivalent). Add 0.2 ml of prehybridization solution for each square centimeter of nitrocellulose filter or nylon membrane. Squeeze as much air as possible from the bag. Seal the open end of the bag with the heat sealer. Incubate for 1—2 h submerged at the appropriate temperature (68°C for aqueous solvents, 42°C for solvents containing 50% formamide).

If the radiolabeled probe is double-stranded, denature it by heating for 5 min at 100°C. Single-stranded probe need not be denatured. Chill the probe rapidly in ice water. Alternatively, the probe may be denatured by adding 0.1 volume of 3 N NaOH. After 5 min at room temperature, transfer the probe to ice water and add

0.05 volume of 1 M Tris. Cl (pH 7.2) and 0.1 volume of 3 N HCl. Store the probe in ice water until it is needed.

For Southern hybridization of mammalian genomic DNA where each lane of the gel contains 10 μg of DNA, 10—20 ng/ml radiolabeled probe (sp. act. = 10^9 cpm/μg or greater) should be sued. For Southern hybridization of fragments of cloned DNA where each band of the restriction digest contains 10 ng of DNA or more, much less probe is required. Typically, hybridization is carried out for 6—8 h using 1—2 ng/ml radiolabeled probe (sp. act. = 10^9 cpm/μg or greater).

Quickly remove the bag containing the filter from the water bath. Open the bag by cutting off one corner with scissors. Add the denatured probe to the prehybridization solution, and then squeeze as much air as possible from the bag. Reseal with the heat sealer so that as few bubbles as possible are trapped in the bag. To avoid radioactive contamination of the water bath, the resealed bag should be sealed inside a second, noncontaminated bag.

When using nylon membranes, the prehybridization solution should be *completely* removed from the bag and immediately replaced with hybridization solution. The probe is then added and the bag is resealed. ***Hybridization solution for nylon membranes*** contains 6 x SSC (or 6 x SSPE); 0.5% SDS; 100 μg/ml denatured, fragmented salmon sperm DNA and 50% formamide (if hybridization is to be carried out at 42°C)

Incubate the bag submerged in a water bath set at the appropriate temperature for the required period of hybridization. Wearing gloves, remove the bag from the water bath and immediately cut off one corner. Pour out the hybridization solution into a container suitable for disposal, and then cut the bag along the length of three sides. Remove the filter and immediately submerge in the a tray containing several hundred ml of 2 x SSC and 0.5% SDS at room temperature. Do not allow the filter to dry out at any stage during the washing procedure.

After 5 min transfer the filter to a fresh tray containing several hundred ml of 2 x SSC and 0.1% SDS and incubate for 15 min at room temperature with occasional gentle agitation. If short oligonucleotides are used as probes, washing should be carried out only for brief periods (1—2 min) at the appropriate temperature. Transfer the filter to a flat-bottom plastic box containing several hundred ml of fresh 0.1 x SSC and 0.5% SDS. Incubate the filter for 30 min to 1 h at 37°C with gentle agitation.

Replace the solution with fresh 0.1 x SSC and 0.5% SDS, and transfer the box to a water bath set at 68°C for an equal period of time. Monitor the amount of radioactivity on the filter using a hand-held minimonitor. The parts of the filter that do not contain DNA should not emit a detectable signal.

Briefly wash the filter with 0.1 x SSC at room temperature. Remove most of the liquid from the filter by placing it on a pad of paper towels. Place the damp filter on a sheet of Saran Wrap. Apply adhesive dot labels marked with radioactive ink to several asymmetric locations on the Saran Wrap. These markers serve to align the autoradiograph with the filter. Cover the labels with Scotch Tape to prevent contamination of the film holder or intensifying screen with the radioactive ink.

Cover the filter with a second sheet of Saran Wrap, and expose the filter to X-ray film (Kodak XAR-2 or equivalent) to obtain an autoradiographic image. Single-copy sequences in mammalian genomic DNA can usually be detected after 16—24 h of exposure at -70°C with an intensifying screen.

Although DNA immobilized within the gel appears to give somewhat stronger hybridization signals than DNA attached to a solid support, it cannot be hybridized sequentially to many different probes. This is a severe disadvantage when the amount of genomic DNA is limited (as is often the case in prenatal diagnosis, for example). We therefore recommend that the genomic DNA be transferred to a nylon membrane such as Nytran (Schleicher and Schuell) or GeneScreen (du Pont) (after Sambrook *et al* 1989).

SECTION B. THE DNA FINGERPRINT

The sequence polymorphism of DNA forms the basis of DNA fingerprinting. Digestion of DNA with specific enzymes, agarose gel electrophoresis, membrane blotting and detection following hybridization with specific oligonucleotides as probes in hybridization have made DNA fingerprinting a very convenient and powerful tool for detection of sequences and their comparison needed in various aspects of genetic analysis.

The analysis of eukaryotic genome has been undoubtedly facilitated by Restriction Fragment Length Polymorphism (RFLP) Analysis, involving enzyme digestion, transfer to membrane and hybridization. However, RFLPs are diallelic, but multiallelic RFLP probes form the very basis of DNA fingerprinting pattern. The hypervariable DNA polymorphism in the genome can be detected by multiallelic analysis by a variable number of tandem repeats of short sequences (10—60 bp), known as *mini*-and *microsatellites*. The *minisatellite* probes are genomic, containing in man, one of the four 33 base long sequences from the first intron of the myoglobin gene. A DNA polycore probe based on myolin minisatellite can detect simultaneously, through hybridization, a large number of dispersed hypervariable loci containing tandem repeats of similar sequences. Tandem - repetitive minisatellites or variable number of tandem repeats (VNTR) show high level of tandem variability. Such a complex fingerprint pattern becomes very specific to individuals, the application of which is enormous. Later, the amplification of the minisatelliates through PCR could be achieved and analysis of minisatellite variant repeat (MVR) is carried out frequently.

The PCR amplification of short tandem repetitive microsatellites has made the test very sensitive, permitting DNA typing of degraded system as well. These are mostly less than 100 bp long. Different alleles of the minisatellite locus could be distinguished along with digital sequencing (Jeffreys *et al*, 1991).

Southern blot hybridization with radioactive probes was the initial step in DNA fingerprinting. However, as synthetic oligonucleotide probes are short and single stranded and can have very high polymorphic information content, their use has greatly accelerated the research on DNA fingerprinting. These probes, complementary to short tandem repeats of genome, are easily synthesized with the aid of information available from Data Bank. They are also obtained from highly variable intron regions or flanking sequences of gene (Ali *et al*, 1989). Repeat copies of $(GATA)_n$ and $(GACA)_n$ from snakes have been shown to hybridize with restriction digested human DNA and polymorphic DNA fingerprint patterns obtained are individual specific (Singh *et al*, 1991; Ali and Eppelen, 1991).

In addition to these, multilocus probes have been obtained in other organisms including *Drosophila* and mouse. Instead of cloning, synthetic oligonucleotides are used.

In the preparation of oligonucleotide probes, the content of the different bases is taken into account. Of the non-Watson-Crick base pairs, CC, AA, TT, GG, CA, CT, GA, GT, the dissociation properties of CA, AA, TT,CT are five times faster whereas GA, GT are 2.5 times faster than Watson - Crick base pairs (Itakura *et al*, 1984).

In addition to VNTR, the new approach of Random Amplification Polymorphic DNA (RAPD) is proving to be very useful. The method uses a ten-base long chain to search for variations in DNA molecule. However, the loss of band often noted following RAPD is not fully accounted for. It may be due to sequence variation or insertion or deletion. At the end, in order to remove the probe from the filter, boiling in 1-5 mM EDTA for 10-15 min or 1 x SSC is necessary. In extreme cases, treatment with 2 x SSC at 60°C for 30 min may be needed.

The oligonucleotides can be synthesized in automatic synthesizers or can be obtained from commercial firms. The purification of oligonucleotides is necessary after synthesis, preferably through polyacrylamide gel electrophoresis, anion exchange chromatography or reverse phase HPLC, as required. After UV detection, the gel is sliced off and oligonucleotides are recovered in STE buffer (150 mM NaCl/10 mM Tris/HCl; pH 8.0; 1 mM EDTA). It can be labelled at the 5′ end by T4 polynucleotide kinase or at the 3′ end with alpha dNTP using terminal transferase.

Isolation of DNA is carried out through various methods described in chapter on Isolation, from different tissues, including blood, urine and solid tissues. DNA is digested through different restriction enzymes *viz*, Alu I, Hae III, Eco R, Hind III, Taq I, following the instruction of the manufacturers. Normally 3-4 units of enzymes are required for digestion of 1 μg of DNA but sometimes more may be needed. As such, initial confirmation in a minigel is preferable.

Agarose gel electrophoresis for DNA separation is normally carried out with TAE buffer (40 mM Tris/HCl, 12 mM/Na acetate, 2 mM EDTA, pH 8.3) using peristaltic pump. For fingerprinting, low voltage (1-2 volts/Cm) run is desirable followed by staining with ethidium bromide and photography under ultraviolet light.

The DNA can be transferred to nitrocellulose or nylon membrane as described for Southern Hybridization or dried in a vacuum dryer for 1h at ambient temperature and 1h at 60°C, followed by direct hybridization. The latter procedure has been found to give better results than membrane hybridization.

The hybridization process with the radio-labelled probe on dried gel or membrane following Southern transfer can be followed. With oligonucleotide probes, prehybridization is not needed. DNA is denatured, annealed and equilibrated prior to hybridization following the usual schedule.

Non-radioactive probes, though slightly less sensitive, are also applied and the oligonucleotides can be labelled with biotin at the 5′ end or digoxigenin - a steroid hapten as reporter. After synthesis in the automatic synthesizer, an aminoethylphosphate linker is added at the 5′ end (Aminolink Applied Biosystems) as described by the supplier. After deprotection, the reactive amino group is then allowed to react with digoxigenin succinimidester (Boehringer) or N- hydroxysuccinimidobiotin (Sigma) for overnight at room temp in Na_2CO_3 buffer (pH 9.0).

The reverse phase HPLC or column is used to purify nonradioactive derivatives. The peaks are listed for biotin or digixogenin by spot blot on nitrocellulose membrane, followed by colorimetric detection with the use of streptavidin - alkaline phosphate (Blue Gene, BRL). The next steps are vacuum drying of positive fraction, dissolving the pellet in redistilled water and determination of the concentration through UV absorbance. Such probes can be hybridized on membranes as well as on dried gels.

Microsatellite probes are observed by ethidium bromide staining following non-denaturing gel electrophoresis. The sensitivity is better with silver staining of native gels (Hearne *et al*, 1992).

The *in situ* hybridization with oligonucleotide probe on chromosomes has also been correlated with DNA fingerprinting in certain respects. The core elements in minisatellites used in DNA fingerprinting have been hypothesized to be hot spots for recombination in the eukaryotic genome (Zischler *el al*). Polymorphic minisatellites have been located at chromosomes ends and subtelomeric segments show high rate of recombination. With minisatellites, correlation is claimed between probe grain and chiasma (Royle *et al*, 1988).

However, hypervariable sequences may not always have core elements as $(GACA)_n$ $(GATA)_n$ and $(GT)_n$. They are still capable of yielding DNA fingerprints due to variation in repeat sequences. In minisatellites, the DNA fingerprinting is not always dependent on core elements. In chromosomes of *Ellobius, in situ* hybridization reveals the absence of GATA/GACA repeat sequences in univalent X chromosome. This fact may indicate that absence of recombination, as with univalent X, is due to the absence of these sequences suggesting also their recombinogenic property (Eppelen *et al*, 1989).

The impact of DNA fingerprinting cuts across different facets of biological sciences. Ranging from identification of susceptible loci involved in common genetic diseases, it extends to the identification of criminals (Chakravorty and Kidd, 1991), resolution of parentage and delineation of status and affinities in evolution.

The protocols followed have been described under *Restriction Fragment Length Polymorphism* and *In situ Hybridization.*

SECTION C: POLYMERASE CHAIN REACTION

The modality of Polymerase Chain Reaction, otherwise abbreviated as PCR, was originally worked out by Kary Mullis and his colleagues at Ehrlich's laboratory of Cetus Corporation, California. It is at present the most convenient method for amplification of gene sequences. It is a technique used for synthesizing millions of copies of a single sequences in a few hours. This enzymatic method thus can synthesize multiple copies of DNA. The entire process needs two synthetic oligodeoxynucleotide primers, a thermostable DNA polymerase and four nucleotide triphosphates functioning on template DNA. On heat denaturation of the template, followed by cooling, the primers anneal to the target ends, one in each strand at two different ends of the single stranded templates. As the primer binds to complementary sequence, the polymerase starts copying the strand. After the duplication of the target sequence, the two strands are again separated by heating and fresh synthesis of strands is initiated with the primers. As the 3′ ends point towards each other repeat cycles of heating and cooling lead to exponential increase in the number of strands bound by the primers. It has been calculated that 20 cycles of heating and cooling lead to the amplification of 10^6 copies of the segment. A single PCR cycle represents denaturation, annealing and extension. This cycle can be repeated several times and within a short period, a huge amount of target DNA can be synthesized. A thermostable bacterium, *Thermus aquaticus* isolated from hot spring, yields the enzyme *Taq polymerase* which continues working indefi-

Polymerase Chain-Reaction

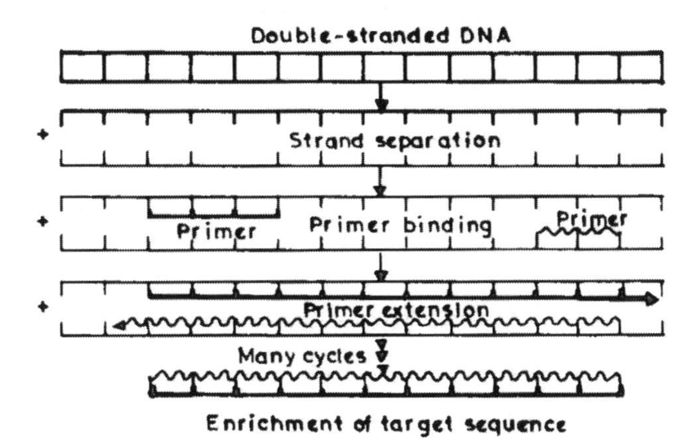

(1) Nature (under Perspective) 21.12, 1989 p.1543 (PCR)

Yeast Artificial Chromosome

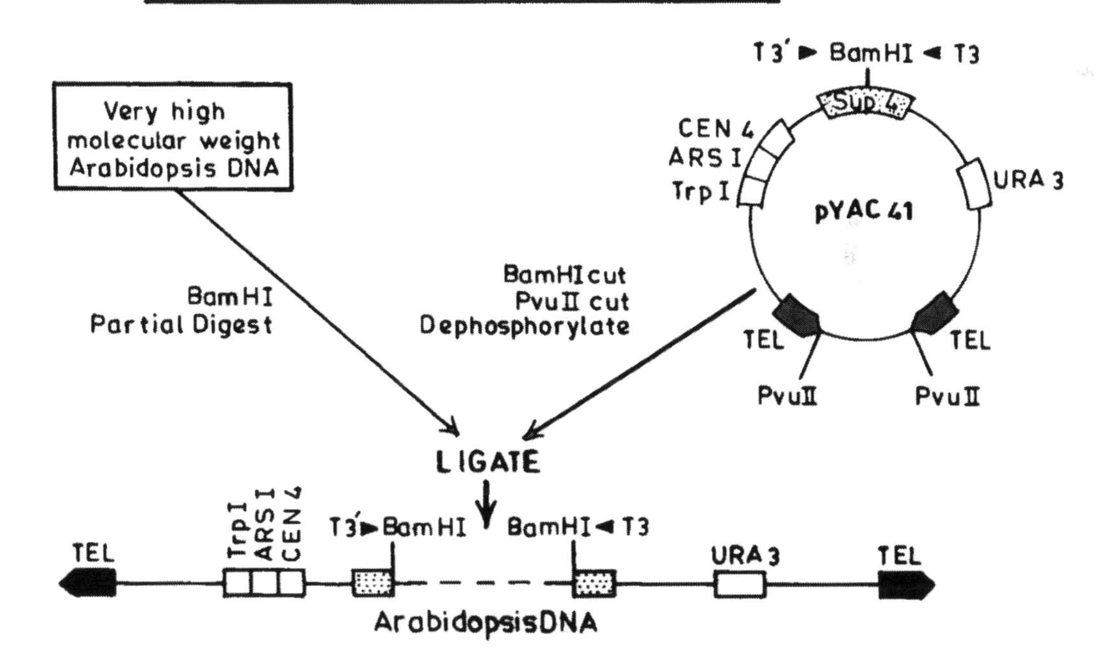

(2) Aust. J. Plant Physiology. 1992, 19, 341-51 p.342

Drawings on (YAC)

Professor R. Schmidt
Cambridge Lab.
AFRC Institute Plant Science Research
John Innes Laboratory
Norwich

DNA Footprint

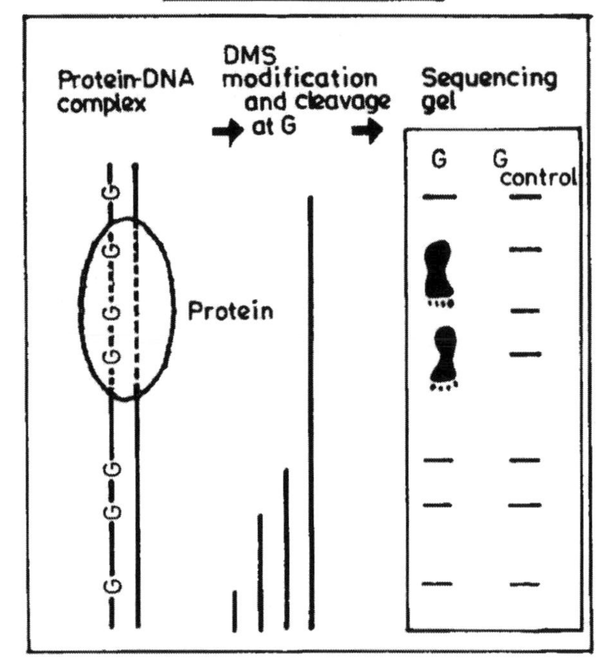

(3) Trends in Genetics - 1991 (DNA Foot Print) 7(7) p.210

Dr. Hans Peter Saluz

Agarose Electrophoresis

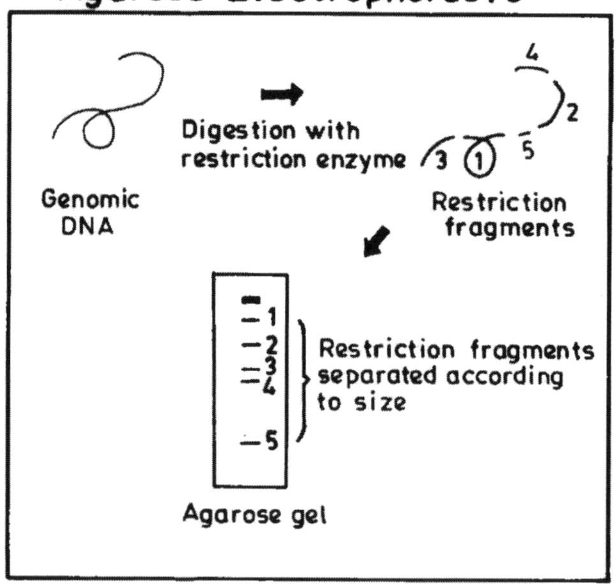

(4) Dr. Gary Kocher
Department of Botany
University of Gergia
Athens, GA 30602

nitely despite the heating needed for denaturation. The technique permits amplification of sequences embodied in a tiny fragment to a large molecule. Fragments of even 10 kb size can be synthesized through Taq polymerase and other thermostable enzymes (see figure).

In fact, 30 cycles can theoretically multiply upto 1 billion. In addition to amplifying known target sequences, the technique has been modified in later years for amplification of flanking unknown sequences with defined primer sites. This method, termed as *Inverse PCR*, permits sequencing of DNA outside the primer sites. Primer molecules are synthesized with their sequences reversed. The target DNA is cut and circularized including sequences outside the primer site. As such, polymerase extends the primer around the circle following thus an opposite direction.

The technique of PCR has greatly facilitated the preparation of probes specially for hybridization *in situ* at the chromosome level, the identification of gene loci, and for chromosome walking. Ranging from identification of loci for genetic disorders in chromosome to sequencing of changes in evolutionary pattern, the PCR technique has emerged as a powerful tool in genetic and chromosome research. The entire process of denaturation, synthesis and reannealing has been automated and PCR system has become an indispensable tool for studies on genetic manipulation.

In the preparation of PCR product, a possible limitation is the contamination caused by carry-over PCR product. The Parkin Elmer Cetus offers a GeneAmp PCR Carryover Prevention Kit for such false amplification. The system utilizes enzymatic and chemical reactions analogous to restriction modification and excision repair system, providing PCR amplification product, without degrading the native DNA template. To make PCR product amenable to degradation, it uses dUTP in place of dTTP in reaction mix and pretreats all subagent PCR reaction mix with uracil-N- glycosylase. After the completion of the final cycle, equal volume of chloroform is added and the sample is soaked at 72°C for overnight. Details of the method are given below (Sambrooke *et al* 1989)

Amplification Reactions

In a sterile 0.5-ml microfuge tube, mix in the following order: Sterile water, 30 µl; 10 × amplification buffer, 10 µl; mixture of four dNTPs, each at a concentration of 1.25 mM, 16 µl; primer 1 (in 5 µl of H_2O) 100 pmoles; primer 2 (in 5 µl of H_2O) 100 pmoles; template DNA (up to 2 µg, depending on the concentration of target sequences); H_2O to a final volume of 100 µl.

10 × Amplification buffer

500 mM KCl; 100 mM Tris. Cl (pH 8.3 at room temperature); 15 mM $MgCl_2$; 0.1% gelatin.

The standard conditions given above work well for a wide range of templates and oligonucleotide primers. The pH of the mixture of dNTPs should 7.0.

For amplification of single sequences from mammalian genomic DNA, use 0.2-2 μg of DNA. The reaction mixture will then contain approximately 0.03-0.3 pg of a target sequence 500 bp in length. For amplification of target DNA cloned in a plasmid vector (e.g., for DNA sequencing), add 20 ng of linearized plasmid DNA. Because the optimal concentration of Mg^{++} in the reaction is quite low, DNAs to be used as templates should be dissolved in 10 mM Tris. Cl (pH 7.6), 0.1 mM EDTA (pH 8.0).

Heat the reaction mixture for 5 min at 94°C to denature the DNA completely. While the mixture is still at 94°C, add 0.5 μl of *Taq* DNA polymerase (5 units/μl; Perkin Elmer Cetus N801-0046).

Taq DNA polymerase is supplied in a storage buffer containing 50% glycerol. This solution is very viscous and is difficult to pipette with accuracy. the best method is to centrifuge the tube containing the enzyme at 12,000g for 10 *s* at 4°C in a microfuge and then to withdraw the required amount of enzyme using a positive displacement pippeting device.

Overlay the reaction mixture with 100 μl of light mineral oil (Sigma M-3516 or equivalent). This prevents evaporation of the sample during repeated cycles of heating and cooling. Carry out amplification as described below. Typical conditions for denaturation, annealing, and polymerization are as follows:

Cycle	*Denaturation*	*Annealing*	*Polymerization*
First cycle	5 min at 94°C	2 min at 50°C	3 min at 72°C
Subsequent cycles	1 min at 94°C	1 min at 50°C	3 min at 72°C
Last cycle	1 min at 94°C	2 min at 50°C	10 min at 72°C

The denaturation step is omitted after the last cycle of amplification and the samples are transferred to –20°C for storage. Polymerase chain reactions can be automated with a thermal cycler (available from Perkin Elmer Cetus or Ericomp).

Withdraw a sample of the amplified DNA from the reaction mixture and analyze it by gel electrophoresis, Southern hybridization, or DNA sequencing. If necessary, the oil can be removed from the sample by extraction with 150 μl of chloroform. The aqueous phase, which contains the amplified DNA, forms a micelle near the meniscus. This micelle can be transferred to a fresh tube with an automatic micropipettor.

With the gradual advances in the amplification protocols in PCR, its application has been extended to a variety of research problems. The characterization of mutation, polymorphisms, evolutionary changes in DNA has all been much simplified through PCR product. It is now widely applied in the study of Human Genome Project, molecular evolution, identification of mutations, gene expression, diagnosis of diseases as well as in forensic studies.

Chromosome Crawling

This procedure is often applied to identify chromosomal sequences adjacent to chromosome segment with known DNA sequences. The method takes advantage of polymerase chain reaction in an inverse direction otherwise termed as ***Inverse PCR***. The procedure involves the complete digestion of DNA with a restriction enzyme which does not have any site of cheavage within the target region. The restriction fragment including the target should not exceed 2-3 kb in length. Then the DNA is diluted and ligated to form a circle. The template DNA is then linearized by cleaving within the target sequence with appropriate enzyme. The sequences flanking the target are then amplified with the aid of oligonucleotide primers complementary to 5 ′ termini of the target. Ultimately, a double stranded linear DNA results with the head to tail arrangement of flanking sequences outside the target DNA molecule. The restriction site used originally to cleave the genomic DNA marks the junction of upstream and down stream sequences. This method of identifying chromosome sequences is termed as *Chromosome Crawling*—an aspect of chromosome *walking*, described in the next section.

SECTION D: CHROMOSOME WALKING

Chromosome Walking involves the localization of gene families within a chromosome domain through the use of overlapping clones. Chromosome walking technique and the mapped markers permit identification of specific loci and isolation of genes for the study of their expression. Large regions of genome can thus be cloned through the isolation of a series of overlapping recombinants. Even a relatively simple family of genes, such as the mammalian globin genes, is spread over at least 70kb of genomic DNA. In this method, a segment of non-repetitive DNA isolated from the end of a recombinant can be used as a probe for identification of clones with adjacent sequences. Because of the slowness of the process, the use of cosmids rather than bacteriophage is preferred. In cosmids, foreign DNA of approx. 45 kb can be inserted (Sambrook *et al.*, 1989).

The primary step in cloning genes through chromosome walking is to secure probes within a few hundred kilobases of the desired locus. In order to achieve this objective, meiotic segregation of the mutants is correlated with restriction fragment length polymorphism. After identification of linked RFLPs, cloning of desired genes entails isolation of cloned DNA fragments, bridging the gap between the RFLP marker and the gene. Chromosome walking through small steps, through cosmid or bacteriophage, is time consuming and laborious.

The later development of the technique for cloning and maintenance of long DNA fragments as inserts in *yeast artificial chromosome vector in yeast cell* has emerged as a powerful tool for chromosome walking. It enables DNA fragments of several hundred kilobases to be cloned. In view of the large size of 600 kb or even more, large genomes can be covered within a few clones. With the YAC system, it is possible to link the genetic map measured in centimorgans to physical map measured in kilobases. In *Arabidopsis*, the conversion of 1 cm for 140 bp has been used to convert the size of YAC inserts to genetic distance (Chang *et al*, 1988). In higher plants, the difficulty of chromosome walk is due to the heavy amount of repetitive sequence present. *Arabidopsis thalliana*, having small genome and low number of interspersed repeats, is therefore, regarded as a model system.

The YAC vector maintained as circular plasmid in *E. coli* has tailored sequences responsible for centromeric, telomeric, autonomously replicated sequences, selective markers and a cloning site maintained through transformation in yeast cells. In addition to functional elements necessary for maintenance in yeast, transposable elements can be inserted in it if necessary, as in *Drosophila*. The generation of YAC clones needs restriction digest of these vectors to release two DNA fragments which become two chromosome arms (Schmid *et al.*, 1992). The next step is to insert high mol. wt. DNA by ligation. Ultimately it is transformed into yeast cells where it remains as a linear artificial chromosome. The YAC clones can have much larger inserts than any other cloning system. The general strategy in the construction of the overlapping YAC library is to hybridize all RFLP markers in the region to at least one of the YAC libraries. All the mapped markers can be identified with corresponding YAC clones. Walk can be initiated to link the clones hybridizing to adjacent markers. This step involves the isolation of end probes from one YAC clone and hybridizing these probes to YAC libraries to identify overlapping clones. The end probes can be generated by inverse PCR and left end rescue.

Yeast artificial chromosome technology permits an increase in genomic cloning size upto 10-fold over cosmids (Bellane-Chantelot *el al*, 1992). Collections of overlapping genomic fragments called "Contigs" can help finger printing and cover all parts of the genome. The human genome so far has been covered in only a few percent by contigs ranging from 200 to 2000 kilobases. The whole human genome can be covered with Yeast Artificial Chromosome cloning.

The YAC clones can be utilized in the preparation of integrated physical, genetical and transcriptional map of the genome. In man (Ross *et al*, 1992), the isolation of 388 YAC clones from a human library with an average insert size of 680 kb has been reported through the hybridization of composite chromosome 21 probe to a high density array of YAC clones. Nearly 50% of these clones hybridize to chromosome 21, as checked by fluorescence *in situ* hybridization. It is possible to subdivide the whole genomic YAC libraries into their chromosomal elements essential for genome mapping. The YAC library has been constructed of *Arabidopsis thalliana* genome to isolate genes by chromosome walking. The hybridization probes from the ends of YAC insert could be produced (Grill and Somerville 1991). In *Arabidopsis,* the complete library of more than 21,000 YAC with an average insert size of 150 kb contains the entire genome. The YAC vector has also been utilized successfully in locating telomere sequences in human chromosome 9 following *in situ* hybridization with biotinylated probe (Guerrini *et al* 1990). The telomere ends help in defining ends of physical and genetic maps of chromosome. Since the sequence is made up of known tandem—rich repeats, cloning of terminal fragments specific for each chromosome is possible.

In order to test the specificity of transcription, inverse PCR is used to amplify vector—insert junction, using the amplified sequence as the template for RNA polymerase. The non- characterized sequences adjoining the vector sequence can also thus be amplified.

Almost 380 RFLP probes have been used (Hwang *et al* 1991, Nam *et al* 1989) for correlation of the physical map with the classical genetic map in *Arabidopsis.*

In maize, nearly 79,000 clones of an average insert size of 145 kb has been prepared with YAC vector (Edwards *et al* 1992). *In vitro* transcripts from phage promoters were successfully employed to secure labelled probes for chromosome walking. RNA probes obtained through *in vitro* transcription of recombinant cosmids carrying promoters of bacteriophage encoded DNA dependent RNA polymerase have also been adopted. In *Drosophila,* nearly 500 clones containing average inserts of 200 kb of *Drosophila* DNA have been assigned positions in salivary gland chromosomes by *in situ* hybridization (Ajioka *et al,* 1991). In different systems, including man, utilization of chromosome specific composite probes has been possible for mapping the genome with high resolution and subdividing the entire genome into specific chromosomal elements.

Preparation of YAC Library for Drosophila genome

Isolate high molecular weight random sheared source DNA of *Drosophila* from the lysed nuclei. Purify the DNA by CsCl density gradient centrifugation. This procedure yields DNA molecules approximately 500 kb long. Repair the DNA ends by bacteriophage T4 DNA polymerase. Follow sucrose velocity gradient centrifugation to separate the fragments by size. Assay collected fraction by Field Inversion Gel Electrophoresis. Pool DNA fragments more than 120 kb in size and concentrate in a collection bag. Use size fractionated DNA as insert ligated to pYACP-1 vector arms. Size fractionate and concentrate again. Use it to transform yeast strain AB 1380 (after Ajoika *et al* 1991).

The vector is a modified version of prototype vector YAC containing centromere, telomeres and selectable markers as well as a transposable P element.

For In Situ Hybridization with YAC Clones

Excise bands containing individual YAC chromosomes from field inversion gels. Purify DNA using sodium iodide—glass powder. Label DNA with biotin dCTP by random primer (hexamer) method. Hybridize on polytene chromosome following usual method. Chromosomes give a brown coloured hybridization signal along the length against a background of dark blue and light blue bands.

Generation of YAC Contig for Chromosome walking

Use probes from the polytene 83.84 region for screening the YAC libraries for clones to construct a YAC contig across the region. Obtain the probes by microdissection at 83 and 84, and amplification by PCR. Generate specific probes by inverse PCR for defining overlapping clones.

Inverse PCR

Use 1 μg of total genomic DNA of yeast and digest with EcoRV (left and YAC circles) Hin II (right end YAC circles) or Alu *I* (left and right end YAC circles). Extract with phenol and precipitate with ethanol. Allow the fragments to ligate at 4°C under very dilute conditions (400 μl) for promoting circularization. Inactivate ligase by heat, precipitate DNA by ethanol, resuspend and divide into aliquots. Digest one aliquot with Nhe*I* (to linearise left and YAC circles) and another aliquot with Ssp*I* (to linearise right end circles). Extract the samples with phenol and pass over a sepharose CL 6B spin column in TE buffer. Follow PCR with the DNA solution (after Schmidt *et al*, 1992).

Preparation of YAC library for plant chromosome

Use vector pYAC41, which is a modified version of pYAC4 with two PruII restriction sites with the adapter and one EcoR1 insert of 64 bp containing two T_3 promoters flanking Bam HI site, in addition to contromeric and telomeric sites (after Grill and Sommerville, 1991).

Isolate high molecular weight Arabidopsis DNA (2 Mb) from leaf protoplasts of 2-3 week old plants. Adopt sucrose gradient contrifugation, phenol purification of DNA and dialyze against 10 mM Tris-HCl (pH 8.0) 0.1 mM EDTA. Digest partially purified DNA with Bam HI to fragments 100-400 kb in size. Ligate to vector DNA treated with 50 fold molar excess of alkaline phosphatase at 14°C overnight. Size select ligated DNA in 0.8% agarose (1:1 mixture of low and ultralow melting agarose) CHEF gels (Bio Rad). Take agarose plugs containing DNA in the range of 120-300 kb and mix with an equal volume of 10 mM Tris-HCl, 1mM EDTA (pH 7.4) and melt at 60°C. Mix DNA solution with an equal volume of 2 M sorbitol and transform yeast sphaeroplast (8 vol).

Select transformants on solidified synthetic complete medium lacking tryptophan and uracil.

Generation of End Probes

Digest total yeast DNA from a YAC containing clone with a series of restriction enzymes producing blunt ends (Alu I, Psa I, Eco RV, Hinc II for the left centromere proximal arm, and Alu I, Nae I, Nru I, Fsp I or Hinc II for the right arm).

Determine the size vector insert, junction fragment by probing a Southern blot with the vector.

Take DNA from individual YACs (0.1 ng) or total yeast DNA (1 μg) from low melting agarose CHEF gels and subject to restriction digestion with an enzyme which cleaves the insert (0.52 kb) from the vector-insert junction. Follow phenol extraction and ethanol precipitation. Resuspend DNA in 0.2 ml ligation mixture. Incubate at 22°C for 3 h. Transfer to 50°C for 30 min to inactivate ligase. Precipitate with ethanol. Digest resuspended DNA with SmaI and FspI to linearize sequences from right and left ends of the insert respectively. Subject aliquots of DNA soln (100 ng of total yeast DNA or 10 ng of pure YAC DNA) to 25 cycles of PCR.

Recut amplified sequences with the same restriction enzymes as used initially. Remove free nucleotides through sephadex spin column, extract with phenol-chloroform and precipitate in phenol and ethanol. Incubate the end fragments with T3 RNA polymerase (50 units BRL) in presence of 0.1 mM α ($^{-32}$P) UTP (100 Ci/mM) and ribonucleoside triphosphates. Follow the usual labelling procedure and hybridization without bovine serum albumin in the hybridization solution. Wash the nylon filters in 5% SDS, 40 mM sodium phosphate (pH 7.2) and 1mM EDTA at 63°C for 20 min each.

YAC Hybridization in situ with Human Chromosome Telomere

Utilizing YAC vector, telomere of the long arm of human chromosome 9 has been cloned and hybridized *in situ* with metaphase chromosome.

The principles underlying the procedures are:

i) Modification of plasmid pYAC to have 5.6 kb and 3.6 kb arms, the former containing autonomous replicating sequence and centromeric sequences; setting free the telomeric ends by Bam HI and generating the cloning site by EcoRI.
ii) Isolation of digested human DNA with EcoRI, ligation with T4 DNA ligase and cloning in YAC vector.
iii) Preparation of telomeric human DNA probe by digestion, FIGE electrophoresis, separation and ligation to 5.6 kb arm (molar ratio 5:1).

iv) Hybridization of DNA with synthetic (TTA GGG)$_7$ and (C_4A_2)$_8$ end labelled oligonucleotides.
v) Biotinylation of positive HG_1 DNA and use as probe for *in situ* hybridization on metaphase chromosome.
vi) Characterization through fluorescent ligand with yellow green signals on red chromosomes counterstained with propidium iodide (Guerrini *et al* 1990).

SECTION E: FOOT PRINT

Genomic foot printing is a method of genome sequencing based on DNA protein interaction. It is based on the premise that a DNA bound protein would not react in a similar way to chemicals or ultraviolet, as compared to unbound DNA bases. For instance, guanine N-7 residue is methylated by dimethyl sulfate (DMS). If the guanine is bound to protein, it would have the altered reaction with DMS and as such may not react with methylation sensitive compounds or enzymes. If living cells are treated with this chemical, and on purification cleared with piperidine at modified sites, the normal agar gel genome sequencing would indicate a gap or window on the sequence ladder as compared to that of cloned control.

Similar genomic footprints can be obtained with UV reaction which differs between protein-free and protein-bound DNA. At the sites of photodamage by UV, the breakage would have two different sets of fragments in the sequencing gel. The region interacting with protein, otherwise termed as the ***Footprint***, would lead to two different strand breakage patterns. This method has been applied in different eukaryotes and the DNA products can be visualized by electroblotting and indirect end labelling or by extension of UV modified templates through radiolabelled primers (Salaz *et al* 1991).

CHAPTER 14

REPRESENTATIVE SCHEDULES FOR DIRECT OBSERVATION OF CHROMOSOMES FROM PLANTS AND ANIMALS *IN VIVO* AND FROM SPECIAL MATERIALS

A. STUDY FROM NORMAL TISSUES

Schedules For The Study of Mitotic Chromosomes

In Plant Materials

In Root Tips

From Paraffin Sections Dehydrated Through Alcohol Chloroform Grades and Stained in Crystal Violet.

Cut fresh root tips of *Allium cepa*, about 1 cm long, wash thoroughly and transfer to a tube containing a mixture of 1% chromic acid and 10% formalin (CF 1:1). Keep for 12-24 h. Wash the roots in a porcelain thimble in running water for 3 h. Transfer the roots to a glass phial containing 30% ethanol and keep for 1 h; then to 50% ethanol, keeping for 1 h; to 70% ethanol, treating overnight; through 80, 90 and 95% ethanol, keeping the roots for 1 h in each. Finally keep overnight in absolute ethanol. Keep in ethanol-chloroform mixtures (3:1, 1:1 and 1:3) successively for 1 h in each. Transfer to pure chloroform and keep for 10 min; then change the used chloroform with pure chloroform, and add small shavings of paraffin. Keep overnight on a hot plate at 35°C, then add a little more wax and keep the phial with contents at 45°C for 2 days and transfer to 60°C. Change the wax with fresh molten wax at intervals of 30 min for 2 h. Pour the molten paraffin with roots into a paper tray and add some more molten wax, then orient the roots in groups of three with their tips pointing to the same side at the same level. After the wax has cooled slightly, plunge the block into cold water. Trim the block and cut transverse sections of the root tips 14 µm thick on the microtome. Cut the ribbons into suitable segments and mount serially in water on a slide previously coated with Mayer's adhesive. For coating a slide, put a tiny drop of Mayer's adhesive on the slide and smear it three-quarters over the surface, then place the slide on a hot plate and help the ribbons to stretch with a pair of needles. Drain off the water and keep the slide overnight on the hot plate to dry. Place the slide with sections in pure xylol grades I and II, keeping in each for 30 min. Transfer the slide to a jar of ethanol-xylol (1:1) and keep for 15 min, then pass through absolute ethanol, 95,90,80,70,50 and 30% ethanol, keeping in the first 3 for 10 min each, and 5 min in each of the rest, and then transfer to water.

For *pre-mordanting*, keep the slide in 1% aqueous chromic acid solution overnight. Wash in running water for 3 h. Stain in 0.5% aqueous crystal violet solution for 20 min. Rinse in water. For *mordanting*, keep in 1% iodine and 1% KI mixture in 80% ethanol for 30-45 *s*, then dip in absolute ethanol for 2 *s*. Pass through three successive grades of absolute ethanol dipping in each for 2 *s*. Transfer to clove oil I, then differentiate under the microscope after keeping in clove oil II for 2 min. Transfer to xylol grade I and keep for 30 min; pass through pure xylol II and III, keeping 1 h in the former and 30 min in the latter. From xylol III, mount in Canada balsam under a coverglass. Allow the slide to dry overnight on the hot plate.

n-Butyl alcohol can be used instead of chloroform in the clearing process. For materials which stain easily, pre- mordanting in 1% chromic acid solution can be omitted; for materials difficult to stain, Navashin's fluid A can be used in pre-mordanting.

From Pre-Treated Squash Preparations

Stained in Feulgen

Treat root tips of *Hemerocallis fulva* in 0.02 oxyquinoline solution at 10-12°C for 3 h. Transfer to acetic ethanol (1:1) mixture and keep for 1 h. Rinse in distilled water. Hydrolyse the root tips in N HCl at 60°C for 12 min. Rinse in water. Transfer root tips to leuco-basic fuchsin solution and keep in it for 30 min to 1 h till the tips are magenta coloured. Transfer each tip to a drop of 45% acetic acid on a slide, cut out the tip region and discard the other tissue. Place a coverslip over the tip and squash it, applying uniform pressure. The preparation can be ringed with paraffin wax and observed. Invert the slide in a closed tray containing glacial acetic acid-ethanol (3:1) mixture. After the cover falls off, pass both slide with material and cover through acetic-ethanol (1:1, 1:3) mixtures, pure ethanol, ethanol -xylol (1:1) mixture and xylol I and II, keeping 5 min in each. Mount in Canada Balsam.

Modifications

Pre-treatment chemicals include *p*-dichlorobenzene, acenaphthene, coumarin, aesculine, isopsoralene, for different materials. Root tips of sugar cane are difficult materials for squash schedules. Pre-treat in sat α-bromonaphthalene soln. in 0.05% saponin for 3h in cold. Fix in acetic-ethanol (1:3) for 48-72h, hydrolyse in N HCl at 60°C for 7-8 min, treat with 3% pectinase soln. in pH 3.6 acetate buffer for 60-90 min and stain in Feulgen. For cereals, soak for 1-2h at 18-20°C in same chemical, soln in 0.05% saponin for 3 h in cold. Fix in acetic-ethanol (1:3) for 48-

72 h, hydrolyse in N.HCl at 60°C for 7-8 min, treat with 3% pectinase sonl. in pH 3.6 acetate buffer for 60-90 min and stain in Feulgen. For cereals, soak for 1-2 h at 18-20°C in same chemical, 1- 2 h in water, and 0.5-1 h at 10-14°C in a mixture of 1% chromic acid 5 ml, 2% osmic acid 1 ml, and aqueous 0.002 M OQ 1 ml. Treat successively in water for 1-2 min, 1% sulphuric acid solution for 10-15 min, water for 1-2 min, 1% chromic acid solution for $\frac{1}{2}$– 1 h, and squash in acetic-carmine.

Treatment in aqueous α-bromonaphthalene followed by successive washing in water and 22% acetic acid, fixation in N HCl: 22% acetic acid mixture (1:12), prior to usual hydrolysis and Feulgen staining schedule can be effectively used for members of Triticinae.

For *Briza*, pre-treat in sat. *p*-dichlorobenzene solution for 18 to 20 h at 4°C, fix in acetic-ethanol (1:3), treat with aqueous pectinase for 2 h at room temperature; hydrolyse at 60°C in N HCl for 8 min and stain in Feulgen for 2 h.

Stained in acetic-orcein solution

Treat fresh root tips of *Aloe vera* in saturated aqueous solution of *p*-dichlorobenzene (PDB) for 2.5-3 h at 12-14°C. Transfer to glacial acetic acid-ethanol mixture (1:2) and keep for 30 min to 2 h, followed by treatment in 45% acetic acid for 15 min. Transfer the root tips to 2% acetic- orcein solution and N HCl mixture in the proportion of 9:1, and heat gently over a flame for 5-10*s*, taking care that the liquid does not boil. Lift a root tip from the mixture and put it in a drop of 1% acetic-orcein solution on a slide and cut off and remove the older part of the tip. Place a coverslip on the tip and squash by applying uniform pressure on the coverslip with the thumb through a piece of blotting paper. Ring with paraffin and observe under the microscope. Invert the slide after squashing in a covered tray containing tertiary butyl alcohol and keep till coverslip is detached. Mount slide and coverslip separately in euparal.

2% propionic-orcein, 1% propionic acid and propionic-ethanol can be used instead of 2% acetic-orcein, 1% acetic-orcein and acetic- ethanol solutions in the respective steps. Treatment in 45% acetic acid is optional.

To intensify stain a drop of aqueous ferric chloride solution can be added to the acetic-orcein-N HCl mixture or the tissue can be kept in the staining mixture for a period extending up to 12 h and then mounted in 45% acetic acid.

Stained in acetic-lacmoid solution

Pre-treatment and fixation are similar to the acetic- orcein schedule. Stain by transferring the root tips to 10 ml standard acetic-lacmoid solution and 1 ml N HCl, and heat for 5- 10 *s* over a flame, taking care not to boil the fluid, then leave for 10 min.

Transfer a root tip to a drop of standard acetic-lacmoid solution on a slide and squash as usual as in previous schedules. Other steps are similar to the acetic-orcein schedule.

In Leaf Tips

Usually squashes made after pre-treatment yield the best results (Sharma and Mookerjea, 1955).

Dissect out very young leaf tips of *Cestrum nocturnum*, wash in water and immerse in saturated aqueous solution of aesculine and keep at 12-14°C for 15 min to 24 h. Fix the materials in acetic-ethanol (1:1) mixture for at least 3 h, the period being extended, if necessary, up to 24 h. Transfer the tips to a mixture of 2% acetic-orcein solution and N HCl (9:1), heat over a flame for 3-4 *s*, then leave the tips in the mixture at 30°C for 30 min.

Squash the tips on a dry slide in a drop of 1% acetic-orcein solution with a coverslip, applying uniform pressure with the help of blotting paper. The next steps are similar to root-tip schedule.

The period of fixation in acetic-ethanol should be increased, if necessary,to remove the chlorophyll completely. Acetic-ethanol mixture(1:2) or (1:3), chilled 80% ethanol or acidulated ethanol (conc. HCl-95% ethanol 1:3) can be used instead of acetic-ethanol (1:1) mixture.

In pollen grains

Dissect out an anther from a flower bud of suitable size, put in a drop of 1% acetic-carmine solution and smear the anther with a clean scalpel, cover with coverslip and observe. In flower buds of a suitable size, mitotic division is observed in the pollen grains. Dissect out the remaining anthers from the flower bud in which pollen grain division was observed. Place the anthers on a clean dry slide, cut off the edges of each anther with a clean scalpel, squeeze out the inner fluid. Reject the empty anther lobes. Then smear the fluid with a clean scalpel. Immediately invert the slides in a covered tray containing Navashin's A and B fixatives, mixed in the proportion 1:1 and keep overnight. Wash the slides in running water for 3 h, then stain in 0.5% aqueous crystal violet for 30 min and rinse in water. The subsequent steps are similar to the technique followed in staining root tip sections cut from paraffin blocks.

Modifications

For wheat pollen treat anthers at 18-20°C in 0.5% aqueous colchicine and 0.002 M aq. oxyquinoline solutions at 10-14°C for 1 h, fix in Carnoy's fluid for 6 h, wash, hydrolyse in HCl, stain in leucobasic fuchsin and smear in 1% acetic-carmine. For *Saccharum* and related genera, pre-treat in 0.5% aqueous colchicine for 1 h, wash, treat in 0.002 M aqueous OQ for 1 h, wash, fix in a mixture of methanol 60 ml; chloroform 30 ml ; water 20 ml; picric acid 1 g and mercuric chloride 1 g, for 24 h. The remaining schedule is similar to that adopted for wheat. For studying chromosomes from herbarium sheets of *Impatiens,* soak anthers overnight in a saturated solution of iron acetate in 45% acetic acid, rinse and smear in dilute acetic-carmine. Heat several times to boiling and seal.

In endosperm

Feulgen squash method

Dissect out very young developing seeds and fix in acetic-ethanol mixture (1:2) for 1-2 h and keep overnight in 95% ethanol. Run the seeds through 70, 50 and 30% ethanol, keeping them for 10 min in each, then wash in running water in porcelain thimble for 10 min. Hydrolyse in N HCl at 60°C for 8-12 min. Rinse in water and stain in leucobasic fuchsin solution for 2 h, then wash for 10 min in two changes of tap water. Dissect out the endosperm on clean dry slide in a drop of 45% acetic acid solution, under a dissecting microscope using tungsten needles pointed in molten $NaNO_2$. Using Mayer's adhesive, film a coverslip and dry it by passing over a flame, then squash the dissected endosperm under the coverslip, exerting strong but uniform pressure. For intensifying the stain, 1% acetic-orcein solution can be used instead of 45% acetic acid as the mounting medium.

Acetic-orcein squash method

Dissect out the very young developing seeds under a dissecting microscope and place immediately in a saturated solution of aesculine, keep at 10-12°C for 3 h. Fix in acetic-ethanol mixture (1:1) at room temperature for 2 h. Heat in a mixture of 2% acetic-orcein solution and N HCl (9:1) over a flame for 9 or 10 *s*, removing the tube at intervals so that the fluid does not boil, keep for 30 min. Transfer each seed to a clean slide in a drop of 1% acetic-orcein solution, cut in into two or three pieces with a scalpel. Squash the whole under a long coverslip exerting strong and uniform pressure and ring with paraffin.

Permanent preparation of endosperm cells flattened in the living stage

Select a suitable ovule and take out the contents of the embryo sac. Press the contents on to a coverslip smeared with a thin layer of a mixture of 0.5% agar, 0.5% gelatin and 3.5 glucose and remove any excess of endosperm fluid. Surface tension aids in flattening the cells. Arrange the preparation in a moist chamber so as to avoid drying before fixation. Fix cover slip with the material in chrome-acetic-formalin fixative diluted in equal parts with distilled water for at least 12 h. Rinse in water and treat in a mixture of decinormal potassium cyanide and 2% magnesium sulphate mixed in equal parts for 1 h. Wash off the cyanide by several changes in water. Hydrolyse the cover slip with material for 4 h at room temperature in a mixture of rectified spirit and concentrated hydrochloric acid(3:1). Dip in water and stain in Feulgen solution for 12 h. Wash in sulphur dioxide water, dehydrate as usual, pass through xylol and mount in balsam.

Study of birefringence in endosperm mitosis

With the same material, follow the same steps as above without using gelatin, and use cover-slip originally ringed with vaseline-paraffin. Place the coverslip with the material on another glucose-agar-coated coverslip and seal with ringed vaseline-paraffin. When the desired flattening of the cells is secure, tilt the preparation to allow excess liquid to drain on the lower coverslip. The excess liquid can also be drained off by inserting a filter paper in between the two coverslips. By proper adjustment, flatten the endosperm cells to obtain the cnromosomes in one plane. Break the liquid contact between the top and bottom coverslips and stabilise the preparation. Study birefringence in a special polarizing microscope with non- rectified coated stain-free 25 × 0.65 n.a. Leitz oil immersion objective in conjunction with 10 × 0.25 n.a. American optical coated objective as condenser. Pure green light(546mm) from a high- pressure mercury arc lamp can be isolated with a multilayer high transmission interference filter. Birefringence in kinetochore region of the chromosome fibril can be observed.

In pollen tube

For the study of mitosis from the division of the generative nucleus in germinating pollen grains, the methods involve, germination of the pollen tube, and treatment for observation of the chromosomes.

Hanging drop culture method

Pollen grain culture. Fit a ring on a slide and smear both rims of the ring with vaseline so that it is attached to the slide at one end. Place a drop of 3% sugar solution on a clean coverslip. Then dust pollen grains from an opened flower into the solution. Invert the cover-slip with the drop of sugar solution on it and attach it to the other vaselined rim of the ring, so that the drop with pollen grains hangs in the closed chamber enclosed by the ring. Growth of the pollen tubes can be noted by observing the ringed slide under the microscope. After about 3 h, remove a few pollen tubes for observation at intervals of 1 h till the optimum time is reached.

Treatment for chromosome study Lift out the pollen tubes with a brush, put in a drop of 1% acetic-orcein solution on a clean slide, warm slightly and squash as usual under a coverslip applying uniform pressure. The slides can be made permanent by the alcohol vapour technique. For accumulating metaphase,0.05% colchicine can be added to the sugar solution. For controlling humidity of the hanging drop chamber, add a drop of water on the slide or place a drop of sugar solution beside the hanging drop.

Coated slide technique and modifications

Weigh out 12 g of lactone, 1.5 g of agar and 0.01 g of colchicine. Heat the lactone and agar with 100 ml distilled water in a double boiler until the agar is dissolved, then add colchicine when the medium has cooled to 80-60°C. Keep the medium at about 60-70°C. Coat slides cleaned in ethanol with a thin layer of egg white. Dip in a beaker containing the medium at 60°C until it warms up, then withdraw the slide, drain off the medium, and wipe the back with a piece of clean cloth. When the medium has set, but is not dry, pick up the pollen with a brush, and dust a thin film on the medium. Immediately place the slide sown with pollen in a moist growing box, the box being a horizontal glass staining dish lined with damp blotting paper on the two sides and top, and keep the temperature between 20 and 25°C. Observe at intervals till the optimum period for maximum division is found. Add a drop of 1% acetic-carmine solution to the pollen tube, squash under a cover-slip and observe. The slides can also be stained by the Feulgen schedule after 12 min hydrolysis in N HCl at 60°C. For species with binucleate pollen, germinate in a medium containing H_3BO_3, 0.01 g; $Ca(NO_3).4H_2O$,0.03 g; $MgSO_4.7H_2O$,0.02 g; KNO_3, 0.01 g; sucrose, 10 g; water, 100 ml, on a slide resting on moist filter paper in a closed petri dish for 24 h. Place crystal of acenaphthene on filter paper. After 24 h, squash pollen tubes in propionic-orcein solution. For convenient handling of germinated pollen during Feulgen and autoradiographic procedure, grow pollen on an autoclaved membrane filter (Millipore AA WP 025 00) in contact with a sterilized medium which contains agar 0.5-1% sucrose 0.1- 0.5% and boric acid 0.01% for 2 h to overnight at 2-4°C on a filter paper with a mixture of OsO_4, 1 g; CrO_3, 1.66 g and water, 233 ml. For *Persica* pollen, add 10% acetic acid. Wash in water, bleach in a mixture of 3% hydrogen peroxide and saturated aqueous ammonium oxalate solution on filter paper. Hydrolyse with 5 n HCL for 18 min at room temperature, stain in Feulgen, wash in three changes of 2% $K_2S_2O_5$ at pH 2.3 (KH_2PO_4, 1.4 g; conc. HCL, 0.35 ml; distilled water, 100 ml) by placing membrane on filter paper wet with the respective fluids. Transfer pollen to glacial acetic acid, squash and process. Sow *Tradescantia* pollen on lactose-agar medium at 38-39°C for 16 h. Fix slides in acetic-ethanol (1:3) for 1-3 h, hydrolyse in N HCl at 60°C, treat with water at 65°C. Delaminate upper layer of medium in cold water. Flatten and fix remaining single layer of pollen tubes to the slide by processing under a coverslip by quick freeze technique, stain in Feulgen and mount as usual. For palm chromosomes, sow pollen in a medium containing H_3BO_3 100 ppm; colchicine, 0.04%; lactose, 12%; gelatin, 5% and egg albumen, 1 drop in 10 ml.

Collodion membrane technique

Mix 1 part of collodion (necol collodion solution from BDH, England) with three parts acetone. Prepare 3% aqueous sugar solution in a petri dish and warm and put a drop of collodion- acetone mixture on it. Cover the open dishes lightly by filter paper and leave in a warm dry place for 3 h till acetone evaporates completely. Dust the pollen directly from the anther on to the smooth areas of the floating membrane towards the centre, replace the lid on the petri dish and transfer it to an incubator. After the pollen tubes have geminated, cut out a piece of the membrane, about 1 cm in diameter, with the pollen tubes, while it is still floating on the sugar solution; lift it with a needle on to a clean dry slide, add a drop of 1% acetic-orcein solution and squash under an albuminized coverslip, applying uniform pressure. Ring with paraffin. Invert the slide in a covered tray containing acetic-ethanol mixture (1:3). After the coverslip is detached, run it through two changes of absolute ethanol-xylol (1:1) mixture and pure xylol, keeping for 10 min in each. Mount the coverslip, pollen side down, in a drop of Canada balsam on a clean slide.

Metaphase arrest technique excludes nutrient medium in the germination of the pollen tube. Line both bottom and cover of a pair of petri dishes with well-moistened filter paper and place a clean slide in the petri dish and dust on it pollen grains from a newly opened flower. Spread 50-100 mg of fine acenaphthene crystal on the filter paper close to the slide and cover and keep at 20-22°C for 24 h. Add a drop of 1% acetic-carmine solution to the pollen tubes on the slide and squash.

In animal material

Squash preparation

The usual sources of material are: larval tails, ganglia, and spermatogonial cells.

Material Urodele larva

Cut off the tail of a growing larva of suitable size immediately behind the anus. Fix in acetic-methanol mixture (1:1) for 2-24 h in a glass phial. Remove the tissue from the phial and place it on a clean slide in a drop of 1% acetic-carmine solution to which a trace of iron has been added. The tissue may be teased by means of two needles before squashing. Squash by placing a cover-slip on the preparation and pressing gently. The slide can be made permanent by any one of the techniques outlined previously. Acetic-lacmoid solution can be used instead of acetic-carmine. For Feulgen staining, after fixation, hydrolyse the material in N HCl at 60°C for 12 min. Modify the treatment if necessary, rinse. Stain in leuco-basic fuchsin solution for 30 min and squash in 45% acetic acid. Carnoy's fluid or Newcomer's fluid can be used as a fixative instead of acetic-ethanol mixture. This method is applicable to all the other tissues. Small ciliates, cultured in petri dishes, may be fixed in San Felice's fluid, attached to slides and stained with Feulgen. Ticks are dissected out in Shen's insect saline solution or 1% sodium extract for 10 min and squashed in 2% aceto-orcein. Entire embryos can be squashed when of suitable size, as for example parthenogenetic psocids(Insecta:Psocoptera).

Material Salivary gland chromosomes of Drosophila

Dissect out third instar larvae in physiological solution. Place the salivary glands in 45% acetic acid for 5-7 s. Fix in 1 N HCl for 30 s. Stain in lactic-acetic-orcein for 30-40 min on a slide. Warm (35°C) the slide for 10 s. Transfer the excess stained gland to lactic-acetic acid (lactic acid: 60% acetic acid, 1:1) on a siliconized slide. Wash thrice in lactic acid and mount in the same solution. Squash under a cover-slip exerting pressure.

Paraffin block preparation

Material Urodele larva

Cut off the growing tail of a larva near the anus and fix in San Felice's fluid (1% chromic acid: 16 ml, 40% formaldehyde: 8 ml, and glacial acetic acid: 1 ml) for 12 h. Transfer the tissue to a porcelain thimble and wash in running water for 3 h. The next steps of dehydration, clearing, infiltration, embedding, section- cutting, bringing down to water are similar to that used in plant material described earlier. Mordant the slide in 4% iron alum for 10-20 min, rinse in water, stain in 0.5% haematoxylin (ripened for 1-2 months) for 5-15 min, rinse in water and destain in saturated picric acid solution for 5-20 min, keep in distilled water containing a few drops of ammonia (0.88%) for 1 min. Dehydrate, differentiate, clear and mount as described for plant material.

Modifications

This technique can also be applied to ganglia and other somatic tisue, except bone marrow cells. For difficult materials, the tissue can be pre-fixed in acetic-methanol (1:3) mixture for 1 h before fixing in San Felice's fluid as usual. Some alternative fixatives which can be used are Flemming's strong fluid, Champy's fluid and Minouchi's fluid. For haematoxylin staining, the more prolonged schedule as followed for plants is also applicable in the case of animals. The Newton's crystal violet staining schedule can also be applied to animal materials.

Air dry preparations from Bone Marrow in vivo

This technique was devised for the bone marrow cells of mammals after pretreatment (Ford and Hamerton, 1956). However, it can also be applied to thymus, spleen, cornea and other tissues.

Inject in mouse 0.04% aqueous colchicine solution intraperitoneally at a dose of 1ml/100mg body weight of the animal. Sacrifice the animal after 1.30 h to 2 h by cervical dislocation. Cut open the abdomen and expose the femur from two sides. Flush out the bone marrow with the help of a clean glass syringe in 8 ml of 1% sodium citrate solution prewarmed at 37°C. Flush the bone marrow thoroughly in the hypotonic solution with a pasteur pipette and keep for 20 min at 37°C. Centrifuge the suspension at 1000rpm for 5 min. Discard supernatant. To the pellet of cells at the bottom add freshly prepared cold fixative (acetic acid-methanol 1:3) and keep for 30 min. Centrifuge and wash in fixative twice. After the third centrifugation, discard supernatant and resuspend cells in a small quantity of fresh fixative.

Draw out the cell suspension by means of a pasteur pipette with a small rubber bulb. Tilt a clean grease-free previously chilled slide at an angle. Drop 4-5 drops of the suspension at the upper end of the slide. Air dry by blowing or flame dry.

For Giemsa staining

Two reliable stains are Gurr's R66 (Hopkin and William) dissolved in methanol. Eastman Kodak Giemsa C8685 (Eastman Organic Chemicals, Rochester NY) - (0.5g powder mixed with 33ml glycerol at 60°C for 2 h, and add 33 ml methanol). Dilute the stock Giemsa to 2% either by adding commercially available buffer tablets (Gurr pH 6.8, one tablet in one 1 water) or with phosphate buffer, pH 6.8. 0.05M phosphate buffer is made by mixing 31.3ml 0.5M KH_2PO_4 with 22.8ml 0.5M Na_2HPO_4 and diluting to 500ml).

Immerse slide with chromosome spreads in the stain for 10 min. Rinse gently in distilled water to remove excess stain. Rinse again in distilled water several times, keeping 1 or 2 min each time. Airdry. Observe slides directly using immersion oil for detailed study. Remove oil by dipping the slide in xylene for 2- 5 min and airdry for future use. The preparation can be mounted after immersion in xylene for a few minutes, using a drop of permount or any other xylene-soluble mounting medium and cover with a coverslip. Keep on hot plate at 40-45°C for 24 h to dry.

For destaining, remove mountant by soaking in xylene and destain by keeping the slide in 95% ethanol for 5 min, rinsing in 70% ethanol and airdrying.

This method has been applied successfully to almost all vertebrates. In certain amphibians, the air-drying or blaze- drying technique has been adapted before staining. In fishes, direct preparations have been made from epithelium from fins, scales, gills and internal organs like kidney, liver and spleen. 3h before sacrifice inject fish with 65μg colchicine/g of fish (stock 0.3% colchicine in 0.9% NaCl).

Prepare chromosomes from kidney tissue immediately after sacrifice. Remove kidney, mince and homogenize with a pasteur pipette in a reaction tube filled with cool (4°C) hypotonic solution (0.4% KCl). Keep in hypotonic solution at room temperature for 50 min. Carefully add 1 ml of chilled fixative (acetic acid-methanol 1:3) carefully onto the hypotonic solution. Centrifuge at 1000rpm for 10 min. Discard supernatant and resuspend twice more. Store in fixative overnight. Prepare airdried spreads.

Epithelial cells can be isolated from rat intestine by placing it *in toto* in a hypotonic solution and subsequently centrifuging and fixing them in chilled acetic-methanol. Other organs have been similarly fixed after pre-treatment in hypotonic solution from *Myxine glutinosa*, sacrificed 3 h after injecting 0.8 ml 0.5% colchicine/100 g body weight. Regenerating tissues of freshwater planarians can be similarly squashed after colchicine treatment and orcein staining as also blastulae of fishes.

In all animal materials, acetic acid-methanol (1:3) is a more effective fixative than acetic acid-ethanol mixture.

Permanent smear preparation

For insects, modified in mammalian chromosomes (also applied to other materials):

Take any dividing tissue, e.g. cornea, bone marrow, liver, testes, etc. If the cells are in suspension as in bone marrow, follow the procedure given before. If the tissue is solid, macerate it by means of a homogenizer in sodium citrate solution (for swelling), incubate the suspension in 37°C and leave for 15 min. Centrifuge gently at about 1000 r/min for 5 min. Decant off the supernatant fluid, add acetic-methanol mixture (1:3) to the sediment and flush it with a pipette, then fix for 15 min or more. If the fixed cells settle down at the bottom, decant; if they remain in the supernatant, gentle centrifuging may also be applied. Add a drop of 45% acetic acid solution and make it a milk emulsion. Place a very small drop on a slide and squash by a coverslip with gentle pressure by means of a thumb (uniform layer). Dry the squashed slide over a small flame (preferably spirit lamp) for about 45 min till the edge of the coverslip appears to be dry. Place the slide in 50% ethanol and allow the coverslip to detach.

Schedules for the Study of Meiotic Chromosomes

In Plant Materials

Study from pollen mother cells

Temporary squash technique

Take flower buds serially from an inflorescence, starting from the smallest and working up to the largest, until the correct bud having divisional stages is found. Dissect out a single anther from a bud with a needle. Place it on a clean slide. With a clean scalpel, smear the entire anther on the slide and add a drop of 1% iron-acetic-carmine solution. Remove the debris. Heat slightly over a flame. Cover with a coverslip and ring with paraffin.

Anthers fixed in acetic-ethanol (1:1) mixture or in Carnoy's fluid, and later stored in 70% ethanol can also be observed following this method. If stored in 70% ethanol, keeping 1 h in each of acetic-ethanol and 45% acetic acid solutions is necessary before smearing in 1% acetic-carmine solution. For materials taking bright stain, treatment in acetic-ethanol can be omitted. Instead of iron-acetic-carmine solution, 1% acetic-carmine solution can be used and a trace of iron added by rubbing a rusty needle in the drop of stain on the slide. 1% acetic-carmine solution can be replaced by acetic-orcein, acetic-lacmoid, nigrosine, etc., solutions. The slides can be kept as such in a refrigerator for a few weeks and then made permanent.

Very small buds in tight inflorescences can be fixed in acetic- ethanol mixture or in Carnoy's fluid for 1 h, hydrolyzed at 60°C for 5-10 min in N HCl, rinsed in water and stained for 1-3 h in leuco-basic fuchsin solution. They should be rinsed in two or three changes of 45% acetic acid solution. Single anthers are to be dissected out, squashed in 45% acetic acid and made permanent as usual.

Maceration in a mixture of 15% chromic acid, 10% nitric acid, 5% HCl (2:1:1) for 5-7 min and hardening in ethanol-propionic acid (1:1), between fixation and staining has been used in *Gossypium* microspores. Restoration of deteriorated temporary acetic carmine preparations involves replacing the acetic-carmine under the coverslip first with 2 N HCl and then with 1% acetic-carmine.

Permanent smear technique

Take flower buds of different sizes. Dissect out a single anther from each bud and observe by squashing in 1% iron-acetic-carmine solution until, in the bud of a particular size, meiotic divisional figures are observed. Dissect out the remaining anthers showing division and place each on a clean slide, cut off one end with a clean scalpel and squeeze out the contents. Discard the empty anther lobe. Quickly draw the fluid into a thin smear on the slide with a clean scalpel and immerse immediately in a tray containing Navashin's fluids A and B, freshly mixed in equal proportion. Keep in the fixative for 3-12 h.

Wash the slide in running water for 1 h. Stain in 0.5% aqueous crystal violet solution for 20 min or more and rinse in water. Mordant in 1% solution of I_2 and KI in 80% ethyl alcohol for 45 *s*. Dehydrate by passing the slide through absolute ethanol grades I, II and III, keeping about 2-3 *s* in each. Transfer the slide to clove oil I, keep for 2-3 min, take out the slide and observe the staining under a microscope. If found satisfactory, transfer the slide to clove oil II and keep for 2-3 min. Pass the slide through xylol grades I, II and III, for 30 min, 1 h and 30 min respectively. Mount in Canada balsam or clarite X.

In animal material

Studies of meiotic stages from testes

In grasshopper testes

Dissect out the testes of a male grasshopper in 0.75% normal saline solution by pulling with two fine forceps at the two ends, separate the head region from the rest of the body, trailing out the intact salivary glands and the anterior portion of the alimentary canal. Remove the glands by breaking off the duct attaching them with a pair of forceps. Take a few lobules on a clean coverslip and remove the excess saline by touching the edge of a filter paper. Cut the tip of the lobules by means of a cataract knife, and the fluid should be spread quickly in uniform layers over the slide. Invert the slide immediately on a tray containing acetic-ethanol solution or any other fixative.

For *acetic-carmine squash*, put testes lobules in a drop of acetic-carmine solution for 5 min and squash. The bulk testes should be fixed in acetic-methanol mixture (1:3) and then put in acetic-carmine solution for 5-15 min.

For ***squash after staining: Feulgen staining***
Bring bulk tissue fixed in acetic-ethanol mixture down through ethanol grades to water. Wash in water and put in cold HCl (N). Hydrolyse in N HCl for 12 min (or adjust accordingly) at 60°C. Rinse in cold HCl and put in Feulgen solution (15 min-1 h). Take a few lobules of the stained tissue in a small drop of 45% acetic acid on a slide and squash gently. Seal and observe. For temporary observation, materials can be sectioned 25 μm thick.

In amphibian testes

Dissect out the testes from the newt, *Triturus* sp. Cut into very small pieces. Fix in acetic-ethanol mixture (1:3) or, in Carnoy's fluid in a glass phial for 2-3 days. Lift a piece of the tissue on a drop of 1% iron-acetic-carmine solution on a clean slide. Squash under a coverslip, warm and seal with paraffin wax.

In Culex testes

Place the pupa on a clean slide in a drop of Ringer solution A (0.65 g of NaCl, 0.025 g of KCl, 0.03 g of $CaCl_2$, 0.02 g of Na_2CO_3 in 100 ml distilled water) and observe under a dissecting microscope. Pull the head and tail of the larva with two needles, breaking it in two. Lift out small translucent testes with a needle. Place the testes in a phial containing acetic-ethanol (1:3) mixture or Carnoy's fluid and fix for 2 min. Transfer to a drop of 1% iron-acetic-carmine solution on a clean slide, warm slightly, cover with a coverslip and squash through blotting paper.

Acetic-carmine can be replaced by 1% acetic-orcein solution. For Feulgen staining, transfer the testes to N HCl after fixation and hydrolyse at 60°C for 4 min. Rinse in water and tranfer to leuco-basic fuchsin solution. After 20 min, squash under a coverslip on a clean slide in 45% acetic acid. The entire operation can be carried out on a slide.

In mammalian testes

Remove the testes entire. Cut into very thin sections. Fix directly in acetic-ethanol mixture (1:3) or 80% chilled ethanol at 10-12°C for 1 h. Lift up a section, put it in 1% iron- acetic-carmine solution, tease out the tubules; warm slightly, and squash under a coverslip and ring with paraffin wax. This method is not very satisfactory. More effective methods are given in the chapter on mammalian chromosomes.
In avian and reptilian testes the techniques are similar to those for mammals.

Testes of trematodes are squashed after collection 37 days after inoculation into eyes of chick.

In silkworm testes:

Dissect out larval sex organs in 0.01% colchicine in 0.45% sodium nitrate solution. Change the solution with fresh hypotonic and keep at room temperature for 30-45 min. Transfer tissue to dry slide, add a drop of 60% acetic-ethanol (1:3), dissect, and squash for 1 min. Freeze on dry ice 2-3 min, melt, keep in glacial acetic acid for 30 *s* and air-dry. Stain in Giemsa diluted 30 times in Sörensen's phosphate buffer (pH 6.8) for 7-10 min at room temperature.

A similar technique has been employed for ants and a slightly modified one for gonads and other parts of Diptera.

In pulmonate land snails

Inject 0.1 ml 10^{-3} M colchicine/g body weight. Remove ovotestis and cut it up in hypotonic solution. Fix and stain in lacto- aceto-orcein, lacto-propiono-orcein or Sudan black B. Mount and squash.

Studies of meiotic stages from eggs

For early oogonial divisions, when the ovary is relatively immature, the techniques used for study are similar to those for testes. For studying older ovaries, both squash and paraffin section techniques are followed.

For oocyte division puncture the shell of the egg. Smear and fix the contents in acetic ethanol mixture (1:1) for a few h, hydrolyse in N HCl and stain following the Feulgen schedule. Puncture the wall to bring out the contents and allow them to dry on a slide. Stain the dried eggs with aqueous Bismarck brown. Certain insects can be fed with a mixture of honey and 1% colchicine solution for 24 h. Dissect out the ovary, keep them in 0.5% colchicine solution for 5-10 min and squash in 1% acetic- carmine solution.

Fluorescence studies from oocyte DNA

Isolate oocytes by dissection or aspiration into PBS from follicles. Wash in PBS thrice. Dissolve zona pellucida by briefly incubating in pronase (Boehringer - Mannheim, 0.5% v/v in PBS, room temperature) for 3 to 12 min. Place oocytes in clean glass slide in a 5μl drop of PBS. Air dry. Fix in acetic acid:methanol (1:3) for 24 h. Prepare stains in the concentrations required in McIlvaine's buffer, pH 7.0. After fixation, air - dry the slides and stain with Chromomycin A_3 (CMA_3, 0.1 mg/ml, Sigma, St. Louis, MD) for 40 min, rinse with distilled water and stain with distamycin (Sigma, 0.1 mg/ml) for one min. Rinse with distilled water. Stain slide again with DAPI (Sigma, 2 μg/ml) for 15 min. Rinse the slide, air dry and cover with a solution of glycerine and McIlvaine's buffer (1:1, pH 7.0). Mount with a coverslip using rubber cement (Fixogum Marabu). Incubate the slides for 72 h at 37°C and then again for 72 h at room temperature. Observe with a Reichert - Polyvar microscope at wavelengths of 450-495 nm (CA_3) and 330-380 nm (Distamycin A - DAPI). Stain in 2% Giemsa following the usual procedure for light microscopy.

For marine invertebrate eggs

Place the eggs in a drop of water on an albuminized slide. Add a few drops of fixative (absolute ethanol 1.5, tertiary butyl alcohol 1, acetic acid 1) to the eggs, allow it to flow over the eggs and drain off. Similarly add 70 and 50% ethanol and distilled water, treating in each for 10 min and then drain off. Hydrolyse at room temperature with three changes of HCl—1 N HCl for 5 min, 5 N HCl for 15 min and again 1 N HCl for 3-5 min. Stain in a jar containing Gomori's chromalum haematoxylin solution at 60°C for 30 min. Mordant in 1 N HCl for 3-5 min. Keep in water for 10 min. Rinse in SO_2 water for 5 min. Treat with chilled 45% acetic acid for 5 min. Wash in water for 5 min. Treat in 1% aqueous papain solution for 10 min. This causes the tubules to shrink. Wash in water and transfer to 60% acetic acid. The cells swell to greater than the original size. Place the coverslip on the material and squash under blotting paper, applying uniform pressure. The slides can be made permanent. Change through 30,50,70,95% and absolute ethanol, keeping 10 min in each. Treat in pure xylol for 30 min and mount in Canada balsam.

Air-Drying schedule for pachytene chromosome study in insects

Dissect out ovaries of fourth and sixth instar larvae, fix for 30 min in Carnoy's fluid and transfer to dry slide. Add 60% acetic acid, tear ovary into small bits with tungsten needles. Keep for 30 *s* at 45°C, move drop a few mm along slide; repeat three to five times. Draw up extra liquid with pipette and reject. Air-dry, stain and mount in lacto-aceto-orcein. Seal.

Detailed descriptions of methods for study of vertebrate chromosomes have been given under the relevant chapters.

B. STUDY FROM SPECIAL MATERIAL

1. Chromosome Segments

Centromere

Under the light microscope, the centromeric apparatus is now seen to consist of several chromomeres attached to each other and to the chromatid arms by interchromomeric threads. These centromeric bodies show the same divisional cycle as the chromatids, and in order to study the structure of the centromere, the gap should be increased. For this purpose, the material is first treated at cold temperature in a pre-treatment chemical, fixed and stained. The most satisfactory pretreatment chemical has been found to be 8-oxyquinoline (OQ) solution.

Cut fresh root tips and treat in aqueous solution of 8- oxyquinoline (0.002 M) at 12-18°C for 3 h. Heat the root tips in a mixture of 2% acetic-orcein solution and N HCl (9:1) for 3-4 *s*. Squash the tips in 1% acetic-orcein solution, seal and observe.

For the study of *centromere in flower buds*, fix the buds in acetic-ethanol mixture (1:4) for 7-9 h. Transfer to 95% ethanol (overnight) and store in 70% ethanol for 2-3 days. Cut the anthers and squeeze out the pollen mother cells into a drop of 45% acetic acid on a slide. Squash by placing a coverslip over the pollen mother cells. Gently heat the slide once over a flame. Invert the slide in 1% aqueous acetic acid. After the coverslip falls off, keep both slide and coverslip in Belling's modification of Navashin's fixative for 3 days. Wash in distilled water for 1 h. Immerse in N HCl at 60°C for 6 min. Wash in distilled water (1-2 min) and immerse in fuchsin sulphurous acid for $3\frac{1}{2}$ h. Pass through fresh SO_2 water, three changes of 10 min each. Rinse in distilled water, pass through 30, 50, 70, 95% and absolute ethanol, keeping in each for 3 min. Mount separately in euparal and seal.

The chromosomes take up the characteristic violet red colour. The centromeric chromomeres, being intercalated among longer fibrillae than the chromomeres of the arms, can be clearly distinguished.

Secondary Constriction

Secondary constrictions lack a centromere, are heterochromatic in nature and do not generally exhibit allocycly. They are the loci where nucleoli are organised, special staining schedules for this region are given under N-banding in the chapter 4 on banding patterns.

Heterochromatin

Heterochromatic segments may be defined as those segments which differ in any respect from euchromatic ones. They may show allocycly (e.g., prochromosomes), may be positively heteropyknotic throughout the division cycle (sex chromosomes) or may be negatively heteropyknotic in all the stages. They are of a heterogeneous nature and include the centromeric heterochromatin, intercalary heterochromatin, telomeric heterochromatin and also the heterochromatin in the secondary constriction regions. Heterochromatic regions exhibit, in general, extreme susceptibility to external conditions, like cold; heteropyknosity and allocycly, in most cases; property of heterochromatizing adjacent euchromatic segments; temporary genetic inactivity; absence of template activity *in vitro*; late or early DNA replication in S phase; and a high content of repetitive DNA.

The techniques for *Q, G, R, O* and *C* banding, discussed in the Chapter 4 on banding patterns are based on the presence of heterochromatic segments. They appear as positively staining bands in metaphase with *Q, G* and *O* techniques and as negatively staining bands with *R*- banding. The *C* band regions correspond with the centromeric type of heterochromatin, which may be located, as in certain plants, in regions other than centromere as well. *G, O* and *Q* bands are correlated mostly with the intercalary heterochromatin.

In this chapter, only the methods for demonstrating the heterochromatic segments of a chromosome, other than the chromocenters and secondary constriction regions, are described here.

Treatment with trichloracetic acid is based on the initial digestion of the tissue in a particular concentration of TCA for a definite period, resulting in differential removal of DNA from the chromosomes, followed by staining (Sharma, 1951).

Fix the cut tissue in acetic-ethanol mixture (1:1) or 80% ethanol in cold (10°C) for 2 h to overnight. Treat in 0.25 M TCA at 60 °C for 40 min. Wash in distilled water. Keep in leucofuchsin solution for 30 min, and squash in 45% acetic acid. The heterochromatin, particularly that of the centromeric and telomeric regions, stains sharply in leucofuchsin, the rest of the chromosome remaining unstained.

Treatment with mercuric nitrate: The tissue is initially treated in an inorganic salt for a period in cold temperature followed by fixation and staining, when the chromosome arms show differential staining in the heterochromatic regions. Wash the tissue and pre-fix in 0.005 mol aqueous mercuric nitrate solution for 4 h at 10-12°C. Transfer to a mixture of equal parts of Navashin A and B solutions, mixed just before use, and keep overnight. Wash, embed in paraffin, cut longitudinal sections and stain and mount following the usual crystal violet staining schedule.

2. Special Chromosomes

Salivary gland chromosomes

Balbiani in 1881 first observed giant chromosomes in the salivary gland cells of some Dipteran species. These chromosomes, also found in some other gland tissue, remain in a state of permanent prophase. Structurally each giant chromosome shows a 'polytene'or 'multiple'nature. The chromosome remains in a permanent prophase condition, in which the chromomeres and chromonemata continue to duplicate without separation, the result being a multi-stranded structure, visible under a hand lens when stained. The aggregation of the homologous chromosomes produces the transverse chromatic bands, and the numerous chromonemata. The salivary gland nuclei are held to be polytenic due to the giant size of the chromosomes, their multiple nature and their total DNA content, which is many times higher than that of ordinary nuclei.For the study of these chromosomes, the salivary glands of different Dipteran larvae are the most suitable material, although they also occur in the cells of malphigian tubules, fat bodies, ovarian nurse cells and gut epithella. Old fat larvae, bred on rich yeast food at 16-20°C, show the largest chromosomes. The diet varies with the type of larva used. The number of the cells in a salivary gland ranges from 28-32 in ***Sciara***, 28-44 in ***Chironomus*** to 100-120 in ***Drosophila***, all in different stages of development.

Place the full-grown larva on a slide in a drop of Ringer's solution or 0.73% isotonic salt solution. Cut off the head with a scalpel and press the body with a needle at the same time. Remove the pressure. The salivary glands float out. Transfer them to a clean slide with the dissecting fluid using a pipette. The material is now ready for staining. For storage, add a few drops of paraffin oil to the material and place a coverslip over the whole. It will remain fresh for 24 h. Staining may be done in either acetic-carmine or acetic-orcein solution. For *Drosophila*, the most effective stain is 2% acetic-orcein in 70% acetic acid; while 1% acetic- orcein in 45% acetic acid, plus 1 ml chloroform may be used for *Sciara*; and 2% orcein in 50% acetic acid for *Chironomus*. Transfer the tissue to a drop of the stain on a clean slide with a pipette. Leave for 5-10 min, depending on the material. Prepare the coverslip by smearing thinly with Mayer's albumin and dry over a flame for 1-3 s. Add the coverslip on the tissue, drain off the excess stain with filter paper, blot the preparation with filter paper, applying uniform pressure to flatten out the chromosomes and remove more stain. If the glands do not rupture, press with a blunt needle over the coverslip in a closed trough, lined with filter paper soaked in 95% ethanol. After 24 h remove the cover-slip and mount in euparal. Alternatively, the slides may be inverted in n-butanol until the cover-slip is detached, and mounted separately. Modification of the schedule given above were suggested by many workers.

Lamp brush chromosomes

In a large number of vertebrates, within the developing oocytes, during diplotene, the chromosomes undergo a phase called 'lamp brush', characterized by a great increase in length and the formation of numerous side loops radiating from the chromosomes. These loops grow in number and size up to diplotene but increase and disappear before metaphase. Lamp brush chromosomes are very elastic and may be pulled to many times their length without injury. However the lateral coils, though elastic, are more fragile. The main axes and loops are DNA-positive and resistant to several chemical treatments. Obviously, the long lateral loops are formed at definite loci due metabolic activity of genes at these regions. Dissect out the ovary or several oocytes, making a small incision at the ventral side of the animal anaesthetized with light ether or 0.1% MS 222 solution for 15 min, and place it in a dry watch-glass containing the coelome with adequate moisture, seal and keep at 4°C for up to two days. Transfer a piece of oocyte or ovary to a mixture of 5 parts 0.1 M KCl and 1 part of 0.1 M NaCl buffered to pH 0.8-7.2 with 0.01 M phosphate. This ratio may vary in different species.

With a pair of fine micro-forceps, break the oocyte and take out the translucent nucleus. Puncturing the cell, followed by squeezing with the needle also results in ejection of the nucleus. In both cases, the nucleus starts swelling on isolation. Transfer the nucleus immediately to the observation chamber filled with the same 5:1 solution. For preparation of the observation chamber, bore a 6.4 mm hole through the centre of a 76.2 × 25.4 mm slide and place a coverslip across the hole and seal with paraffin wax. Fill it with S11 solution before transferring the nucleus. In the observation chamber, remove the nuclear membrane carefully, under a dissecting microscope with the help of a pair of very fine forceps and a tungsten needle pointed by dipping in molten sodium nitrate. The chromosomes are then liberated and confined in the hole in the slide on the coverslip forming its bottom.

It can be kept in the cold for observation for a day in the unfixed condition. For fixation, it is preferable to keep the slide 3-5 min in formaldehyde chamber or to add a drop of concentrated formalin. After fixation, the preparations can be kept for several weeks.

Cover the slide on the other slide with another cover-slip and seal. Invert the optical train of the phase contrast microscope and observe the preparation. The chromosomes will be observed as coiled loops in a state of continuous movement. Care must be taken to handle the preparation very gently. Micro photographs are taken in flashlight.

For making permanent preparations

Use a slide instead of a coverslip at the bottom of the observation chamber so that the chromosome can be isolated on the slide or alternatively, the cover slip may be used. After isolation and clearing of the nucleus in 5:1 medium (calcium free), pass the nucleus quickly through a watch glass containing 0.1 M 511 and 0.5×10^{-4} M $CaCl_2$ and transfer to the observation chamber again. Remove the nuclear membrane, isolate the chromosome as above and transfer to a moist chamber for an hour, which latter on becomes transformed to a formaldehyde chamber till the chromosome gets attached to the slide (in about 15 min). The pH may be lowered to facilitate attachment. Transfer the set-up to a jar containing 2-4% of 5-10% formalin, buffered to pH 4-5, and keep for 1 h. With a scalpel, separate the slide from the observation chamber. Dehydrate in ethanol, removing paraffin or vaseline with which the slide was attached and follow the usual procedure for staining in Feulgen solution for DNA and Azure B at pH 4 for RNA.

In securing autoradiographs, lamp brush chromosomes are convenient materials, since the incorporation of precursors is very rapid. Specific ratioactive precursors, such as, tritium-uridine (100-300 mCi specific activity 0.5-5.0 Ci/mmol) may be injected into the body cavity. If the explanting is carried out at 48-72 h, profuse labelling indicating synthesis of RNA or DNA loops can be obtained. But for short term and rapid incorporation, explanted ovaries can be treated in concentrated radioactive precursors taken in a watch glass. Both stripping film and emulsion coating can be performed as outlined in the chapter 5 on autoradiography.

For electron microscope studies, lamp brush chromosomes are not convenient materials due to the thickness of the chromosomes and difficulty in interpreting the isolated material. However, chromosomes attached to coverslips can be dehydrated and then polymerized, after inverting over a gelatin capsule filled with plastic monomer and the cover slip separated out, using the freezing technique. In whole mounts, mounting directly on steel or platinum grids coated with carbon can be performed. (Please see chapter 6 on electron microscopy).

Pachytene chromosome

Due to their configuration during the pachytene stage in pollen mother cells, the chromosomes of a large number of plants present very good material for the study of their individual structure and the nature of pairing.

For B chromosome study in rye fix a complete spike of rye with awns cut close in acetic-ethanol mixture (1:4) for 3-4 h. Keep overnight in 95% ethanol and pass on to 70% ethanol. Dissect out the three anthers of each flower. Transfer one anther to a drop of iron-acetic-carmine solution (two drops of iron acetate in 10 ml acetic-carmine solution). Under a widefield binocular microscope, cut the anther into two pieces, press each half gently to squeeze out the pollen mother cells, and remove all pieces of anther wall and tapetum with a needle, leaving only the pollen mother cells. Observe under the microscope if the nuclei are in the pachytene stage. If so, add another drop of stain, place a coverslip on the material and heat gently over a flame three or four times. Apply uniform pressure vertically with a U- shaped needle-point, blotting off excess stain from the slide, check under a microscope. If stronger pressure is wanted, add more stain, heat again and press. Heat again and invert the slide whilst still warm in 10% aqueous acetic acid and after the coverslip falls off, pass both slide and coverslip through acetic-ethanol and ethanol-xylol grades and mount in Canada balsam.

For mammalian materials, the squash method involves the following steps: Place a piece of testicular tissue in 0.3% aqueous sodium nitrate solution for 1 to 6 h. depending on the size. Dissect out and observe a few tubules in stain to ascertain the presence of the required stage. Keep tissue in 3 M gluconate- deltalactone solution for 2 h for softening; stain in 1% acetic or propionic-carmine for 10-12 h; wash in four changes of 70% ethanol; and transfer to acetic-ethanol (1:1) mixture. Mince into a thick suspension; filter through several layers of cheesecloth. Centrifuge at 250 r/min for 15 min and pipette off the fluid between the supernatant and coarse precipitate. To a drop of fluid containing the cell suspension, add a drop of water soluble mounting medium, cover with coverslip, blot; warm and squash, exerting uniform pressure.

For air-dry preparations take fresh tissue after biopsy in tissue culture medium containing Eagle's basal amino acids and vitamins at double concentration in Eagle's BSS adjusted to pH 7.0 with 7.5% $NaHCO_3$ with glutamine 2mM, penicillin 100 units/ml, streptomycin 100 μg/ml, phenol red 7μg/ml, new born agammaglobin bovine serum 15% and USP heparin sodium 20000 units. Mince tissue in BSS at 37°C; centrifuge suspension at 100 g for a few min. Centrifuge, supernatant at 150 g for 4 min; discard supernatant and resuspend pellet in excess 0.125 M KCl with heparin added (20000 units/1) and incubate at 37°C for 1 h. Centrifuge discard supernatant and resuspend pellet in acetic-methanol (1:3) for 10-15 min. Take a drop on cooled slide and allow the fixative to evaporate in *warm* air as the drop flows down. Stain the dried slide for 2 h in 1% solution of orcein in 60% acetic acid and mount with Diaphane.

For a general analysis of the chromosomes of barley, fix the anthers inthree parts 95% ethanol; one part acetic acid. Tease out the pollen mother cells and place in iron-acetic-carmine solution. Place the slides over a hot water bath, and before final squashing, tap gently on the coverslip to separate the pollen mother cells. The slides are made permanent by dry ice technique and mounted in balsam.

Pollen grains

The pollen grains of Angiosperms are utilised for studying (a) mitotic division in pollen grains; (b) fertility of the grains, and (c) mitotic division in the generative cell inside the growing pollen tube. *For the study of the first mitotic divisions in the pollen grains* the methods followed are similar to those adopted for the study of meiosis in pollen mother cells as described earlier. *In order to study the apparent fertility of pollen grains* two sets of methods are available; (a) based on staining the contents of the grains, and (b) based on the germination of the pollen tube.

To study mitotic division in generative cells inside pollen tubes the practice is to study the mitotic division of the generative nucleus into the two sperm nuclei, which usually takes place between 2 and 48 h after germination of the pollen tube. The methods generally include two main stages, (a) germination of the pollen tube and (b) study of the chromosome sturucture. The chief factors necessary for the artificial germiantion of pollen tubes are temperatue, humidity and cultue media.

The principal methods have been described earlier in this chapter.

Embryosac Mother Cells

Meiotic division can also be studied from the embryosac mother cells; it usually takes place after the meiotic division in anthers, sometimes as long as three weeks afterwards.

The preliminary step is to expose the ovules to the treatment fluids, having first dissected them out. Ovule squash technique, the small ovules being detached from the placenta just before squashing, involves maceration in N HCl and staining with Feulgen reaction or by the acetic-orcein, acetic-carmine or acetic-lacmoid schedule. The former technique is more useful for larger ovules while the latter can be used for both small and large ones, the unwanted tissue being dissected out before final squashing.

For squash preparations, the most effective fixative is a modification of Carnoy's fluid having four parts chloroform, three parts absolute ethanol and one part glacial acetic acid. The high proportion of chloroform is necesssary to keep the pliability of cell structures. The period of fixation may vary from two days to three weeks. Whole mounts can be prepared by fixing in bulk for 24 h followed by hydrolysis and staining in Feulgen solution. The ovules are then dissected out.

Nucleolus

Staining of nucleolonema (a) Silver impregnation procedures and iron-pyrogallic stain for fixed material, and (b) phase contrast microscopy, dark-field illumination and oblique transillumination for fresh material have been used for studying the filamentous structures within the nucleolus.

Acetic-carmine staining Treat tissues with rectified spirit, 2; formalin, 1;5% glacial acetic acid, 1; hydrolyse with HCl at 60°C for fixed periods and squash in 1% acetic-carmine solution.

Feulgen-light green schedule Bring down section fixed in a fixative without acetic acid to water. Treat for 2-3 h in 75% ethanol. Wash in distilled water. Hydrolyse for 10 min in N HCl at 60°C. Stain for 2 h in leuco-basic fuchsin solution. Wash in two changes of SO_2 water, keep for 10 min in each. Rinse successively in distilled water, 50 and 70% ethanol. Mordant for 1 h in 80% ethanol saturated with Na_2CO_3. Dip in 80 and 95% ethanol. Stain for 20-25 min in filtered saturated alcoholic solution of light green with 2-3 drops of aniline oil. Drain. Rinse in a saturated solution of Na_2CO_3 in 80% ethanol, 10 ml and 90 ml 80% ethanol. Differentiate in 95% ethanol. Dehydrate through absolute ethanol, ethanol-xylol and xylol grades and mount in balsam. For smears and squashes, the initial treatment in 75% ethanol is omitted. In squashes, squash cells after Feulgen staining in 45% acetic acid. Separate coverslip in 40% ethanol and proceed as usual.

Toluidine blue-molydate method based on the principle of gradual blocking of— NH_2 group of nucleoprotein and unmasking of phosphate groups of nucleic acid binding cationic dyes.

The dyes giving satisfactory results are: Coleman Bell CU-3, National Aniline NU-2 and NU-17, Harleco NU-14, Matheson-Coleman Bell CU-9 and Biological Stain Commission NU-19. The dyes are dissolved in McIlvaine's buffer at pH 3.0 at 20°C. Wash tissue culture in 0.85% saline for 10 s before fixation. Treat two slides in 5% aqueous trichloracetic acid for 10 min. Wash in distilled water. Fix one slide for 5 min and another for 10 min in formal sublimate, containing 40% formaldehyde, 1 part, and 6% aqueous mercuric chloride, 9 parts. Rinse in tap water and treat with Lugol's iodine for 5 min. Immerse the slides in 5% sodium thiosulphate for 5 min and rinse in water. Stain for 30 min in toluidine blue. Treat with 4% aqucous ammonium molybdate solution for 15 min. Wash in tap water. Dehydrate in tertiary butyl alcohol, clear in xylol and mount in a synthetic resin. The *pars amorpha* of the nucleolus takes up bluish-green stain while the nucleolini are bright purple.

The later techniques with special *N*-banding have been included in the chapter 4 on Banding.

3. Special Groups

Algae

Chromosome structure in algae is of particular interest since the group presents a variable pattern. In most of the forms, including a majority of Chlorophyceae, chromosome morphology is comparable to that of higher plants, whereas in others, like Conjugales, Euglenophyceae, and especially Dinophyceae, the structure demonstrates a number of unusual characteristics. In Conjugales the chromosomes are devoid of localised centromeres, comparable to some extent with the structure reported in *Luzula* of angiosperms.

Euglenophyceae show certain features of special interest like the absence of a typical equatorial plate and centromere and the persistence of RNA-containing endosome throughout mitosis, as well as different types of chromosome aggregation and movement from higher plants.

In Dinoflagellates the cytochemical and ultrastructural data have shown the chromosomes to be sausage-shaped structures composed of continuous fine fibrils of DNA only, without any basic protein.

The methods of chromosome study in algae principally include two separate stages culture to obtain suitable divisional figures, and processing to observe the chromosomes. Culture is usually necessary, in the case of most forms of algae, for acquiring a sufficient number of metaphase plates in polar view for chromosome analysis. In the simplest form, the alga is grown in its natural medium, but progressively complex media with different proportions have been devised for obtaining better growth and synchronization of mitosis. The other major factor is light period and in most cases, artificial light and dark photoperiods are provided. For successful culture, the bulk material after collection is teased out under dissecting compound microscopes to separate pure filaments from mixed ones. Media ranging from water of the habitat to different types of synthetic media have been devised.

In Chlorophyceae and certain other algae, synchrony in division can be obtained even in culture without any special treatment. Godward suggested the possibility of synchronization of division through temperature control in continuous light with bubbled air enriched with 4% carbon dioxide and by continuous culture in medium which is continuously renewed. She has also stressed the possibility of the evolution of variant races in cultures maintained for a long time by mitotic changes or mutations.

Smaller members of marine Dinophyceae can be cultured in sea water, to which nitrate, phosphate and soil extracts have been added (50 ml soil extract; 0.2 g sodium nitrate and 0.03 M sodium phosphate, both in 10 ml of water added per litre of sea water.

Both soil extract and sea water have been replaced by artificial components by some workers. The cultures are kept under fluorescent lighting between 1076-2152 lx at a temperature of 15- 25°C for 12-18 h. Cultures of fresh water forms are not so successful. Culture of gametophytes and sporophytes of the Laminariales is influenced by nutrient, temperature and light. Sea water enriched by various nutrients has been used. The most suitable temperature is between 10-16°C and an indirect light intensity of 538-3228 lx, but never direct sunlight. Two commonly used nutrients are:

Nutrient A 0.01 M KH_2PO_4, 0.5 ml and 0.01 M KNO_3, 0.5 ml, added to 25 ml of sea water every 14 days.

Nutrient B 10^{-5} M KH_2PO_4 ($\frac{1}{20}$th of nutrient A stock); 5 x 10^{-5}M KNO_3 ($\frac{1}{4}$th of nutrient A stock) 0.5 ml of each added to 25 ml sea water every 14 days.

For a number of varieties belonging to the Phaeophyceae large pieces of the fruiting region are collected, washed in jets of boiled sea water and immersed in boiled sea water. After the zoospores are released, the suspension is poured in a flat dish containing sterilized slides or coverslips. After 30 min, the coverslips are removed and placed in a culture chamber. Coverslips may be placed back to back in cellophane covered glass pots. Rapid growth is ensured by long hours of daylight or by permanent illumination with three 80 W fluorescent tubes at about 15°C. In the Rhodophyta, culture methods have only a limited application. The habitat medium, with frequent changes, is sufficient in most cases. Complex media are often not necessary for cultures required for cytological study alone.

Schedules for studying the chromosomes of algae involve mainly fixation, staining and mounting. Pre-treatment is given in specific cases, as are also different maceration techniques. Decolorization of the pigment is an important factor in the choice of a fixative. The more common ones are:

Formalin-acetic-ethanol Also used in different formulae, the one by Westbrook (1935) being glacial acetic acid, 2.5 ml; 40% formaldehyde, 6.5 ml, and 50% ethanol, 100 ml. It is recommended for immediate use in Rhodophyceae since storage may cause shrinkage of the nucleus and later disintegration of the material.

Amongst the metallic fixatives Chromic acid-acetic acid mixtures have been used in a large range of variations, a very common one being a modification of Karpechenko's fluid: Solution A containing chromic acid, 1 g; glacial acetic acid, 5 ml, sea water, 65 ml and Solution B with 40% formaldehyde, 40 ml in 35 ml sea water. The two solutions are mixed immediately before use. Sea water is replaced by distilled water for fresh-water forms and the use of formaldehyde is optional. Osmic acid vapour, nitric acid vapour, 2% osmic acid solution. Belling's Navashin solution enriched with osmic acid, iodine water, bromine water and chromic acid solution have all been successfully used on different members of

Chlorophyceae. Fixation in osmic acid must be a brief one and the osmium must be washed out before it forms a black deposit.

Acetic acid: ethanol and acetic acid: methanol mixtures with different concentrations and proportions of the constituents are widely used. Mixtures of glacial acetic acid and 95% ethanol (1:1, 1:2, 1:3) are effective for Rhodophyceae, the two former giving better results.

Within Phaeophyceae, filamentous forms can be fixed in acetic acid-ethanol mixture (1:3) for up to 24 h, preferably changing the fixative after 1h. Lithium chloride may be used for large parenchymatous plants after fixation as a softening agent. Decolorization is usually satisfactory after 24 h.

Acetic-ethanol (1:3) is satisfactory for members of the Dinophyceae which are large and easy to handle. Acetic acid-methanol, in different proportions, gives very good results. Fixation is carried out in centrifuge tubes and the cells collected by centrifugation. Addition of a few drops of saturated ferric acetate (in acetic acid) to the fixative 1 h before staining is beneficial for certain species.

Various modifications have been used in different types of green algae. Mixtures of glacial acetic acid and 95% ethanol (1:1, 1:2, 1:3) and also ethanol or methanol alone, gave good results in different forms (*see* Godward, 1966).

Processing whole mounts is possible for uniseriate thalli in the Rhodophyceae as also for various members of the Dinophyceae and Chlorophyceae. Serial sections of paraffin- embedded material are utilized only in cases of certain Rhodophyceae following the usual block preparation schedule. Squash preparations are most commonly used in studying algal chromosomes. The fixed material can be squashed directly or it may require a softening process, depending upon the hardness. Dilute solutions of acid or alkali in water or ethanol are used for softening, the more common ones being HCl or NaOH in concentrations ranging from 1 to 50%.

The stains commonly applied to higher plants have also been used for algae, like acetic-carmine, Feulgen, haematoxylin, brazilin and methyl green-pyronin. The *acetic-carmine* stain is the one most widely applied following the iron alum-acetic-carmine method developed by Godward.

Members of the Dinophyceae are stained after fixation in a drop of acetic-carmine on a slide. Acetic-orcein is equally effective. For Laminariales, acetic-carmine, containing 1 drop of saturated ferric acetate solution per 25 ml, is added to the material *after* squashing, followed by alternate heating and cooling. Large parenchymatous brown algae require fixation in acetic-ethanol (1:3), enriched by a few drops of ferric acetate. The material is washed in 70% ethanol, hand sections are cut and mounted in 6% Na_2CO_3. After squashing, distilled water and later the stain, are added and the slide gently boiled. The iron- acetic-carmine schedule has also been used successfully for the Rhodophyceae after slight modifications.

Feulgen staining schedule has been applied in almost all forms of algae. The method of handling is similar to that for acetic-carmine staining, both for Chlorophyceae and Dinophyceae. Gametophyte materials of Laminariales take up bright stain. For large parenchymatous brown algae, fixation in Karpechenko's fluid is followed by washing in running water and bleaching for 3-4 h in 20% aqueous H_2O_2 solution and again washing in running water. The material is heated to 60°C in distilled water, hydrolyzed in N HCl at 60°C for 7-10 min. Without bleaching, hydrolysis has to be done for 15-30 min. The material is transferred to cold distilled water, washed in running water for 10 min, hand sections are cut and squashed in SO_2 water (N HCl, 5 ml; 10 % $K_2S_2O_3$, 5 ml and water, 100 ml).

Heidenhain's haematoxylin schedule and its modifications have been used in the red algae. Haematoxylin in acetic acid stains the nuclei in green algae but it is not permanent.

Brazilin was extensively used in red algae. Both the stain and its mordant require a long period of ripening in the dark to give satisfactory preparations. Sections of squash preparations are treated in a 2% ferric ammonium sulphate solution in 70% ethanol for 1h, washed in 70% ethanol and stained for 12-16 h in 0.5% brazilin in 70% ethanol. Finally they are washed twice in 70% ethanol and dehydrated through ethanol grades.

Methyl green pyronin staining schedule has been applied in the green algae, using a BDH dye mixture. A very small amount is dissolved in water and the material, after acetic fixation, is mounted in it.

Acetic-carmine preparations in Laminariales. Fix in acetic- ethanol (1:3) for 12-18 h, wash in running water; immerse in 1 M lithium chloride solution for 15 min; keep in water for 15 min; dissect out material and squash under a coverslip; insert a few drops of acetic-carmine solution with a trace of ferric acetate at the side of the coverslip, heat at intervals for 30 *s.* without boiling, squash again and blot excess stain. If over- stained, de-stain by heating in a mixture of acetic-carmine solution and glacial acetic acid (1:5) for 15 *s.*

In filamentous Phaeophyta, fixation is carried out for 24 h and a few drops of saturated ferric acetate in 45% acetic acid is added to the fixative. After washing in acetic-ethanol (1:3), add stain, cover, boil and squash.

Feulgen schedule for Phaeophyta Fix coverslips with growing gametophytes in acetic-ethanol (1:3), wash, hydrolyse in N HCl at 60°C for 8-10 min, transfer to cold water, and then keep in decolorised Schiff's reagent for 8 h at room temperature, bleach in three changes of SO_2 water, squash in SO_2 water and later dehydrate through ethanol grades and mount in euparal.

For softening and isolating female conceptacles split the receptacles longitudinally and fix for 12 h, followed by treatment for 20 min in a mixture of saturated ammonium oxalate solution and 20 vol. H_2O_2 (1:1), wash for 15 min before staining. Alternatively, for male material, macerate 30 g for 4 min in 150 ml of fixative in a Waring Blendor; transfer to a graduated cylinder; pipette off middle layer with sex organs and stain as usual.

Fungi

Even though belonging to Eukaryota in its nuclear constitution, this group is distinctive in its extremely simple thalloid constitution and absence of such pigments that are universal for higher plants. The meagre cytological data so far available have also indicated the occurrence of unusual features in certain groups, such as the existence of nuclear membrane during division and the controversial double reduction division during meiosis.

Chromosome analysis in fungi is not only important from a fundamental standpoint, including biochemical mutagenesis in *Neurospora crassa,* but also because of their implications in utilitarian research. The genetic basis of the fermenting capacity of the *Saccharomyces* complex, the medicinal values of scores of fungi including species of *Aspergillus* and *Penicillium* and the infecting capacity of fungi, need solutions in which chromosome research may play a significant role. Some methods are outlined here.

Fixatives include Helly's modified fluid (mercuric chloride, 5 g; potassium dichromate 3 g; distilled water, 100 ml; add 5 ml formalin just before use) which is very successful as a fixative in a wide range of fungi. Other fixatives used are osmium tetroxide vapour and acetic-ethanol.

Different chromosomal stains have been tried out.

Giemsa staining gives effective results for quite a large number of filamentous fungi and yeasts. The preparations can be stained directly or after hydrolysis in N HCl. A stain used for yeast chromosomes contains 16 drops of Gurr's Giemsa R66 dissolved in 10-12 ml of Gurr's Giemsa buffer at pH 6.9. If hydrolysis is required, fixed cells are usually extracted for $1\frac{1}{2}$ h with 1% NaCl at 60°C, the time, concentration and temperatures depending on the material; then treated for 10 min with N HCl at 60°C, rinsed with tap water and kept in the stain for several hours.

Haematoxylin staining has been advocated for squash preparations, following fixation in Lu's BAC fixative.

Feulgen staining is utilized according to the usual schedule of acid hydrolysis followed by staining. Schiff reagent prepared with Diamant Fuchsin is used for yeast chromosomes. The smears can be mounted in water or acetic-carmine.

Representative schedules

For yeast place a loopful of slimy growth from a 2-5 day- old plate culture on a No. 1 coverslip. Place another coverslip on top with corners turned away at an angle of 5 degrees to the former. Allow the drop to spread, pull apart the coverslips so that a smear is formed on each and immerse in Helly's modified fixative for 10 min. Rinse thoroughly in 70% ethanol, transfer to Newcomer's fixative for preservation in the cold. The film must not be permitted to dry prior to fixation. The yeast culture may be exposed briefly to formalin vapour before the smear is drawn for better preparation. In Giemsa staining, extract fixed cells for $1\frac{1}{2}$ h with 1% NaCl at 60°C, treat for 10 min in N HCl at 60°C, rinse in tap water and stain for several hours in Giemsa solution. Differentiate by moving the smears repeatedly for 10-12 *s* at a time in 40 ml distilled water to which a few drops of acetic acid have been added (pH 4.2). Observe under a water immersion lens to control extraction of excess stain. When the nuclei are brightly differentiated, mount the coverslip on a slide in a drop of buffer containing 2-3 drops of Giemsa per 10 ml. Blot excess medium, squash with uniform pressure, seal and observe. Squashing without pressure may also be done.

In Feulgen schedule, hydrolyse the smear for 10 min, rinse with tap water, stain in Schiff's reagent for 3.5 h, rinse quickly in 10 changes of SO_2 water (tap water 90 ml; N HCl 5 ml; 10% sodium metabisulphite 5 ml), keep in running water for 20 min and mount in water or acetic-carmine.

For mitosis in basidiomycetes centrifuge the culture and suspend in distilled water. Homogenise, spread in a thin film on slide, air dry, fix in acetic-lactic-ethanol (6:1:1) for 10 min, pass through 95% and 70% ethanol and rinse in water. Hydrolyse in N HCl at room temperature for 5 min, then at 60°C for 6 min, wash thoroughly in distilled water and suspend for 5 min in a phosphate buffer (pH 7.2). Stain for 25 min in Giemsa's stain, rinse successively in water and buffer, drain, flood with Abopon and cover with a coverslip.

For general mitotic preparations, Pour a mixture of BAC fixative (*n*-butanol, acetic acid and 10% aqueous chromic acid solution, 9:6:2) over the culture in a petri dish and store under partial vauum at 0-6°C for 1-5 days. Dissect out fruit bodies in a mixture of conc. HCl and 95% ethanol (1:1), heat gently, wash with Carnoy's fixative for 1-2 min, heat in propionic-carmine solution with a trace of iron, apply coverslip, allow to stand, heat to boiling and press the coverslip. Add a drop of 45% acetic acid to all corners, apply a drop of glycerin- acetic acid mixture to one corner, blot and seal.

For Pezizales, fix in Carnoy's fluid for 24-72 h, followed by direct squash in acetic carmine.

For basidia of agarics fix bits of hymenial tissue in Newcomer's fixative for 1-12 h, hydrolyse and mordant in aqueous N HCl containing 2% aluminium alum, 2% chrome alum and 2% iodic acid for 5 min at room temperature, then at 60°C for 10-15 min. Wash in three changes of distilled water and keep in Wittmann's acetic-iron-haematoxylin for 2 h. Mount in a drop of stain, cover with a coverslip, press, heat to just below boiling and seal.

For chromosomes of Neurospora

Dip strips of agar containing perithecia, after four days of crossing, in ethanol : acetic acid : 85% lactic acid mixture (6 : 1 : 1) in closed vials and store in deep freeze. Prepare stain by mixing a stock solution of 47 ml acetic acid and 20 ml lactic acid solution (1 ml 85% lactic acid in 24 ml distilled water) with 5 ml N HCl and 28 ml distilled water at room temperature. Mix 5 ml of this fluid with 100 mg Natural Green (Gurr) and reflux for 4 min after boiling over a low flame, in a small beaker covered by a petri plate bottom containing two ice cubes. Tease out the asci in a drop of stain, mount after separation in another drop under a coverglass, heat slightly and seal with dental wax.

4. Components-Nucleic acids

Pyronin-methyl green technique

Of all the methods employed for the differential localisation of RNA and DNA, the universally accepted standard technique is that of Brachet based on Pappenheim and Unna's methyl green-pyronin G mixture. In this schedule, methyl green imparts a green colour to chromatin or, more precisely, DNA, whereas RNA present in the cytoplasm, nucleolus, etc., appears pinkish red with pyronin, in both cases the reaction involving the phosphoric groups of the nucleic acid moiety. Methyl green is correctly represented as 'Methyl green OO', a basic dye of the triphenyl methane series. Balbiani (1881) first utilized methyl green to secure green colour in salivary gland chromosomes of *Chironomus*, followed later by Pappenheim and Unna, who employed it in combination with pyronin. In Unna's modification, phenol was a constituent of the mixture which evidently adjusted the pH for optimum staining DNA by methyl green was first demonstrated by Brachet (1942). Kurnick (1947) claimed that selectivity of DNA staining by methyl green is dependent on the polymerised nature of the DNA molecule, and depolymerised DNA could not be satisfactorily stained with methyl green, and also that histones competed with the dye for nuclei acid. The constant stoichiometry of the staining showed the formation of the chemical compound with the phosphoric group during staining. Kurnick and Mirksy (1950) asserted that one dye molecule combines with ten phosphoric groups of DNA, basing their observation on the stoichiometry of the reaction by dialysis, precipitation of stain-nucleic acid mixtures and staining of nucleic of known DNA content. This ratio of 1:10 was given by heptamethyl pararosaniline (CI 684) and the ratio 1:13 by hexamethyl pararosaniline (CI 685) (Pearse, 1960, 1972). Kurnick (1950a, b) and Errera (1951) claimed that stable binding of methyl green should involve at least two amino groups of the dye. The stainability of methyl green is therefore controlled largely by different agents which under certain conditions may bring about depolymerization. Goldstein (1961) claimed that the explanation so far provided for methyl green staining of DNA is unnecessary, the chief factor being the weight of the dye cation, and he has suggested that dye cations, having a combined atomic weight between 350 and 500, can penetrate and adhere to nuclear DNA whereas smaller cations adhere to denser molecules such as cytoplasmic RNA. Methyl green, having a cationic weight of 387, is specific for DNA. Heat or other agents may bring about alterations in the structure of DNA in such a way that it becomes denser and, as such, becomes stainable with pyronin.

Pyronin is a basic dye of the xanthene group and is available in three different forms, 'pyronin B', 'pyronin Y' and 'pyronin G'. Pyronin Y (Michrome No. 339, Gurr) is a tetramethyl whereas pyronin B (Michrome No. 44, Gurr) is a tetraethyl compound. Pyronin preferentially stains low polymers of nucleic acid, but stoichiometric studies did not reveal constancy in the binding of depolymerized DNA by pyronin. To secure selective staining, pyronin G or Y is suitable as pyronin B results in non-specific cytoplasmic staining. Evidently, methylation may have some connection with the staining property of pyronin.

Schedules for pyronin-methyl green staining

Schedule A (Brachet 1942)

The recommended fixation time is for 4-16 h at pH 7.0 in 10% aqueous formalin.
Reagents are:
Methyl green : Before use, wash methyl green repeatedly with chloroform or amyl alcohol to dissolve and remove the traces of methyl violet which are formed when methyl green is exposed to the atmosphere. Filter off the residual methyl green and dry it. Alternatively, shake an aqueous solution of methyl green with an excess of chloroform or amyl alcohol. The violet component is to be gradually removed. Allow to stand for 2-3 days. Remove the aqueous supernatant liquid for use.

Pyronin Y or pyronin G: In general, the bluish shades are more satisfactory.

Ribonuclease Ribonuclease solution can be prepared by Brachet's method, as follows: Mince finely 0.5-1 kg of ox pancreas by passing through mincer. Pound the minced meat into a smooth paste with mortar and pestle. Suspend the paste in an equal volume N/10 acetic acid for 24 h. Boil for 10 min, cool and filter. Bring the pH to 6.0. Filter again. Add a few thymol or camphor crystals as preservative and store in cold. The enzyme retains its activity for several months.

Pyronin-methyl green mixture contains methyl green (washed in chloroform and dried), 0.15 g; pyronin Y, 0.25g; 90% ethanol, 2.5 ml; M/5 acetate buffer pH 4.7, 97.5 ml.

In an alternative method, the solutions used are:

Solution (1): 5% aq. pyronin solution, 17.5 ml; 2% aq. methyl green solution (chloroform-washed), 10 ml; dist. water, 250 ml.

Solution (2) M/5 acetate buffer pH 4.8. Mix equal quantities of (1) and (2) in a staining jar. Do not use mixture after keeping for a week.

Ribonuclease 0.1% solution in dist. water adjusted to pH 6.0.

Procedure

Fix sliced tissue in Carnoy's or Zenker's fluid. Wash. Dehydrate as usual. Embed in paraffin, Cut paraffin sections. Deparaffinize in toluene and bring the sections down through alcohol grades to distilled water. Stain the sections in pyronin-methyl green mixture for 20 min. Wash them rapidly with distilled water. Differentiate in 95% ethanol for 5-10 min. Dehydrate in absolute ethanol, clear in toluene and mount in DPX. For control experiments: (a) Keep one set of slides, marked 'A', in distilled water pH 6.0 in an oven at 37°C for 1 h. Then stain, diferentiate, dehydrate and mount as described before; (b) keep a second set of slides, marked 'B', in ribonuclease solution in an oven at 37°C for 1 h and then follow the staining schedule described before.

In the normally stained tissue, RNA in nucleolus and cytoplasm takes up red colour, while DNA appears as green particles in the nuclear chromatin. In control experiment (a) no green colour is observed, showing that DNA has been removed by warm water. In control experiment (b), RNA is removed by ribonuclease, shown by the complete absence of red colour.

Schedule B (Kurnick 1955b)

The recommended fixative is Carnoy's fluid or freeze-drying. Dissolve 2 g of pyronin Y in 100 ml of distilled water. Add chloroform and shake the mixture in a separating funnel till the layer of chloroform becomes colourless. Separate the dissolved dye. Similarly prepare a 2% solution of methyl green and extract it with chloroform. The solutions can be kept as stock. For use, mix together 12.5 ml of pyronin Y solution and 7.5 ml of methyl green solution and add 30 ml of distilled water.

Fix and embed the tissue. Bring down the paraffin sections to distilled water. Immerse the sections in the staining mixture for 6 min. Freeze-dried sections can be stained directly. Remove excess stain by blotting with filter paper. Transfer to *n*- butanol, keeping for 5 min and then treat in a further change of *n*-butanol for 5 min. Transfer to xylol and keep for 5 min. Transfer to cedarwood oil and keep for 5 min. Mount, preferably in Permount. Chromatin is stained green while cytoplasm and nucleoli are bright red.

In an alternative schedule, increase the period of staining up to 10-30 min, rinse in distilled water, drain and blot, keep in two changes of *n*-butanol for 5 min each and mount directly in euparal.

C. Modification of Brachet's Schedule (1955)

Mix together: 0.5% aq. pyronin G solution 37 ml; 0.5% aq. extracted methyl green solution 13 ml; 0.2 M acetate buffer 50 ml.

This mixture retains its capacity for at least four months.

Bring down the paraffin sections to distilled water and blot to remove moisture. Immerse for 30 min in the buffered staining mixture. Wash in distilled water for a few seconds and blot again. Keep in pure acetone for 1 min. Transfer to acetone-xylol mixture (1:1) and then to pure xylol, keeping for 1 min in each. Mount in neutral balsam.

D. Modified Schedule for Plant Tissues

Prepare staining mixture by adding together:

2% aq. pyronin solution 50 ml; 2% aq. methyl green solution 30 ml; Dist. water 10 ml; Chloroform 10 ml.

Procedure

Bring paraffin sections of tissues, previously fixed in Carnoy's fluid, down to distilled water. Shake the staining mixture before use. Then immerse the slides in it for 15 min at 20°C. Blot excess stain with a filter paper and dip the slide in 50% ethanol and pure *n*-butanol mixture (1:1) for 2 *s*. Immerse in *n*-butanol for 10 min. Again immerse in *n*-butanol for 5 min. Treat in toluol for 1 h. Mount in neutral balsam. Chromosomes and nuclear chromatin stain bright green while nucleolar and cytoplasmic RNA stain pink to deep rosy red.

Alkaline Phosphatase

The technique for the demonstration of alkaline phosphatase, a term loosely applied to phosphomonoesterase, is one of the most important ones for localization of specific enzyme sites on chromosomes. The demonstration of the enzyme activity in the cell, and especially in the chromosome, can be carried out principally through two different methods: (a) calcium phosphate deposition; and (b) azo dye methods.

Calcium Phosphate Method

The calcium phosphate precipitation technique was first developed independently by Gomori (1939) and Takamatsu (1939), and was based on the principle that if a tissue containing the enzyme is incubated at 37°C in a medium containing phosphate salt as the principal constituent at an alkaline pH (9.4), the liberated phosphoric acid can be deposited at the site of the enzyme as an insoluble phosphate precipitate if a calcium salt is present as one of the components. The visualization of calcium phosphate can be done by converting it into silver sulphate and then to metallic silver or into cobalt phosphate through a cobalt salt and finally to a black precipitate of cobalt sulphide through ammonium sulphide.

Several phosphate salts have been used as substrates, such as α, β-glycerophosphate, fructose diphosphate, adenosine triphosphate, sodium dihydrogen phosphate, etc., but the most commonly used one is sodium-β-glycerophosphate. As the enzyme is effective in an alkaline medium, the maintenance of proper pH (9.2-9.8) is necessary.

Gomori (1950) suggested that the alkaline phosphatase technique can be applied for quantitative estimation. Danielli (1950) stated that quantitative results can be estimated on the principle that the sites of highest activity appear before those of less activity. Several methods for the quantitative estimation of alkaline phosphatase activity, with the aid of spectrophotometry, radioactivity as well as interferometry, have been developed

Azo dye method

The azo dye technique is based on the principle of precipitating the alcoholic part of the phosphate ester instead of the phosphoric acid. The method was originally developed by Menten, Junge and Green (1944). They used β-naphthyl phosphate as the substrate and the liberated β-napthol, after hydrolysis, is reacted upon *in situ* by diazotised β-naphthyl amine at pH 9.4, whereby a red precipitate is obtained. It can be applied to both fresh and formalin-fixed tissues. Danielli (1946) suggested the use of phenyl phosphate and β-naphthyl phosphate to obtain the reaction. Mannheimer and Seligman (1948) used magnesium ion as activator and modified the procedure and used α-naphthyl diazoniumnaphthalene-1,5-disulphonate for coupling. The purplish red dye, which precipitates out, has been found to be insoluble. This method has the added advantage over the original technique in that the coupling reagent used here is stable as compared to the diazonium salt used by them. In Gomori's modification of the original method, sodium α-naphthyl phosphate has been used instead of calcium β-naphthyl phosphate, as the sodium salt is relatively less soluble.

Calcium cobalt method:

Incubation mixture contains 3% aq. sodium - β - glycerophosphate soln 10 ml; 2% aq. sodium diethylbarbiturate 10 ml; Dist. water 5 ml; 2% aq. calcium chloride solution 20 ml; 5% aq. magnesium sulphate solution 1 ml; 2% aqueous cobalt nitrate or acetate solution. Aqueous dilute yellow ammonium sulphide solution contains about 50 drops of concentrated liquid in 20 ml water. Fix in cold acetone at 4°C with three changes for 24 h. Transfer through progressive alochol grades to absolute ethanol, keeping 30 min in each grade. Treat with ethanol ether mixture (1:1) for 1h and transfer to 1 % celloidin. Decant excess celloidin and treat successively in chloroform and benezene, keeping 1 h in each. Embed in paraffin. Cut sections 5 μm thick and mount on albuminised slides. Dry at 37°C and store at 4°C. Bring down the slides in distilled water through immersion successively in petroleum ether and absolute acetone. Incubate in the incubating mixture at 37°C for 30 min to 16 h, 4 h being the optimum period. Wash thoroughly in distilled water after bringing down the slides to room temperature. Immerse in 2 % cobalt nitrate solution for 3-5 min. Wash in distilled water. Immerse in yellow ammonium sulphide solution for 1-2 min. Rinse in distilled water. Stain with 1% eosin solution for 5 min if required. Dehydrate and clear through ethanol and xylol grades and mount in neutral balsam. A deposition of a black precipitate at the sites of alkaline phosphatase activity is seen.

Alternatives For frozen sections, cut very thin sections and mount on slides without albumin. Dry at room temperature for 1-2 h. The remaining steps are similar to paraffin sections. In a later schedule by Gomori(1952), the incubation mixture contains: 3% aq. sodium glycerophosphate solution 20 ml; 2% aq. sodium diethylbarbiturate solution 30 ml; 2% aq. calcium chloride solution 4 ml; 2% aq. magnesium sulphate solution 2 ml; Dist. water 30 ml. Cobalt chloride can be used instead of cobalt nitrate or acetate. Non- metallic fixatives give quite good results. The preparation of paraffin blocks can be done through the usual alcohol-chloroform grades.

Caution : In no step, from fixation to mounting, should the temperature exceed 56°C. The incubation mixture should be freshly prepared before use.

α-Napthyl phosphate method

Incubation mixture contains:

Sodium-α-naphthyl phosphate 0.05 g; 5% aq. borax solution 10 ml; 10% aq. magnesium chloride or sulphate solution 0.5 ml; Cold dist. water (20°C) 100 ml; A stabilised diazonium salt 0.25 mg. Either tetrazotised-*o*-dianisidine (michrome blue salt 250), 3-nitroanisole-4-diazonium chloride (michrome red salt 612) or 3-nitrotoluene-4-diazonium naphthalene-1, 5-disulphonate (michrome scarlet salt 618) can be used. Also needed are haematoxylin solution in water, 70% ethanol 99 ml; glacial acetic acid 1 ml.
Mix together.

Procedure

For paraffin block preparations, fix thin cold slices of tissue at 4°C for 24 h in three or four changes of acetone. Clear in xylol. Embed in paraffin rapidly. Cut sections and attach to albuminized slides. Deparaffinize as usual. Rinse three times in pure acetone and then in three changes of distilled water. Incubate in the incubation mixture at 37°C for 10-30 min or more. Remove a slide at intervals and observe under the microscope until the correct brightness of colour is attained. Stir the mixture mechanically during incubation. Rinse thoroughly in distilled water after bringing the slides to room temperature. Immerse in haematoxylin solution for counterstaining. Treat with acetic-ethanol mixture for 5-10 min. Rinse thoroughly in water. Mount in glycerine jelly or dehydrate, clear and mount in neutral balsam.

Observations

The sites of alkaline phosphate activity stain purplish black with michrome blue salt; 250 purplish brown with michrome red salt 606 and reddish brown with michrome red salts 612 and 618.

Alternative

Sodium β-naphthylphosphate may be used instead of sodium -α- naphthyl-phosphate but α-salts yield a more specific and non- diffusible precipitate.

A modified coupling azo dye method for alkaline phosphatase

For frozen sections

Fix thin slices of tissue in 10% neutral formalin in the cold for 10-16 h, or use fresh frozen cold microtome sections, mounted on cover slips. Cut frozen sections 10-15 μm and mount on slides without adhesive. Dry in air for 1-3 h. Dissolve 10-20 mg sodium- naphthyl phosphate in 20 ml 0.1 M stock tris buffer (pH 10). Add to it, with stirring, 20 mg of the stable diazotate of 5-chloro-*o*-toluidine. Cover the sections on the slides with the filtered solution and incubate at room temperature for 15-60 min. Rinse in running water for 1-3 min. Counterstain for 1-2 min in Mayer's haemalum. Rinse in running water for 30-60 min and mount in glycerine jelly. The sites of alkaline phosphatase activity appear brown with Fast Red TR and Fast Violet B or black with Fast Black B. The nuclei are dark blue.

For paraffin sections of material fixed in cold acetone

Bring down the sections to water after passing successively through light petroleum and absolute acetone. Cover with freshly prepared filtrate fo substrate-diazonium salt mixture as in previous schedule. Incubate for 30 min to 4 h for salt Fast Blue RR or for up to 2 h for salt Fast Red RC or for up to 12 h for salt Fast Red TR. Rinse in water, counterstain with haemalum as given in previous schedule, wash in running water and mount in glycerine jelly. Salt Fast Red TR gives the best results, the sites of alkaline phosphatase activity appearing reddish brown and the nuclei blue.

REFERENCES

PREFACE

Brown, T. A. (ed.) (1991) *Essential Molecular Biology* vol. I and II. IRL Press, Oxford.

Sambrook, J., Fritsch, F., Maniatis, T. (1989) *Molecular cloning — a laboratory manual* 2nd ed. **3** 2-58.

CHAPTER 1

Baker, J. R. (1966) *Cytological technique*. 5th ed. Methuen, London.

Caspersson, T., Farber, S., Foley, G. E., Kudynowski, J., Modest, E. J., Simonsen, E., Wagh U. (1968) Chemical differentiation along metaphase chromosomes. *Exptl. Cell Res.* **49** 219-222.

Casselman, W. G. B. (1959) *Histochemical Technique*. Methuen, London.

Conn, H. J. (1953) *Biological stains*. Biotech. Publ., Geneva, New York.

Chaudhuri, M., Chakravarty, D. P., Sharma, A. K. (1962) Isopsoralene and its use in karyotype analysis. *Stain Techn.* **37** 95-97.

Darlington, C. D., La Cour, L. F. (1968) *The handling of chromosomes*. Allen & Unwin, London.

Feulgen R., Rossenbeck, H. (1924) Microscopisch chemisches Nachweis einer Nucleinsaure von Typus der Thymonuclein Saure. *Zelfs. Physiol. Chemie* **135** 203-248.

Gurr, E. (1960) *Encyclopedia of microscopic stains*. Leonard Hill, London.

La Cour, L. F. (1941) Acetic orcein. *Stain Techn.* **16** 169-174.

Lassek, A. M., Lunetta, S. (1950) The hydrogen ion concentration of primary fixing fluids. *Stain Techn.* **25** 45-47.

Manna, G. K., Ray Chaudhuri, S. P. (1953) A study of the somatic chromosome complex in normal endometrium and in malignant tumour of the cervix uteri with an improved technique. *40th Ind. Sc. Congr. Abst.* 181-182.

Pearse, A. G. E. (1972) *Histochemistry*. Churchill, London.

Pelc, S. R. (1956) Effect of X rays on the metabolism of cell nuclei of nondividing tissues. *Nature* **178** 359-361.

Sharma, A. K., Bal, A. (1953) Coumarin in chromosome analysis. *Stain Techn.* **28** 255-257.

Sharma, A. K., Chaudhuri, M. (1962) An investigation on the viscosity changes in the cell caused by coumarin and its derivatives. *Nucleus* **5** 137-142.

Sharma, A. K., Roy, M. (1955) Orcein staining and the study of the effect of chemicals on chromosomes. *Chromosoma* **7** 275- 280.

Sharma, A. K., Sharma, A. (1957) Permanent smears of leaf tips for the study of chromosomes. *Stain Techn.* **32** 167-164.

Sharma, A. K., Sharma, A. (1980) *Chromosome Techniques — Theory and Practice.* 3rd ed. Butterworth, London.

Tjio, J. H., Levan, A. (1950) The use of oxyquinoline in chromosome analysis. *Anal. Estac. Exptl. de Aula dei* **2** 21- 64.

Weisblum, B., Hasseth, P. L. (1972) Quinacrine, a chromosome stain specific for deoxyadenylate - deoxythymidylate rich regions in DNA. *Proc. Natl. Acad. Sci. US* **69** 629-632.

Wittman, W. (1962) Aceto-iron haematoxylin for staining chromosomes in squashes of plant material. *Stain Techn.* **37** 27-30.

CHAPTER 2

Berlyn, G. P., Miksche, J. P. (1976) *Botanical microtechnique and cytochemistry.* Iowa State University Press, Iowa.

Bennett, M. D., Smith, J. B., Heslop-Harrison, J. S. (1982) Nuclear DNA amounts in Angiosperms. *Proc. Roy. Soc. London* **B 216** 179-195.

Caspersson, T., Zech, L. (1973) Chromosome identification: techniques and applications in biology and medicine. *Nobel Symposium* **23** Academic Press, New York.

Dolezel, J., Binarova, P., Lucretti, S. (1989) Analysis of nuclear DNA content in plant cells by flow cytometry. *Biol. Plant.* **31** 113-120.

Fakan, S., Fakan, J. (1989) Autoradiography of spread molecular complexes. In **Somerville, J., Scheer, U.** (eds.) *Electron microscope in molecular biology : A practical approach.* IRL Press, Oxford, 201-204.

Fakan, S., Hughes, M. E. (1989) Fine structural ribonucleoprotein components of the cell nucleus visualized after spreading and high resolution autoradiography. *Chromosoma* **98** 242-249.

Fukui, K. (1986) Standardization of karyotyping plant chromosomes by a newly developed chromosome image analysing system CHIAS. *Theor. Appl. Genet.* **72** 27-32.

Fukui, K., Kakeda, K. (1990) Quantitative karyotyping of barley chromosomes by image analysis methods. *Genome* **33** 450-458.

Gosalvez, J., Sumner, A. T., Lopez-Fernandez, C., Rossino, R., Goyanes, V., Mezzanotte, R. (1990) Electron microscopy and biochemical analysis of mouse metaphase chromosomes after digestion with restriction endonucleases. *Chromosoma* **99** 36-43.

Hamkalo, B. A., Miller, O. L. Jr (1973) Electron microscopy of genetic activity. *Annu. Rev. Biochem.* **42** 379-396.

Harrison, C. J., Jack, E. M., Allen, T. D. (1987) Light and scanning electron microscopy of the same metaphase chromosomes. In **Hayat, M. A.** (ed.) *Correlative microscopy in biology : instrumentation and methods*. Academic Press, New York, 189-248.

Hodgson, J. (1991) Microimaging *in vivo. Biotechnology* **9** 353-356.

Kamisugi, Y., Fukui, K. (1990) Automatic karyotyping of plant chromosomes by imaging techniques. *Biotechniques* **8** 290- 295.

Leitch, A. R., Schwarzacher, T., Wang, M. Z., Moore, G., Heslop-Harrison, J. S. (1991) Flow cytometry of cereal chromosomes. *Cytometry* Suppl. **5** 39.

Lucas, J. N., Mullikin, J. C., Gray, J. W. (1991) Dicentric chromosome frequency analysis using slit scan flow cytometry. *Cytometry* **12** 316-322.

Needham, G. H. (1958) *The use of microscope, including photomicrography.* Thomas, Springfield, Illinois.

Oliver, C.W. (1947) *The intelligent use of the microscope*. Chapman & Hall, London.

Oster, G. (1950) *Progress in Biophysics* **1**, 73. Butterworth-Springer, London.

Oster, G. (1955) In *Physical techniques in biological research* **1**, 439, 1st ed. Academic Press, New York.

Passwater, R. A. (1970) *Guide to fluorescence literature* **2**. Plenum, New York.

Patau, K. (1952) Absorption microphotometry of irregular shaped objects. *Chromosoma* **5** 341-362.

Pollister, A. W., Swift, H., Rasch, E. M. (1969) In *Physical Techniques for biological research* **31** 201. Academic Press, New York.

Razavi, L. (1968) In *Nucleic acids in immunology*, 248. Springer-Verlag, Berlin .

Reid, N. (1974) In *Practical methods in Electron Microscopy* **Glauert, A. M.** (ed.) North Holland, Amsterdam.

Richards, B. M., Davies, H. G. (1958) *General cytochemical methods* **1**. Academic Press, New York.

Richards, O. W. (1955) In *Analytical cytology*, 501. McGraw-Hill, New York.

Ris, H. (1962) In *The interpretation of ultrastructure, Symp. Intern. Soc. Cell Biol.* Academic Press, New York.

Ross, K. F. A. (1954) *Quart. J. Microscop. Sci.* **95** 425.

Rost, F. W. D. (1972) In *Histochemistry — theoretical and applied* **Pearse, A. G. E.** (ed.) **2** 11-71. Churchill, London.

Ruch, F. (1966) In *Physical techniques for biological research* **3A** 57, 2nd ed. Academic Press, New York.

Sjöstrand, F. S. (1969) Electron microscopy of cells and tissues. In *Physical techniques in biological research* **3C**. Academic Press, New York .

Sommerville, J., Scheer, U. (1987) *Electron microscopy in molecular biology — practical approach.* IRL Press, Oxford, 215- 232.

Steiner, R. F., Beers, R. F. Jr. (1961) *Polynucleotides,* 301. Elsevier, Amsterdam.

Stevens, G. W. (1957) *Microphotography at extreme resolution.* Wiley, New York.

Strugger, S. (1949) *Fluoreszenmikroskopie und Mikrobiologie*. M. H. Schaper, Hannover.

Sumner, A. T. (1991) Scanning electron microscopy of mammalian chromosomes from prophase to telophase. *Chromosoma* **100** 410-418.

Swift, H. (1962) In *Interpretation of Ultrastructure, Symp. Intern. Soc. Cell Biol.* **1**. Academic Press, New York.

Turner B. M., Keohane, A. (1987) Antibody labelling and flow cytometric analysis of metaphase chromosomes reveals two discrete structural forms. *Chromosoma* 95 263-270.

Wanner, G., Formanek, H., Martin, R., Herrmann, R.G. (1991) High resolution scanning electron microscopy of plant chromosomes. *Chromosoma* **100** 103-109.

West, S. S. (1969) In *Physical techniques in biological research* **3C** 253, 2nd ed. Academic Press, New York.

Wilson, G. B., Morrison, J. H. (1961) *Cytology*. Reinhold, New York.

Zhao, J., Hao, S., Xing, M. (1991) The fine structure of the mitotic chromosome core (scaffold) of *Trilophidia annulata. Chromosoma* **100** 323-329.

CHAPTER 3

Bahr, G. F. (1973) Tutorial proceedings. *Int. Academy of Cytology* **2** 58.

Babu, A., Verma, R. S. (1986) The heteromorphic marker on chromosome 18 using restriction endonuclease Alu I. *Am. J. Hum. Genet.* **38** 549-554.

Bianchi, N. O., Bianchi, M. S., Cleaver, J. E.(1984) The action of ultraviolet light on the patterns of banding induced by restriction endonucleases in human chromosomes. *Chromosoma* **90** 133-138.

Burkholder, G. D., Schmidt, G. J. (1986) Endonuclease banding in isolated mammalian metaphase chromosomes. *Exptl. Cell Res.* **164** 379-384.

Chattopadhyaya, D., Sharma, A. K. (1988) A new technique for orcein banding with acid treatment. *Stain Techn.* **63** 283-289.

De la Torre, J., Mitchell, A. R., Sumner, A. T. (1991) Restriction endonuclease/nick translation of fixed mouse chromosomes. *Chromosoma* **100** 203-211.

De Carvalho, C. R., Saraiva, L. S. (1993) A new heterochromatin banding pattern revealed by modified HKG banding technique in maize chromosomes. *Heredity* **70** 515-519.

Dolezel, J., Cihalikova, J., Lucretti, S. (1992) A high yield procedure for isolation of metaphase chromosomes from root tip of *Vicia faba* L. *Planta* **188** 93-98.

Endo, T. R. (1986) Complete identification of wheat chromosomes by means of a C-banding technique. *Jap. J. Genet.* **61** 89-93.

Endo, T. R., Gill, B. S. (1984) Somatic karyotype, heterochromatin distribution and nature of chromosome differentiation in common wheat, *Triticum aestivum* L. *Chromosoma* **89** 361-369.

Ferrucci, L., Romano, E., De Stefano, G. F. (1987) Alu I induced bands in great apes and man: implication for heterochromatin characterization and satellite DNA distribution. *Cytogenet . Cell Genet.* **44** 53-57.

Frediani, M., Mezzanotte, R., Vanni, R. *et al.* (1987) The biochemical and cytochemical characterisation of Vicia faba DNA by means of Mbo I, Alu I and Bam H I restriction endonucleases. *Theor. Appl. Genet.* **75** 46-50.

Gerlach, W. L. (1977) N-banded karyotypes of wheat species. *Chromosoma* **62** 49-56.

Gill, B. S., Friebe, B., Endo, T. R. (1991) Standard karyotype and nomenclature system for description of chromosome bands and structural aberrations in wheat (*Triticum aestivum*). *Genome* **34** 830-839.

Giraldez, R., Cermeno, M. C., Orellana, J. (1979) Comparison of the C-banding pattern in the chromosomes of inbred lines and open pollinated varieties of rye. *Z. Pflanzenzücht.* **83** 40-48.

Gosalvez, J., Goyanes, V. (1988) Selective digestion of mouse chromosomes with restriction endonucleases I. Scaffold-like structures and bands by electron microscopy. *Cytogenet. Cell Genet.* **48** 198-200.

Gosalvez, J., Lopez-Fernandez, C., Ferrucci, L., Mezzanotte, R. (1989) DNA base sequence is not the only factor for restriction nuclease activity on metaphase chromosomes : evidence using isoschizomers. *Cytogenet. Cell Genet.* **50** 142-144 (See also Chapter 3).

Haaf, L., Muller, H., Schmid, M. (1986) Distamycin A/DAPI staining of heterochromatin in male meiosis of man. *Genetica* **70** 179-185.

Hederman, U., Schurmann, M., Schwinger, E. (1988) The effect of restriction enzyme digestion of human metaphase chromosomes on C-band variants of chromosomes 1 and 9. *Genome* **30** 652-655.

Hizume, M., Tanaka, A., Yonezawa, Y., Tanaka, R. (1980) A technique for C-banding in Vicia faba chromosomes. *Jap. J. Genet.* **55** 301-305.

Holmquist, G. (1975) Hoechst 33258 fluorescent staining of Drosophila chromosomes. *Chromosoma* **49** 333-356.

Ikushima, T. (1990) Bimodal induction of sister chromatid exchanges by luminol during the S-phase of the cell cycle. *Chromosoma* **99** 360-364.

Jewell, D. C. (1981) Recognition of two types of positive staining chromosomal material by manipulation of critical steps in the N-banding technique. *Stain Techn.* **56** 227-234.

Kakeda, K., Fukui, K., Yamagata, H. (1991) Heterochromatin differentiation in barley chromosomes revealed by C and N banding techniques. *Theor. Appl. Genet.* **81** 144-150.

Khachaturov, E. N., Barsky, V. E., Galkina, I. G., Stonova, N. S., Maksimaldo, Y. B. (1975) *Izvestia Akad. Nauk SSR - Biol* **6** 873.

Kihlman, B. A., Kronborg, D. (1975) Sister chromatid exchanges in Vicia faba. *Chromosoma* **51** 1-10.

Kenton, A. (1991) Heterochromatin accumulation, disposition and diversity in *Gibasis*. *Chromosoma* **100** 467-478.

Korenberg, J. R., Freedlander, E. F. (1974) Giemsa technique for the detection of sister chromatid exchanges. *Chromosoma* **48** 355-360.

Latt, S. A. (1981) Sister chromatid exchange formation. *Annu. Rev. Genet.* **15** 11-15.

Lavania, U. C., Sharma, A. K. (1979) Trypsin orcein banding in plant chromosomes. Stain Techn. **54** 261-263.

Lica, L., Hamkalo, B. (1983) Preparation of centromeric heterochromatin by restriction endonuclease digestion of mouse L929 cells. *Chromosoma* **88** 42-49.

Ludena, P., Sentis, C., Fernandez - Peralta, A. M., Gonzalez - Aquilera, J., Fernandez - Piqueras, J. (1990) New distinctions between regions of centromeric heterochromatin in human chromosomes by treatments with the isoschizomers NdeII-Sau3AI. *Genome* **33** 785-788.

Mezzanotte, R., Bianchi, U., Vanni, R., Ferrucci, L. (1983) Chromatin organization and restriction endonuclease activity on human metaphase chromosomes. *Cytogenet. Cell Genet.* **36** 562-566.

Matsui, S. I., Sasaki, M. S. (1975) The mechanism of Giemsa banding of mammalian chromosomes with special attention to the role of non histone proteins. *Jap. J. Genet.* **50** 189-204.

Prantera, G., Ferraro, M. (1990) Analysis of methylation and distribution of C_pG sequences on human active and inactive chromosomes. *Chromosoma* **99** 18-23.

Perry, P., Wolff, S. (1974) New Giemsa method for the differential staining of sister chromatids. *Nature* **251** 156-158.

Scheres, J. M. J. C., Hustin, T. W. J., Batten, F. J., Merk, G. F. M. (1977) *Helsinki chromosome conference*, 56.

Schnedl, W. (1973) Identification of chromosome segments. In *Nobel Symposia* **Caspersson, T., Zech, L.,** (eds.) **23** 342 Academic Press, New York.

Schubert, I. (1990) Restriction endonuclease (RE-) banding of plant chromosomes. *Caryologia* **43** 117-130.

Shang, M. X., Jackson, R. C., Nguyen, H. T. (1988) Heterochromatin diversity and chromosome morphology in wheats analyzed by the HKG banding technique. *Genome* **30** 956-965.

Sharma, A. K. (1975) Orcein banding and repeated DNA. *J. Ind. Bot. Soc.* **54** 1-6.

Takayama, S. (1976) Configurational changes in chromatids from helical to banded structures. *Chromosoma* **56** 47-54.

Tagarro, I., Fernandez - Peralta, A. M., Gonzalez - Aquilera, J. J. (1992a) Distribution of satellite DNA fractions within major heterochromatic regions of human chromosomes as revealed by PleI and TfiI digestion. *Cytogenet. Cell Genet.* **60** 102-106.

Tagarro, I., Gonzalez-Aquilera, J. J., Fernandez-Peralta, A. M. (1992b) Induction of R-bands on human chromosomes by TfiI as the consequence of local differences on target richness. *Cytogenet. Cell Genet.* **60** 154-156.

Tagarro, I., Gonzalez-Aquilera, J. J., Fernandez-Peralta, A. M. (1990) TaqI digestion reveals fractions of satellite DNAs on human chromosomes. *Genome* **34** 251-254.

Taylor, J. H. (1958) Sister chromatid exchanges in tritium labelled chromosomes. *Genetics* **43** 515-529.

Wagner, R. P., Maguire, M. P., Stallings, R. L. (1993) *Chromosomes. A synthesis.* Wiley-Liss, New York.

Wolff, S. (1977) Sister chromatid exchange. *Ann. Rev. Genet.* **11** 183-201.

Wray, W. (1975) In *Methods in Cell Biology* **Prescott, D. M.** (ed.) **15** 111. Academic Press, New York.

CHAPTER 4

Caro, L. G. (1964) High Resolution Autoradiography. In *Methods in Cell Physiology* **1**. Academic Press, New York.

Liquier-Milward, J. (1956) Electron microscopy and radioautography as coupled techniques in tracer experiments. *Nature* **177** 619.

Moorhead, P. S. (1960) *Exptl. Cell Res.* **9** 474.

Moorhead, P. S., Nowell, P. C, Mellman, W. S., Battips, D. M., Hungerford, D. A. (1964) The blood technique and human chromosomes. In *Symposium on Mammalian Tissue Culture Cytology*, Sao Paulo, 1962. Academic Press, New York.

Reid, N. (1974) In *Practical methods in electron microscopy* **Glauert, A. M.** (ed.) North Holland, Amsterdam.

Salpeter, M. M. (1966) General area of autoradiography at the electron microscope level. In *Methods in Cell Physiology* **2**. Academic Press, New York.

Stevens, A. R. (1966) High resolution autoradiography. In *Methods in Cell Physiology* **2**. Academic Press, New York.

CHAPTER 5

Albini, S. M., Jones, G. H. (1987) Synaptonemal complex spreading in *Allium cepa* and *A. fistulosum. Chromosoma* **95** 324-338.

Costas, E., Gyanes, V. J. (1987) Ultrastructure and division behaviour of dinoflagellate chromosomes. *Chromosoma* **95** 435-441.

Dille, J. E., Bittel, D. C., Ross, K., Gustafson, J. P. (1990) Preparing plant chromosomes for scanning electron microscopy. *Genome* **33** 333-339.

Dollin, A. E., Murray, J. D., Gillies, C. B. (1989) Synaptonemal complex analysis of hybrid cattle I. *Genome* **32** 856-864.

Dresser, M. E., Moses, M. J. (1980) Synaptonemal complex karyotyping in spermatocytes of chinese hamster. *Chromosoma* **76** 1-22.

Drouin, R., Messier, P. E., Richer, C. L. (1989) Dynamic G- and R-banding of human chromosomes for electron microscopy. *Chromosoma* **98** 40-48.

Fawcett, D. C. (1964) In *Histology and Cytology in modern developments in Electron Microscopy.* Academic Press, New York.

Gay, H., Anderson, T. F. (1954) Serial sections for electron microscopy. *Science* **120** 1071-1073.

Gillies, C. B. (1983) Spreading plant synaptonemal complexes for electron microscopy, *Kew Chromosome Conference* II, 115-122 **Brandham, P. E., Bennett, M. D.** (eds.) George Allen Unwin, London.

Haapala, O. (1985) Chromosome Axis In *Advances in chromosome and Cell Genetics* 173-201. **Sharma, A. K., Sharma, A.** (eds.) Gordon and Breach, London.

Hamkalo, B. A., Narayanaswami, S. (1985) *In situ* hybidisation at the EM level. In *Advances in chromosome and Cell Genetics* 203-218. **Sharma, A. K., Sharma, A.** (eds.) Gordon and Breach, London.

Harauz, G., Borland, L., Bahr, G. F., Zeitler, E., Van Heel, M. (1987) Three-dimensional reconstruction of a human metaphase chromosome from electron micrographs. *Chromosoma* **95** 366-374.

Harrison, C. J., Jack, E. M., Allen, T. D. (1987) Light and scanning electron microscopy of same metaphase chromosomes. In *Correlative microscopy in Biology : instrumentation and methods* 189-248. Academic Press, New York.

Jagiello, G., Sung, W. K., Van't Hof, J. (1983) Fiber DNA studies on premeiotic spermatogenesis. *Exptl. Cell Res.* **146** 281-287.

Jeffrey, J. Y., Geneix, A. (1974) *Humangenetik* **25** 119.

Kornberg, R. (1981) The location of nucleosomes is chromatin specific or statistical. *Nature* **292** 579-580.

Laskey, R. A., Earnshaw, W. C. (1980) Nucleosome assembly. *Nature* **286** 763-767.

Lewis, C. D., Laemmli, U. K. (1982) Higher order metaphase chromosome structure, evidence for metalloprotein interactions. *Cell* **29** 171.

Messier, P. E., Jean, P., Richer, C-L. (1986) Easy transfer of selected mitosis for light to electron microscopy. *Cytogenet. Cell Genet.* **43** 207-210.

Miller, O. L., Jr., Bakken, A. H. (1972) Morphological studies of transcription. *Acta Endocrinol.* Suppl. **168** 155-177.

Olins, A. L., Olins, D. E. (1974) Spheroid chromatin units (v Bodies). *Science* **143** 330-332.

Olins, A. L., Olins, D. E. (1990) Identification of 10nm non-chromatin filaments in the macronucleus of *Euplotes eurystomus. Chromosoma* **99** 205-211.

Paulson, J. R., Laemmli, U. K. (1977) The structure of histone depleted chromosomes. *Cell* **12** 817.

Peachey, L. D. (1958) *J. Biophys. Cytol.* **4** 322.

Stubblefeld, V., Wray, W. (1971) Architecture of the chinese hamster metaphase chromosomes. *Chromosoma* **22** 262.

Sumner, A. T. (1991) Scanning Electron Microscopy of mammalian chromosomes from prophase to telophase. *Chromosoma* **100** 400- 418.

Wanner, G., Formanek, H., Martin, R., Herrmann, R. G. (1991) High resolution scanning electron microscopy of plant chromosomes. *Chromosoma* **100** 103-109.

Zhao, J., Hao, S., Xing, M. (1991) The fine structure of mitotic core (scaffold) of *Trilophidia annulata. Chromosoma* **100** 323-329.

CHAPTER 6

Bales, G. W., Husenkampf, C. A. (1985) Culture of plant somatic hybrids following electrical fusion. *Theor. Appl. Genet.* **70** 227-233.

Bales, G. W., Biastuch, W., Riggs, C. D., Rabussay, D. (1988) Electroporation for DNA delivery to plant protoplasts. *Plant Cell Tissue and Organ Culture* **12** 213-218.

Carlson, P. S., Smith, H. H., Dearings, P. D. (1972) *Proc. Natl. Acad. Sci. US* **66** 2292.

Climelius, G. (1988) Potentials of protoplast fusion in plant breeding programme. *Plant Cell Tissue and Organ Culture* **12** 163-172.

DeFaat, A. D. M. M., Blaas, A. (1989) An improved method for protoplast microinjection suitable for transfer of entire plant chromosomes. *Plant Science* **50** 161-169.

Doyle, A., Griffiths, J. B., Newell, D. G. (eds.) (1993) *Cell and tissue laboratory procedures*, Wiley, New York.

Dudits, D. (1988) Asymmetric cell hybridization and isolated chromosome transfer. *Plant Cell Tissue and Organ Culture* **12** 205-211.

Galun, E., Aviv, D. (1986) In *Plant Molecular Biology Methods in Enzymology* **Weissbach, A., Weissbach, H.** (eds.) **118** 595- 611.

Giles, K. L. (1974) *Plant Cell Physiol.* **15** 281.

Grierson, D. (ed.) (1991) *Plant Genetic Engineering*, Chapman and Hall, London.

Griesbach, R. J., Malmberg, R. L., Carlson, P. S. (1982) Uptake of isolated lily chromosomes by tobacco protoplasts. *J. Hered.* **73** 151-152.

Griesbach, R. J. (1985) Advances in microinjection of higher plant cells. *Biotechnique* **3** 348-349.

Horsch, R. B., Fry, J., Hoffmann, N., Neidermeyer, J., Rogers, S. G., Fraley, R. T. (1988) Leaf disc transformation. *Plant molecular biology manual* **A5** 1-9.

Ichikawa, H., Tanno-Suenage, L., Imamwea, J. (1988) Formation of carrot cybrids based on metabolic complementation by fusion between X-irradiated cells and iodoacetamide treated cells. *Plant Cell Tissue and Organ Culture* **12** 201-204.

Jones, L. E., Hildebrandt, A. C., Riker, A. J., Wu, J. H. (1960) Growth of somatic tobacco cells in microculture. *Amer. Jour. Bot.* **47** 468.

Kao, K. N., Michayluk, M. R. (1974) A method for high frequency intergenetic fusion of plant protoplasts. *Planta* **115** 335-367.

Kung, S-D., Arntzen, C. J. (eds.) (1989) *Plant Biotechnology*, Butterworth, London.

Keller, W. A., Melchers, G. (1973) The effect of high pH and Calcium on tobacco leaf protoplast fusion. *Z. Naturforsch.* **280** 737.

Lee, M., Phillips, R. L. (1988) The chromosomal basis of somaclonal variation. *Ann. Rev. Plant Physiol. Plant Mol. Biol.* **39** 413-437.

Lee, C. H., Power, J. B. (1988) Inter and intra somatic gametosomatic hybridization within the genus *Petunia*. *Plant Cell Tissue and Organ Culture* **12** 197-200.

Murashige, T., Skoog, F. (1962) A revised medium for rapid growth and bioassays with tobacco tissue culture. *Physiologia Plant.* **15** 473-497.

Murata, M. (1983) Staining airdried protoplasts for study of plant chromosomes. *Stain Techn.* **58** 101-106.

Oard, J. H. (1991) Physical methods for the transformation of plant cells. *Biotech. Adv.* **9** 1-11.

Pen, W. H., Houben, A., Schlegel, R. (1993) Highly effective cell synchronization in plant roots by hydroxyurea, amiprophosphomethyl and colchicine, *Genome* (in press).

Paszkowshi, J., Schillito, R. D., Saul, M., Mandad, V., John, T., Hohn, V., Potrykus, J. (1984) Directed plasmid uptake into protoplasts through chemical and physical treatments. *EMBO J.* **3** 2717-2722.

Power, J. B., Cummins, S. E., Cocking, E. C. (1970) Fusion of isolated plant protoplasts. *Nature* **225** 1016-1018.

Reinert, J., Bajaj, Y. P. S. (1977) *Plant Cell, Tissue and Organ Culture.* Springer-Verlag, Berlin.

Samford, J. C., Devitt, M. J., Russel, J. A., Smith, F. D. *et al.* (1991) An improved helium-driven biolistic device. *J. Methods Cell Mol. Biol.* **3** 3-16.

Schieder, O. (1984) Isolation and culture of protoplasts (Datura). In *Cell culture and somatic cell genetics in plants* **Vasil, I. K.**, (ed.) 350-355. Academic Press, New York.

Sharma, A. K., Sharma, A. (1980) *Chromosome Techniques - Theory and Practice.* Butterworth, London.

CHAPTER 7

Adolph, R. W. (ed.) (1988) *Chromosome and Chromatin Structure* **2**. CRC Press, Boca Raton.

Ambrose, E. J., Easty, D. M., Wylie, J. A. H. (1967) *The Cancer cell in vitro.* Butterworth, London.

Bannasch, P. (1968) *Recent results in cancer research; the cytoplasm of hepatocytes during carcinogenesis.* Springer-Verlag, Berlin.

Barr, M. L. (1965) *Sex chromatin techniques.* In *Human chromosome methodology.* Academic Press, New York.

Barr, M. L., Bertram, E. G. (1949) A morphological distinction between neurons of the male and female and the behaviour of the nucleolar satellite during accelerated nucleoprotein synthesis. *Nature* **163** 676-677.

Bjorklund, B., Bjorklund, V., Paulsson, J. E. (1961) *Proc. Soc. Exp. Biol. Med.* **108**, 385.

Caspersson, T., Hulten, M., Lindsten, J., Zech, L. (1971) Identification of chromosome bivalents in human male meiosis by quinacrine mustard fluorescence analysis. *Hereditas* **67** 147-150.

Court-Brown, W. (1967) *Human Population Cytogenetics*. North-Holland, Amsterdam.

De Bruyn, W. M. (1956) Jaarboek van kankeronderzoek in kanker bestrijding in Nederland, 50.

De Bruyn, W. M., Hampe, J. F. (1961) Jaarbock van kanker onderzoek in Kankerbestriding in Nederland, 107.

Denton, T. E. (1973) *Fish chromosome methodology*. Thomas, Springfield, Illinois.

Eagle, H. (1955) Nutrition needs of mammalian cells in tissue culture. *Science* **122** 501.

Edwards, R. G. (1962) Meiosis in ovarian oocytes of adult mammals. *Nature* **196** 446-450.

Emery, A. E. H. (ed.) (1975) *Modern Trends in Human Genetics* **2**, 21. Butterworths, London.

Evans, E. P., Breckson, G., Ford, C. E. (1964) *Cytogenetics* **3**, 289.

Evans, H. J. (1984) Structure and organization of the human genome. In *Mutations in man*. **Obe, G.** (ed.) Springer-Verlag, Berlin.

Ford, C. E. (1961) Human Cytogenetics. *Brit. Med. Bull.* **17** 179.

Ford, C. E., Jacobs, P. A., Lajtha, L. G. (1958) Human somatic chromosomes. *Nature* **181** 565.

Ford, C. E., Evans, E. P. (1969) In *Comparative mammalian cytogenetics* **461**. **Benirschke, K.** (ed.) Springer-Verlag, Berlin.

Gardner, M. H., Punnett, H. H. (1964) An improved squash technique for human male meiotic chromosome softening, and concentration of cells, mounting in Hayer's medium. *Stain Techn.* **39** 245-248.

Griesbach, R. J. (1987) Advances in the microinjection of higher plants. *Biotechniques* **3** 348-350.

Hamerton, J. L. (1971) *Human Cytogenetics*. Academic Press, New York.

Harnden, D. G., Brunton, S. (1965) The skin culture technique. In *Human Chromosome Methodology*. Academic Press, New York.

Harris, H., Miller, O. J., Klein, G., Worst, P., Tachibana, T. (1969) Suppression of malignancy by cell fusion. *Nature* **223** 363-369.

Healy, G. M., Fisher, D. C., Parker, R. C. (1955) Nutrition of animal cells in tissue culture. Synthetic medium No. 858. *Proc. Soc. Exp. Biol. Med.* **89** 71-79.

Hsu, T. C. (1952) Mammalian chromosomes in vitro : the karyotype of man. *J. Hered.* **43** 167-172.

Hsu, T. C. (1965) In *Cells and tissues in culture* **1** 397. Academic Press, New York.

Hueper, W. C., Conway, W. D. (1964) *Chemical carcinogenesis and cancers.* Thomas, Springfield, Illinois.

Hungerford, D. A. (1965) Leucocytes cultured from small inocula of whole blood and the preparation of metaphase chromosomes by treatment with hypotonic KCl. *Stain Techn.* **40** 333.

Hungerford, D. A., Donnelly, A. J., Nowell, P. C., Beck, S. (1959) *Amer. J. Human Genet.* **11** 215.

Jagiello, G., Karnicki, J., Ryan, R. (1968) Cytogenetic observation in mammalian oocyte. *Adv. Human Genet.* **8** 347- 438, Plenum, New York.

Klein, G. (1966) *Viruses inducing cancer, implications for therapy,* Univ. of Utah Press, Salt Lake City.

Kligerman, A. D., Bloom, S. E. (1977) Rapid chromosome preparations from solid tissues of fishes. *J. Fish Res. Board Can.* **34** 266-269.

Koller, P. C. (1960) In *Cell physiology of neoplasia — 14th Sym. Fund. Cancer Res.* University of Texas Press, Texas.

Kroeger, H. (1966) In *Methods in Cell Physiology* **2** 61. Academic Press, New York.

Lampert, F. (1971) In *Advances in Cell and molecular biology* **1**. Academic Press, New York.

Lee, M. R. (1969) A widely applicable technique for direct processing of bone marrow for chromosomes of vertebrates. *Stain Techn.* **44** 155-158.

Lumsden, C. E. (1963) In *Pathology of tumours of the nervous system,* 281, 2nd ed. Arnold, London.

Makino, S. (1957) The chromosome cytology of the ascites tumors of rats with special reference to the concept of the stemline cell. *Int. Rev. Cytol.* **6** 25-84.

Mellman, W. J. (1965) Human peripheral blood leucocyte. In *Human Chromosome Methodology*. Academic Press, New York.

Moorhead, P. S., Nowell, P. C., Mellman, W. J., Battips, D. M., Hungerford, D. A. (1960) Chromosome preparation of leucocytes cultured from human peripheral blood. *Exptl. Cell Res.* **20** 613.

Nitsch, J. P., Nitsch, C. (1956) Auxin dependent growth of excised *Helianthus tuberosus* tissues. *Amer. Jour. Bot.* **43** 839-851.

Nowell, P. C., Hungerford, D. A., Brooks, C. D. (1958) *Proc. Amer. Assoc. Cancer Res.* **2** 331.

Ohno, S. (1965) Direct banding of germ cells. In *Human chromosome methodology* . Academic Press, New York.

Paul, J. (1959) *Cell and Tissue Culture*. Livingstone, Edinburgh.

Paulson, J. R., Laemmli, U. K. (1977) The structure of histone depleted metaphase chromosomes. *Cell* **12** 817.

Priest, J. H. (1969) *Cytogenetics*. Lea and Febiger, Philadelphia.

Raven, R. W., Roe, F. J. C. (eds.) (1967) *The Prevention of cancer*. Butterworths, London.

Rothfels, K. H., Siminovitch, L. (1958) An air drying technique for flattening chromosomes in mammalian cells grown *in vitro. Stain Techn.* **33** 73-77.

Sasaki, M. S., Makino, S. (1965) The meiotic chromosomes of man. *Chromosoma* **16** 637-651.

Sharma, A., Talukder, G. (1974) *Laboratory Procedures in Human Genetics, I. Chromosome methodology*. The Nucleus, Calcutta.

Talukder, G. (1985) In *Laboratory Manual, Centre of Advanced Study in Cell and Chromosome Research*, 73. Dept. of Botany, University of Calcutta.

Tarkowski, A. K. (1966) An airdrying method for chromosome preparations from mouse eggs. *Cytogenetics* **5** 394-400.

Tips, R. L., Smith, G. S., Meyer, D. L., Ushijima, R. N. (1963) *Texas Rep. Biol. Med.* **21** 581.

Turpin, R., Lejeune, J. (1969) *Human afflictions and chromosomal aberrations*. Pergamon Press, Paris.

Wachiel, A. W., Gettner, M. E., Ornstein, L. (1966) In *Physical techniques in biological research* **3A**, 173. Academic Press, New York.

Wada, B. (1966) Analysis of mitosis. *Cytologia* **30** suppl.

Waymouth, C. (1956) *J. Nat. Cancer Inst.* **17** 315.

Weber, B., Schempp, W., Wiesner, H. (1986) An evolutionarily conserved early replicating segment on the sex chromosomes of man and the great apes. *Cytogenet. Cell Genet.* **43** 72-78.

White, P. R. (1954) *The cultivation of animal and plant cells*. Ronald Press, New York.

White, P. R. (1959) In *The Cell* **1** 291. Academic Press, New York.

Yunis, J. J. (ed.) (1965 and 1974) *Human chromosome methodology*. Academic Press, New York.

CHAPTER 8

Barski, G., Sorieul, S., Cornefort, F. (1960) Production dans des cultures *in vitro* de deux souches cellulaires an association de cellules de caractere hybride. *Cr. hebd Seanc. Acad. Sci. Paris* **251** 18-25.

Dubbs, D. R., Kit, S. (1968) *Exptl. Cell Res.* **33** 19.

Harris, H. (1970) *Cell Fusion*. Clarendon Press, Oxford.

Vig, B. K., Athwal, R. S. (1989) Sequence of centromere separation: separation in a quasi-stable mouse-human somatic cell hybrid. *Chromosoma* **98** 167-173.

Weiss, M. C., Ephrussi, B., Scaletta. L. J. (1966/68) *Proc. Natl. Acad. Sci.* U.S. **59** 1132.

CHAPTER 9

Alink, G. M., Houdtvan, J. J. (1985) In *Mutagenicity Testing in Environmental Pollution* **Zimmerman, F. K., Taylor-Mater, R. E.** (eds.) Ellis Horwood, Chicester.

Auerbach, C., Robson, J. M. (1946) Chemical production of mutations and methods of inducing doubling of chromosomes in plants. *Nature* **157** 302.

Blakeslee, A. F., Avery, A. (1937) *J. Hered.* **28** 392.

Berg, K. (ed.) (1979) *Genetic damage in man caused by environmental agents*. Academic Press, New York, 510.

Brusick, D. J. (1987) Implications of treatment condition-induced genotoxicity for chemical screening and data interpretation. *Mutat. Res.* **189** 1-6.

Degrassi, F., Rizzoni, M. (1982) Micronucleus test in *Vicia faba* root-tips to detect mutagen damage in freshwater pollution. *Mutat. Res.* **97** 19-33.

Fishbein, L. (1984) Mutagens and carcinogens in the environment. In *Genetics: New Frontiers* **Chopra, V. L., Joshi, B. C., Sharma, R. P., Bansal, H. C.** (eds.) 3-42. Oxford and IBH, New Delhi.

Goodman, D. R., James, R. C., Herbison, R. D. (1985) Assessment of mutagenicity using germ cells and the application of test results. In *Reproductive toxicology* **Dixon, R. L.**, (ed.) Raven Press, New York.

Grant, W. F. (1982) Chromosome aberration assays in *Allium* — a report of the US Environmental Agency Gene Tox Programme. *Mutat. Res.* **99** 273-291.

Green, S., Auletta, A., Fabricant, J., Kapp, R., Manandhar, M., Chin-In-Sheu, Springer, J., Whitfield, B. (1985) Current status of bioassays in genetic toxicology. *Mutat. Res.* **154** 59-67.

Hsu, T. C. (ed.) (1982) *Cytogenetic assays of environmental mutagens.* Allanheld, Osmun, Totowa, New Jersey, 430.

Huskins, C. L., Steinitz, L. M. (1948) The nucleus in differentiation and development. *J. Hered.* **39** 34-77.

Kihlman, B. A. (1975) Root-tips of Vicia faba for the study of the induction of chromosomal aberrations. *Mutat. Res.* **31** 401-412.

Latt, S. A., Schreck, R. R., Sahar, E., Parks, I. J., Kittral, C. (1982) In *Environmental Mutagens and Carcinogens.* University of Tokyo Press, Tokyo and Alan R. Liss, New York, 331.

Levan, A. (1949) The influence on chromosomes and mitosis of chemicals, as studied by the Allium test. 8th International Congr. Genetics. *Hereditas* Suppl. vol. 325- 337.

Naismith, R. W. (ed.) (1987) Guidelines for minimal criteria of acceptability for selected short-term assays for genotoxicity. *Mutat. Res.* **189** 81-183.

Oehlkers, F. (1943) Die Auslosung von chromosomenmutationen in der meiosisdurde. Einwerkung von chemikaleen. *Zl. A. V.* **81** 313-341.

Oftedal, P., Brögger, A. (eds.) (1986) *Risk and reasons: risk assessment in relation to environmental mutagens and carcinogens.* Alan R. Liss, New York, 189.

Ooka, T. (1976) In *Methods in Cell Biology* **Prescott, D. M.** (ed.) **14** 287. Academic Press, New York.

Parry, J. M. (1985) In *Mutagenicity Testing in Environmental Pollution Control.* Ellis Horwood, Chicester.

Plewa, M. J. (1982) Maize as a monitor for environmental mutagens. In *Environmental mutagens and carcinogens* **Sugimura, T., Kondo, S., Takebe, H.** (eds.) 411-419. University of Tokyo Press, Tokyo.

Preston, R. J., San Sebastian, J. R., McFee, A. F. (1987) The *in vitro* lymphocyte assay for assessing the clastogenicity of chemical agents. *Mutat. Res.* **189** 175-183.

Schvartzman, J. B. (1987) Sister chromatid exchanges in higher plant cells: past and perspectives. *Mutat. Res.* **182** 127- 145.

Sharma, A. (1984) *Environmental chemical mutagenesis.* Persp. Rep. Ser. 6, Golden Jubilee Publ. Indian National Science Academy, New Delhi, 53.

Sharma, A. (1986) Higher plants as cytogenetic monitors for chemical agents. *J. Ind. Bot. Soc.* **64** 9-16.

Sharma, A., Talukder, G. (1987) Effects of metals on chromosomes of higher organisms. *Environmental Mutagenesis* **9** 191-226.

Sen, S. (1970) *Res. Bull.* 2. Dept of Botany, University of Calcutta, Calcutta.

Sharma, A. K. (1978) Change in chromosome concept. *Proc. Ind. Acad. Sci.* **87B** 161-190.

Sharma, A. K. (1981) *Impact of the development of science and technology on environment.* Address by General President, 66th Ind. Sci. Congress, Varanasi, 1-41.

Sharma, A. K. (1987) Genetic Toxicology. *Indo-US Seminar on genetic toxicology, Calcutta,* Nov. 16-18. *The Nucleus* **30** 159-160.

Sharma, A. K., Mookerjea, A. (1954) Induction of division in cells - a study of the causal factors involved. *Bull. Bot. Soc. Bengal* **8** 24-100.

Sharma, A. K., Roy, M. (1956) Chemical constitution and enzyme activity of chromosomes and related structures. *La Cellule* **58** 109-133.

Sharma, A. K., Sen, S. (1954) Study of the effect of water on nuclear constituents. *Genet. Iber.* **6** 19-32.

Sharma, A. K., Sharma, A. (1960) Spontaneous and chemically induced chromosome breakage. *Internat. Rev. Cytol.* **10** 101- 136.

Sharma, A. K., Sharma, A. (1980) *Chromosome Techniques — Theory and Practice,* 3rd ed. Butterworth, London.

Snell, G. D. (1963) In *Conceptual advances in immunology and oncology,* 323. Harper, New York.

Sobels, F. H. (1987) *In vitro* testing as a step in the evaluation of *in vivo* toxicity. *Mutat. Res.* **18** 7-10.

Sorsa, M., Hemminki, K., Vainio, H. (1982) Biological monitoring of exposure to chemical mutagens in the occupational environment. *Teratogenesis, Carcinogenesis and Mutagenesis* **2** 137-150.

Sugimura, T., Kondo, S., Takebe, H. (eds.) (1982) *Environmental mutagens and carcinogens.* Alan R. Liss, New York.

Torrey, J. C. (1961/67) *Phys. Pl.* **20** 265.

Valentine, R. C., Horne, R. W. (1962) In *Symposia for the Society of Cell Biology* **1**. Academic Press, New York.

Vant Hof, J., Schaire, L. A. (1982) Tradescantia assay system for gaseous mutagens US : EPA Gentox Programme. *Mutat. Res.* **99** 293-302.

Walk, R. A., Jenderney, J., Rohrborn, G., Hackenberg, U. (1987) Chromosomal abnormalities and sister chromatid exchanges in bone marrow cells of mice and chinese hamster after inhalation and intraperitoneal administration. *Mutat. Res.* **182** 333-342.

Wolf, S. (1974) Cytogenetics and environmental mutagens. In *Genetics: New Frontiers* **Chopra, V. L.** *et al.* (eds.) Oxford and IBH, New Delhi, 83.

Zimmerman, F. K., Taylor-Meyer, R. E. (1985) *Mutagenicity testing in environment pollution control.* Ellis Horwood, Chicester, 47-68.

CHAPTER 10

Boyd, J. B. (1975) In *Methods in Cell Biology* **Prescott, D. M.** (ed.) **10** 135. Academic Press, New York.

Brown, T. A. (ed.) (1991) *Essentials in molecular biology — a practical approach* vols. **1** and **2**. IRL Press, Oxford.

Busch, H. (1967) In *Methods in Enzymology* **Grossman, L., Moldev, K.** (eds.) **12** 439. Academic Press, New York.

Davis, R. W., Thomas, M., Cameron, J., St John, T. P., Scherer, S., Padgett, R. A. (1980) Rapid DNA isolation for enzymatic and hybridization analysis. *Methods in Enzymology* **65** 404-411.

Dellaporta, S. L., Wood, J., Hicks, J. B. (1983) A plant DNA minipreparation : version II. *Plant Molecular Biology Reporter* **1**(4) 19-21.

Doyle, J. (1991) *DNA protocols for plants* NATO ASI series H57. In *Molecular Techniques in Taxonomy*, **Hewitt, G. M. et al**, (eds.) Springer-Verlag, Berlin.

Dolezel, J., Cihalikova, J., Lucretti, S. (1992) A high yield procedure for isolation of metaphase chromosomes from root tips of *Vicia faba*. *Planta* **188** 93-98.

Edstrom, J. E., Kaiser, R., Rohme, D. (1987) Microcloning of mammalian metaphase chromosomes. *Methods Enzymol.* **151** 503-516.

Galun, E., Aviv, D. (1986) In *Plant Molecular Biology : Methods in Enzymology* **Weissbach, A., Weissbach, H.** (eds.) **188** 595-611, Academic Press, New York.

Gillies, C. B. (1983) Spreading plant synaptonemal complexes for electron microscopy. *Kew Chromosome Conference*. **Brandham, P. E., Bennett, M. D.** (ed.) George Allen & Unwin, London 115-122.

Griesbach, R. J., Sink, K. (1983) Evacuolation of mesophyll protoplasts. *Plant Sci. Lett.* **30** 29-30.

Griesbach, R. J. (1985) Advances in microinjection of higher plant cells. *Biotechnique* 3 348-349.

Griesbach, R. J., Malmberg, R. L., Carlson, P. S. (1982) An improved technique for the isolation of higher plant chromosomes. *Plant Science Letters* **24** 55-60.

Hadlaczky, G. (1984) Isolation of organelles : Chromosomes. In *Cell culture and somatic cell genetics of plants* vol. **1**, Chapter 53, 461-470, Academic Press, New York.

Hadlaczky, G., Bisztray, G., Praznovsky, T., Dudits, D. (1983) Mass isolation of plant chromosomes and nuclei. *Planta* **157** 278-285.

Hadlaczky, Gy., Went, M., Ringertz, N. R. (1986) Direct evidence of the non-random localization of mammalian chromosomes in the interphase nucleus. *Exptl. Cell Res.* **167** 1-15.

Hagag, N. G., Viola, M. V. (1993) *Chromosome microdissection and cloning: a practical guide*. Academic Press, San Diego, CA, USA.

Maio, J. J., Schildkraut, C. L. (1966) In *Methods in Cell Physiology* **2** Academic Press, New York.

Maniatis, T., Fritsch, E. F., Sambrook, J. (1982) *Molecular cloning : a laboratory manual*. Cold Spring Harbor Laboratory Press, New York.

Samford, J. C., Devitt, M. J., Russel, J. A., Smith, F. D., Haepending, R. R., Ray, M. K., Johnston, S. A. (1991) An improved helium driven biolistic device. *J. Methods Cell Mol. Biol.* **3** 3-16.

Schubert, I., Dolezel, J., Houben, A., Scherthan, H., Wanner, G. (1993) Refined examination of plant metaphase chromosome structure at different levels made feasible by new isolation methods. *Chromosoma* **102** 96-101.

Sonnebichler, J., Machikao, F., Zetl, I. (1977) In *Methods in Cell Biology* **Prescott, D. M.**, (ed.) **15** 150. Academic Press, New York.

Travaglini, E. C. (1973) In *Methods in Cell Biology* **Prescott, D. M.** (ed.) Academic Press, New York.

Wanner, G., Formanek, H., Martin, R., Herrmann, R. G. (1991) High resolution electron microscopy of plant chromosomes. *Chromosoma* **100** 103-109.

Wray, W. (1977) In *Methods in Cell Biology* **Prescott, D. M.**, (ed.) **15** 111. Academic Press, New York.

Wray, W. (1973) In *Methods in Cell Biology* **Prescott, D. M.**, (ed.) **6** 283. Academic Press, New York.

Wray, W. (1975) Parallel isolation procedures for metaphase chromosomes, mitotic apparatus and nuclei. *Methods Enzymol.* **40** 75-89.

Wu, H. K., Chung, M. C., Wu, T., Ning, C. N., Wu, R. (1991) Localisation of specific repetitive DNA sequences in individual rice chromosomes. *Chromosoma* **100** 330-338.

Wu, T. Y., Wu, R. (1987) A rice repetitive DNA shows sequence homology to both 5S RNA and tRNA. *Nucleic Acids Res.* **15** 5913-5923.

Xie, Y., Wu, R. (1989) Rice alcohol dehydrogenase genes : anaerobic induction, organ specific expression and characterisation of cDNA clones. *Plant Mol. Biol.* **13** 53-68.

Zhao, X., Wu, T., Xie, Y., Wu, R. (1989) Genome-specific repetitive sequences in the genus Oryza. *Theor. Appl. Genet.* **78** 201-209.

CHAPTER 11

Fukui, K., Minezawa, M., Kamisugi, Y., Yanagisawa, T., Fujishita, M., Sakai, F. (1991) Microdissection of barley chromosomes by cell work station. *Barley Genet.* **6** 272-276.

Fukui, K., Minezawa, M., Kamisugi, Y., Ishikawa, M., Ohmido, N., Yanagisawa, T., Fujishita, M., Sakai, F. (1992) Microdissection of plant chromosomes by argon ion laser beam. *Theor. Appl. Genet.* **84** 787-791.

Hearne, C. H., Ghosh, S., Todd, J. A. (1992) Microsatellites for linkage analysis of genetic traits. *Theor. Appl. Genet.* **8** 288-293.

Kamisugi, Y., Sakai, F., Minezawa, M., Fujishita, M., Fukui, K. (1993) Recovery of dissected C-band regions in *Crepis* chromosomes. *Theor. Appl. Genet.* **85** 825-828.

Kao, F. T., Yu, J. W. (1991) Chromosome microdissection and cloning in human genome and genetic disease analysis. *Proc. Natl. Acad. Sci. USA* **88** 1844-1848.

Meltzer, P. S., Guan, X. Y., Burgess, A., Trent, J. M. (1990) Rapid generation of region specific probes by chromosome microdissection and their application. *Nature Genetics* **1** 24-28.

Monajembashi, S., Cramer, C., Cramer, T., Wolfram, J., Greulich, K. O. (1986) Microdissection of human chromosomes by a laser micro beam. *Exptl. Cell Res.* **167** 262-265.

Neuhaus, G., Spangenberg, G., Scheid, O. M., Schweiger, H. G. (1987) Transgenic rapeseed plants obtained by the microinjection of DNA into microspore derived embryoids. *Theor. Appl. Genet.* **75** 30-36.

Ponelles, N., Bautz, E.K.F., Monajembashi, S., Wolfram, J., Greulich, K. O. (1989) Telomeric sequence derived from laser microdissected polytene chromosomes. *Chromosoma* **98** 351-357.

Scalenghe, F., Tureo, E., Edstrom, J. E., Pirotte, V. (1981) Microdissection and cloning of DNA from a specific region of *Drosophila melanogaster* polytene chromosomes. *Chromosoma* **82** 205-216.

Schondelmaier, J., Martin, R., Jahoor, A., Houben, A., Graner, A., Koop - H. U., Herrmann, R. G., Jung, C. (1993) Microdissection and microcloning of the barley chromosome 1 HS. *Theor. Appl. Genet.* **86** 629-636.

Spangenberg, G., Koop, H. U. (1992) Low density cultures, microdroplets and single cell nurse cultures. In *Plant tissue culture manual* Suppl. **1A Lindsey, K.** (ed.) Kluwer Academic, Boston.

CHAPTER 12

Amabis, J. M., Amabis, D. C., Kaburaki, J., Stollar, B. D. (1990) The presence of an antigen reactive with a human autoantibody in *Trichosia pubescence* (Diptera: Sciaridae) and its association with certain transcriptionally active regions of the genome. *Chromosoma* **99** 102-110.

Appels, R., Dvorak, J. (1982) The wheat ribosomal spacer region: its structure and variation in populations among species. *Theor. Appl. Genet.* **63** 337-348.

Attenburg, L. C., Getz, M. J., Saunders, G. F. (1975) In *Methods in Cell Biology* **Prescott, D. M.,** (ed.) **10** 325. Academic Press, New York.

Bedbrook, J. R., Jones, J., O'Dell, M., Thompson, R. J., Flavell, R. B. (1980) A molecular description of telomeric heterochromatin in *Secale* species. *Cell* **19** 545-560.

Bennett, S. T. (1993) DNA-DNA *in situ* hybridisation protocol (personal communication).

Bennett, S. T., Kenton, A. Y., Bennett, M. D. (1992) Genomic *in situ* hybridization reveals the allopolyploid nature of *Milium montianum* (Gramineae). *Chromosoma* **101** 420-424.

Brown, T. A. (ed.) (1991) *Essential molecular biology — a practical approach*, vols. **1** and **2**. IRL Press, Oxford.

Burkholder, G. D., Latimer, L. J. P., Lee, J. S. (1988) Immunofluorescent staining of mammalian nuclei and chromosomes with a monoclonal antibody to triplex DNA. *Chromosoma* **97** 185-192.

Chumakov, I. M., LeGull, I., Billault, A., Ongen, P., Soularne, P., Guillon, S. ***et al.*** (1992) Isolation of chromosome 21- specific YAC from a total human complement. *Nature Genetics* **1** 222-225.

Degrassi, F., Tanzarella, C. (1988) Immunofluorescent staining of kinetochores in micronuclei: a new assay for the detection of aneuploidy. *Mutat. Res.* **203** 339-345.

Drouin, R., Messier, P. E., Richer, C. L. (1989a) DNA denaturation for ultrastructural banding and the mechanism underlying the fluorochrome-photolysis-Giemsa technique studied with anti-5- bromodeoxyuridine antibodies. *Chromosoma* **98** 174-180.

Drouin, R., Messier, P. E., Richer, C. L. (1989b) Dynamic G and R banding of human chromosomes for electron microscopy. *Chromosoma* **98** 40-48.

Ehrlich, H. A., Gelfand, D., Sninsky, J. J. (1991) Recent advances in the polymerase chain reaction. *Science* **252** 1643-1651.

Earnshaw, W. C., Heck, M. M. S. (1985) Localisation of topoisomerase II in mitotic chromosomes. *J. Cell Biol.* **100** 1716-1725.

Fenech, M., Morley, A. A. (1989) Kinetochore detection in micronuclei : an alternative method for measuring chromosome loss. *Mutagenesis* **4** 98-104.

Gall, J. C., Pardue, M. L. (1971) Nucleic acid hybridization in cytological preparations. *Meth. Enzymol* **21** 470-480.

Gimenez-Martin, G., Lopez-Saez, J. F., Moreno, P., Gonzales-Fernandes, A. (1968) On the triggering of mitosis and the division cycle of polynucleate cell. *Chromosoma* **25** 282-296.

Goday, C., Pimpinelle, S. (1989) Centromere organization in meiotic chromosomes of *Parascaris univalens. Chromosoma* **98** 160-166.

Gustafson, J. P., Butler, E., McIntyre, C. L. (1990) Physical mapping of a low copy DNA sequence in rye. *Proc. Natl. Acad. Sci. USA* **87** 1899-1902.

Haaf, A., Muller, H., Schmid, M. (1986) Distamyin A (DA)/DAPI staining of heterochromatin in male meiosis of man. *Genetica* **70** 179-185.

Hadlaczky, G., Went, M., Ringertz, N. R. (1986) Special article. *Exptl. Cell Res.* **167** 1-5.

Hadlacky, G., Praznovszky, T., Rasko, I., Kereso, J. (1989) Centromere proteins, I. *Chromosoma* **97** 282-288.

Hennig, U. G. G., Rudd, N. L., Hoar, D. I. (1988) Kinetochore immuno-fluorescence in micronuclei: a rapid method for the *in situ* detection of aneuploidy and chromosome breakage in human fibroblasts. *Mutat. Res.* **203** 405-414.

Heslop-Harrison, J. S., Schwarzacher, T., Anamthawat-Jonsson, K., Leitch, A. R., Shi, M., Leitch, I. J. (1991) *In situ* hybridisation with automated chromosome denaturation. *Technique-a journal of methods in Cell and Molecular Biology* 3 109-116.

Hodgson, J. (1991) Micro-imaging: *in vivo veritas. Biotechnology* 9 353-356.

Howes, N. K., Lukow, O. M., Dawood, M. R., Bushuk, W. (1989) A rapid detection of the 1BL/1BS chromosome translocation in hexaploid wheats by monoclonal antibodies. *J. Cereal Sci.* **10** 1-4.

Huang, P. L., Hahlbrock, K., Somssich, I. E. (1988) Detection of a single copy gene on plant chromosomes by *in situ* hybridization. *Mol. Gen. Genet.* **211** 143-147.

Hutchinson, J., Flavell, R. B., Jones, J. (1981) Physical mapping of plant chromosomes by *in situ* hybridisation. In *Genetic Engineering - Principles and Methods* 3 202-221. Plenum, New York.

Ikushima, T. (1990) Bimodal induction of SCE by luminol during the S phase of cell cycle. *Chromosoma* 99 360-364.

Kakeda, K., Fukui, K., Yamagata, H. (1991) Heterochromatic differentiation in barley chromosomes revealed by C and N banding techniques. *Theor. Appl. Genet.* **81** 144-150.

Kakeda, K., Yamagata, H., Fukui, K., Ohno, M., Fukui, K., Wei, Z. Z., Zhu, F. S. (1990) High resolution bands in maize chromosomes by G banding methods. *Theor. Appl. Genet.* **80** 265-272.

Kamisugi, Y., Ikeda, Y., Ohno, M., Minezawa, M., Fukui, K. (1992) *In situ* digestion of barley chromosomes with restriction endonucleases. *Genome* **35** 793-798.

Kamisugi, Y., Fukui, K. (1990) Automatic karyotyping of plant chromosomes by imaging techniques. *Bio Techniques* **8** 290- 295.

Kingwell, B., Rattner, J. B. (1987) Mammalian kinetochore centromere composition of 50 kDA antigen is present in the mammalian kinetochore centromere. *Chromosoma* **95** 403-407.

Koch, J. E., Kolvraa, S., Petersen, K. B., Gregersen, N., Bolund, L. (1989) Oligonucleotide priming methods for the chromosome- specific labelling of alpha satellite DNA *in situ. Chromosoma* **98** 259-265.

Knox, R. E., Howes, N. K., Aung, T. (1992) Application of chromosome specific monoclonal antibodies in wheat genetics. *Genome* **35** 831-837.

Lawrence, J. B., Villnave, C. A., Singer, R. H. (1988) Sensitive, high resolution chromatin and chromosome mapping *in situ*: presence and orientation of two closely integrated copies of EBV in a lymphoma line. *Cell* **52** 51-61.

Leitch, I. J., Leitch, A. R., Heslop-Harrison, J. S. (1991) Physical mapping of plant DNA sequences by simultaneous *in situ* hybridization of two differently labelled fluorescent probes. *Genome* **34** 329-333.

Leitch, I. J., Leitch, A. R., Schwarzacher, T., Maluszynska, J., Anamthawat-Jonsson, K., Shi, M., Harrison, G., Heslop-Harrison, J. S. (1992) Two colour mapping of plant DNA sequences using digoxigenin and biotin. *Update* brochure.

Leitch, A. R., Mosgoller, W., Schwarzacher, T., Bennett, M. D., Heslop-Harrison, J. S. (1990) Genomic *in situ* hybridization to sectioned nuclei shows chromosome domains in grass hybrids. *J. Cell Science* **95** 335-341.

Lozano, R., Sentis, C., Fernandez-Piqueras, J., Ruiz Rejon, M. (1991) *In situ* digestion of satellite DNA of *Scilla sibirica. Chromosoma* **100** 439-442.

Lucas, J. N., Mullikin, J. C., Gray, J. W. (1991) Dicentric chromosome frequency analysis using slit-scan flow cytometry. *Cytometry* **12** 316-322.

Maniatis, T., Fritsch, E. F., Sambrook, J. (1986) *Molecular cloning: a laboratory manual.* Cold Spring Harbor Laboratory, Cold Spring Harbor, New York.

Manuelidis, L., Langer-Safer, P. R., Ward, D. C. (1982) High resolution mapping of satellite DNA using biotin-labeled DNA probes. *J. Cell Biol.* **95** 619-625.

Meltzer, P. S., Guan, X. Y., Burgess, A., Trent, J. M. (1992) Rapid generation of region specific probes by chromosome microdissection and their application. *Nature Genetics* **1** 24-28.

Miller, D. A. Choi, Y. C., Miller, O. J. (1983) Chromosome localisation of highly repetitive human DNAs and amplified ribosomal DNA with restriction enzymes. *Science* **219** 395-397.

Miller, O. L. Jr., Bakken, A. H. (1972) Morphological studies of transcription. *Acta Endocrinol. Suppl.* **168** 155-177.

Moens, P. B., Earnshaw, W. C. (1989) Anti-topoisomerase II recognises meiotic chromosome cores. *Chromosoma* **98** 317- 322.

Moens, P. B., Pearlman, R. E. (1990) Telomere and centromere DNA are associated with the cores of meiotic prophase chromosomes. *Chromosoma* **100** 8-14.

Moyzis, R. K., Albright, K. L., Bartholdi, M. F. *et al.* (1987) Human chromosome specific repetitive DNA sequences: novel markers for genetic analysis. *Chromosoma* **95** 375-386.

Mukai, Y., Endo, T. R., Gill, B. S. (1990) Physical mapping of 5s rRNA multigene family in common wheat. *J. Hered.* **81** 290- 295.

Murata, M. (1983) Staining air dried protoplasts for study of plant chromosomes. *Stain Techn.* **58** 101-106.

Murray, V., Martin, R. F. (1987) Nucleotide sequences of human alpha DNA repeats. *Gene* **57** 255-259.

Ohnuki, Y. (1968) Structure of chromosomes, I. Morphological study of the spiral structure of human somatic chromosomes. *Chromosoma* **25** 402-428.

Palevitz, B. A. (1990) Kinetochore behaviour during generative cell division in *Tradescantia virginiana. Protoplasma* **157** 120-127.

Palmer, D. K., O'Day, K., Margolis, R. L. (1990) The centromere specific histone CENPA is selectively retained in discrete foci in mammalian sperm nuclei. *Chromosoma* **100** 32-36.

Pankov, R., Lemieux, M., Hancock, R. (1990) An antigen located in the kinetochore region in metaphase and on polar microtubule ends in the midbody region in anaphase, characterised using a monoclonal antibody. *Chromosoma* **99** 95-101.

Pardue, M. L., Gall, J. G. (1975) In *Methods in Cell Biology* **Prescott, D. M.** (ed.) **10** Academic Press, New York.

Pinkel, D., Straume, T., Gray, J. W. (1986) Cytogenetic analysis using quantitative, high sensitivity fluorescence hybridization. *Proc. Natl. Acad. Sci. USA* **83** 2934-2938.

Prantera, G., Ferraro, M. (1990) Analysis of methylation and distribution of CPG sequences on human active and inactive X chromosomes by *in situ* nick translation. *Chromosoma* **99** 18-23.

Rawlins, D. J., Highett, M. I., Shaw, P. J. (1991) Localization of telomeres in plant interphase nuclei by *in situ* hybridization and 3D confocal microscopy. *Chromosoma* **100** 424-431.

Rayburn, A. L., Gill, B. S. (1985) Use of biotin-labeled probes to map specific DNA sequences on wheat chromosomes. *J. Heredity* **76** 78-81.

Rigby, P. W. J., Dieckmann, M., Rhodes, C., Berg, P. (1977) Labeling deoxyribonucleic acid to high specific activity *in vitro* by nick translation with DNA polymerase I. *J. Mol. Biol.* **113** 237-251.

Sabbath, M., Anderson, B. (1977) In *Methods in Cell Biology* **Prescott, D. M.** (ed.) **15** 435. Academic Press, New York.

Saiki, R. K., Scharf, S., Fallona, F., Mullis, K. D., Horn, G. T., Erlich, H. A., Arnheim, N. (1988) *Science* **239** 481-491.

Sambrook, J., Fritsch, E. F., Maniatis, T. (1989) *Molecular Cloning — a laboratory manual* **1**. Cold Spring Harbor Press, New York.

Santoz, C. S. J., Fernandez-Piqueras, J. (1989) *In situ* methylation of insect chromosomes with methylase HpaII. *Chromosoma* **98** 105-108.

Schellander, K., Mayr, B., Kalat, M., Fuhrer, F., Keefer, C. L., Brackett, B. G. (1990) A DNA-specific technique for pig and cattle oocytes using counter-stain-enhanced fluorescence. *Caryologia* **48** 117-130.

Schmid, M., Steinlein, C., Friedl, R., De Almeida, C. E., Haaf, T., Hillis, D. M., Duellman, W. E. (1990) Chromosome banding in amphibia, xv. Two types of Y chromosomes and heterochromatin hypervariability in *Gastrotheca pseustes* (Anura, Hylidae). *Chromosoma* **99** 413-423.

Schwarzacher, H. G. (1976) *Chromosomes* Springer - Verlag, Berlin

Schwarzacher, T., Anamthawat-Jonsson, K., Harrison, G. E., Islam, A. K. M. R., Jia, J. Z., King, I. P., Leitch, A. R., Miller, T. E., Reader, S. M., Rogers, W. J., Shi, M., Heslop-Harrison, J. S. (1992) Genomic *in situ* hybridization to identify alien chromosomes and chromosome segments in wheat. *Theor. Appl. Genet.* **84** 18.

Schwarzacher, T., Leitch, A. R., Bennett, M. D., Heslop-Harrison, J. S. (1989) *In situ* localisation of parental genomes in a wide hybrid. *Ann. Bot.* **64** 315-324.

Schweizer, D. (1980) Simultaneous fluorescent staining of R-band and specific heterochromatin regions (DA/DAPI) bands in human chromosomes. *Cytogenet. Cell Genet.* **27** 190-193.

Shen, D., Wang, Z., Wu, M. (1987) Gene mapping on maize pachytene chromosomes by *in situ* hybridization. *Chromosoma* **95** 311-314.

Steffenson, D. M., Wimber, D. E. (1972) In *Nucleic Acid Hybridization in the Study of Cell Differentiation* **Ursprung, H.**, (ed.) **3**, 47. Springer - Verlag , Berlin.

Sumner, A. T. (1987) Immunocytochemical demonstration of kinetochores in human sperm heads. *Exptl. Cell Res.* **171** 250-253.

Tagle, D. A., Swaroop, M., Lovett, A. M., Collins, F. S. (1993) Magnetic bead capture of expressed sequences encoded within large genomic segments. *Nature* **361** 751-753.

Taniguchi, K., Tanaka, R. (1991) Visualization of replicating bands in plant chromosomes with a monoclonal anti-BrdU-antibody method. *Japan. J. Genet.* **66** 485-489.

Trask, B. J. (1991) Fluorescence *in situ* hybridization: applications in cytogenetics and gene mapping. *Trends in Genetics* **7** 149-154.

Turner, B. M., Keohane, A. (1987) Antibody labelling and flow cytometric analysis of metaphase chromosomes reveals two discrete structural forms. *Chromosoma* **95** 263-270.

Ueda, T., Kato, Y., Irie, S. (1990) Regional differences in immunostainability of isolated metaphase chromosomes of Indian muntjac with anti-Z DNA antibody. *Chromosoma* **99** 161-168.

Vig, B. K., Athwal, R. S. (1989) Sequence of centromere separation : separation in a quasi-stable mouse-human somatic cell hybrid. *Chromosoma* **98** 167-173.

Visa, N., Marfany, G., Vilageliu, L., Albalat, R., Atrian, S., Gonzalez-Duarte, R. (1991) The Adh in Drosophila: Chromosomal location and restriction analysis in species with different phylogenetic relationships. *Chromosoma* **100** 315-322.

Vogel, W., Autenrieth, M., Mehnert, K. (1989) Analysis of chromosome replication by a BrdU antibody technique. *Chromosoma* **98** 335-341.

Vogel, W., Autenrieth, M., Speit, G. (1986) Detection of bromodeoxyuridine incorporation in mammalian chromosomes by a bromodeoxyuridine - antibody. *Human Genet.* 72 129-132.

Weier, H. U. G., Lucas, J. N., Poggensee, M., Segraves, R., Pinkel, D., Gray, J.W. (1991) Two color hybridization with high complexity chromosome-specific probes and a degenerate alpha satellite probe DNA allows unambiguous discrimination between symmetrical and asymmetrical translocations. *Chromosoma* **100** 371-376.

Willard, H. F., Waye, J. S. (1987) Hierarchial order in chromosome specific human alpha satellite DNA. *Trends in Genetics* **3** 192- 198.

Yoneda, A., Yoneda, Y., Kaneda, Y., Hayes, H., Uchida, T., Okada,Y. (1991) Monoclonal antibodies specific for human chromosome 5 obtained with a monochromosomal hybrid can be used to sort out cells containing the chromosome with a FACS. *Chromosoma* **100** 187-192.

Zhang, F. R., Heilig, R., Thomas, G., Aurias, A. (1990) A one-step efficient and specific non-radioactive non-fluorescent method for *in situ* hybridization of banded chromosomes. *Chromosoma* **99** 436-439.

Zischler, H., Schaefer, R., Eppelen, J. T. (1989) *Nucleic Acids Res.* **17** 1411.

CHAPTER 13

Ajioka, J. W., Smoller, D. A., Jones, R. W., Carnette, J. P., Vellek, A. E. C., Garza Delink, A. J., Duncan, I. W., Hartl, D. L. (1991) *Drosophila* genome project one hit coverage in Yeast Artificial Chromosome. *Chromosoma* **100** 495-509.

Ali, S., Muller, C. R., Eppelen, J. T. (1986) *Hum. Gen.* **74** 239- 243.

Ali, S., Eppelen, J. T. (1991) DNA fingerprinting of eukaryotic genomes by synthetic oligodeoxyribonucleotide probes. *Ind. Jour. Biochem. Biophys.* **28** 1-9.

Ali, S., Wallace, R. B. (1988) *Nucleic Acid Res.* **16** 848-849 and **Ali, S., Wallace, R. B.** (1989) *Anal. Biochem.* **179** 280-283. In **Ali S.** and **Eppelen J. T.** (1991).

Bell, J. (1989) Chromosome crawling in the MHC. *Trends in Genetics* **5** 289-290.

Bellane-Chantellot, C., Lacroix, B., Ougen, P. *et al.* (1992) Mapping the whole human genome by fingerprinting yeast artificial chromosomes. *Cell* **70** 1059-1068.

Bernatzky, R. (1988) Restriction fragment length polymorphism. *Plant Molecular Biology Manual* **c2** 1-18.

Chakravorty, R., Kidd, K. K. (1991) The utility of DNA typing in forensic work. *Science* **254** 1735-1739.

Chang, C., Bowman, J. L., Dejohn, A. W., Lander, E. S., Meyerowitz, E. M. (1988) Restriction fragment length polymorphism linkage map for *Arabidopsis thaliana. Proc. Natl. Acad. Sci. USA* **81** 1991-1994.

Di Franco, C., Pisano, C., Dimitri, P., Gigliotte, S., Junakovik, V. (1989) Genomic distribution of Copia-like transposable elements in somatic tissues and during development in Drosophila melanogaster. *Chromosoma* **98** 402-410.

Edwards, K. J., Thompson, H., Edwards, D., De Saizieu, A., Sparks, C., Thompson, J. A., Greenland, A. J., Yers, M., Schuck, W. (1992) Construction and characterization of a Yeast Artificial Chromosome library containing three haploid maize genome equivalents. *Plant Mol. Biol.* **19** 299-308.

Eppelen, J. T., Kammerbauer, C., Steimle, V., Zischler, H., Albert, E., Andrews, A., Hala, K., Nanda, I., Schmid, M., Riess, O., Weissing, K. (1989) *Electrophoresis Forum* **89** 179.

Farr, C. J., Stevanovic, M., Thomson, E. J., Goodfellow, P. N., Cooke, H. J. (1992) Telomere-associated chromosome fragmentation. *Nature Genetics* **2** 275-282.

Fukui, K. (1986) CHIAS - Standardization of karyotyping plant chromosomes by a newly developed chromosome image analyzing system. *TAG* **72** 27-32.

Fukui, K., Kakeda, K. (1990) Quantitative karyotyping of barley chromosomes by image analysis methods. *Genome* **33** 450-458.

Grill, E., Somerville, S. (1991) Construction and characterization of a Yeast Artificial Chromosome library of *Arabidopsis* which is suitable for chromosome walking. *Mol. Gen. Genet.* **226** 484-490.

Guerrini, A. M., Ascenzioni Pisani, G., Roffazzo, G., Valle, G. D., Donini, P. (1990) Cloning a fragment from the telomere of the long arm of human chromosome 9 in a YAC vector. *Chromosoma* **99** 138-142.

Hwang, I., Kohchi, T., Hauge, B. M., Goodman, H. M. (1991) Identification and map position of YAC clones comprising one third of the Arabidopsis genome. *The Plant Journal* **1** 367- 374.

Itakura, K., Rossi, J. J., Wallace, R. B. (1984) *Am. Rev. Bio. Chem.* **33** 323-356.

Jeffreys, A. J., McLeod, A., Tomaki, K., Neil, D. L., Monclon, D. G. (1991) Minisatellite repeat coding as a digital approach to DNA typing. *Nature* **354** 204-209.

Kamisugi, Y., Fukui, K. (1990) Automatic karyotyping of plant chromosomes by imaging techniques. In *Adv. in Microscopy* IV. *Biotechniques* **8** 290-295.

Kocher, T. D., Wilson, A. C. (1990) DNA amplification by the Polymerase Chain Reaction. In *Lectures on Occupational Medicine* Chap. 7 185-207. Blackwell, London.

Maniatis, T., Jeffrey, A., Kleid, D. G. (1975) Nucleotide sequences of the rightward operator of the phage lambda. *Proc. Natl. Acad. Sci.* **72** 1184-1188.

Nam, H. G., Giraudat, J., den Ber, B., Moonan, F., Loos, W. D. B., Hauge, B. M., Goodman, H. M. (1989) Restriction fragment length polymorphism linkage map of *Arabidopsis thaliana. Plant Cell* **1** 699-705.

Ochman, H., Gerber, A. S., Hartl, D. L. (1988) Genetic applications of Inverse Polymerase Chain Reaction. *Genetics* **120** 621-623.

Ross, M. T., Nizetic, D., Nguyen, C., Kneghts, C., Vatcheva, R., Burden, N., Douglas, C., Zehetner, G., Ward. D. C., Baldini, A., Lehrucht, T. (1992) Selection of a human chromosome 21 enriched YAC sublibrary using a chromosome specific composite probe. *Nature Genetics* **1** 284-290.

Sambrook, J., Fritsch, E. F., Maniatis, T. (1989) *Molecular cloning — a laboratory manual* **1.** Cold Spring Harbor Press, New York.

Salaz, H. P., Wiebaner, K., Wallace, A. (1991) Standing DNA modification and DNA protein interactions *in vivo. TIG* **7** 207-212.

Schmidt, R., Gnops, G., Bancroft, I., Dean, C. (1992) Construction of an overlapping YAC library of the *Arabidopsis thaliana* genome. *Aust. J. Pl. Physiol.* **19** 341-351.

Singh, L., Purdom, L. F., Jones, K. W. (1981) *Cold Spring Harbor. Symp. Quant. Biol.* **45** 805-811.

CHAPTER 14

Balbiani, E. G. (1881) *Zool. Anz.* **4** 662

Brachet, J. (1942, 1958) *Biochemial Cytology*. Academic Press, New York.

Callan, H. J. (1973) Replication of DNA in eukaryotic chromosome. *Brit. Med. Bull.* **29** 192-195.

Conger, A. D. (1953) Culture of pollen tubes for chromosomal analysis at the pollen tube division. *Stain Techn.* **28** 289- 293.

Danielli, J. F. (1950) Studies on the cytochemistry of proteins. *Cold Spring Harbor Symp. Quant. Biol.* **4** 32-39.

Ford, C. E., Hamerton, J. L. (1956) A colchicine hypotonic citrate squash sequence for mammalian chromosomes. *Stain Techn.* **31** 247-251.

Godward, M. B. E. (1966) *The chromosomes of algae*. Edward Arnold, London.

Gomori, G. (1939) Biochemical demonstration of phosphatase in tissue sections. *Proc. Soc. Exp. Biol. Med.* **42** 23-26.

Gomori, G. (1952) *Microscopic histochemistry*. Chicago University Press, Chicago.

Kurnick, J. B. (1950) Methyl Green Pyronin - basis of selective staining of nucleic acids. *Jour. Gen. Physiol.* **33** 243-264.

Meuten, M. L., Junge, J., Green, M. H. (1944) *J. Biol. Chem.* **153** 471.

Pearse, A. G. E. (1972) *Histochemistry - Theoretical and Applied*. Little Brown, Boston.

Ray, V. A., Kier, L. D., Kannan, K. L., Hass, R. T., Auletta, A. E., Wassom, J. S., Nesnow, S., Waters, M. D. C. (1987) An approach to identifying specialized batteries of bioassays for specific classes of chemials: class analysis using mutagenicity and carcinogenicity relationships and phylogenetic concordance and discordance patterns 1. Composition and analysis of overall data base. *Mutat. Res.* **185** 197-241.

Sarma, Y. S. R. K. (1973) Algal Karyology and Evolution Trends. In *Chromosome Evolution on Eukaryotic Groups 1* **Sharma, A. K., Sharma, A.** (eds.) CRC Press, Florida, 1-22.

Sharma, A. K. (1951) Trichloroacetic acid and fuelgen staining. *Nature* **167** 441-442.

THE REFERENCES CITED ALSO INCLUDE PUBLICATIONS WHICH MAY BE OF HELP TO THE READERS, BUT HAVE NOT BEEN INCLUDED IN THE TEXT

INDEX